4th International Conference on Micro Pattern Gaseous Detectors (MPGD 2015)

EPJ Web of Conferences Volume 174 (2018)

Trieste, Italy
12-15 October 2015

Editors:

S. D. Torre
B. Gobbo
S. Levorato
L. Ropelewski
F. Tessarotto

ISBN: 978-1-5108-5876-3

Printed from e-media with permission by:

Curran Associates, Inc.
57 Morehouse Lane
Red Hook, NY 12571

Some format issues inherent in the e-media version may also appear in this print version.

Printed by Curran Associates, Inc. (2018)

For additional information, please contact EDP Sciences – Web of Conferences
at the address below.

EDP Sciences – Web of Conferences
17, Avenue du Hoggar
Parc d'Activité de Courtabœuf
BP 112
F-91944 Les Ulis Cedex A
France

Phone: +33 (0) 1 69 18 75 75
Fax: +33 (0) 1 69 28 84 91

contact-edps@webofconferences.org

Additional copies of this publication are available from:

Curran Associates, Inc.
57 Morehouse Lane
Red Hook, NY 12571 USA
Phone: 845-758-0400
Fax: 845-758-2633
Email: curran@proceedings.com
Web: www.proceedings.com

TABLE OF CONTENTS

PROCCEDEDING OF MPGD2015

The "4th International Conference on MicroPattern Gaseous Detectors – MPGD2015" (web site: http://mpgd2015.ts.infn.it/) has taken place in Trieste from 12 to 17 October 2015, organized by INFN – Sezione di Trieste. Previous conferences in this series have been help in 2009 in Kolympari Greece, in 2011 in Kobe, Japan and in 2013 in Zaragoza, Spain. For the third time the venue is in Europe, where the first seminal Conference took place in 2009, thanks to the enthusiasm of the MPGD-dedicated community and the support of the technological collaboration RD51, CERN-based and dedicated to MPGD developments and dissemination.

These series of meetings, where experts from around the world review both the state of the art and novel perspectives in the sector, have become a reference point in the field of MPGDs. Following this tradition, the Trieste Workshop attracted more than 130 participants with its scientific programme including six invited talks, forty contributed talks and forty posters. The topical sessions were dedicated to "New Developments in MPDGs", "Production techniques", "Material and Ageing Tests", "MPGD Detector Physics", " Simulation and Software", "Electronics" and "Applications". The Workshop also recognized young researchers in the field with the George Charpak Young Scientist Award, where the MPGD 2015 Award winners are: Fabian KUGER for the poster "Electron losses during drift and mesh transit in an ATLAS-like MicroMegas", Maxence VANDERNBROUCKE for the oral presentation: "Resistive Strip Micromegas Tracker for CLAS12 Experiment" and Eran ERDAL for the oral presentation: "Bubble-assisted Liquid Hole-Multipliers: bubble stability and electroluminescence in varying electrode configurations".

The Workshop confirmed the great vitality of the MPGDs: in the current era of high-resolution and high-precision experiments, the domain of applications is becoming larger and larger, and more challenging requirements for the design of new detectors are arising. Nowadays, these detectors play a central role in particle, nuclear and astroparticle physics, where they form key systems in current and future experiments in a variety of research fields, while the panorama of the applications beyond fundamental research is continuously growing.

We would like to thank all the members of the International Organizing Committee, the Local Organizing Committee and the session conveners: their contribution has been fundamental to guarantee the high scientific standard of the Conference.

The participants had benefited of the superb Conference site, in the centre of Trieste, directly on a pier protruded in the Adriatic see, and some relax has been provided by the social programme, offering opportunities to meet Trieste and its surroundings: a welcome lunch, a seminar dedicated to the special relation between Trieste and coffee, a concert performance offered to the Conference participants, the Conference excursion with the visit of the historical town of Cividale followed by the Conference banquet.

We are grateful to all our colleagues at INFN Trieste for their help, contribution and support. Special thanks to Erica Novacco for taking care of the Workshop secretariat, a job she performed with great competence and generosity. MPGD2015 was sponsored by several Italian and other European Institutions, by the Rd51 Collaboration and private companies to which we express our gratitude.

Proceedings Editors

S. Dalla Torre
INFN, Trieste, Italy

B. Gobbo
INFN, Trieste, Italy

S. Levorato
INFN, Trieste, Italy

L. Ropelewski
CERN, Geneva, Switzerland

F. Tessarotto
INFN, Trieste, Italy

edp sciences

EPJ Web of Conferences

Statement of Peer review

In submitting conference proceedings to EPJ Web of Conferences, I certify to the Publisher that I adhere to the **Policy on Publishing Integrity** of the journal in order to safeguard good scientific practice in publishing.

1. All articles have been subjected to peer review administered by the proceedings editors.
2. Reviews have been conducted by expert referees, who have been requested to provide unbiased and constructive comments aimed, whenever possible, at improving the work.
3. Proceedings editors have taken all reasonable steps to ensure the quality of the materials they publish and their decision to accept or reject a paper for publication has been based only on the merits of the work and the relevance to the journal.

Title, date and place of the conference:
4th International Conference on MicroPattern Gaseous Detectors - MPGD2015, 12-15 October 2015, Trieste, Italy

Proceedings editor(s):
S. Dalla Torre, B. Gobbo, S. Levorato, L. Ropelewski, F. Tessarotto

Date and editor's signature: 31 January 2018, Silvia Dalla Torre for the editor team

A Prototype Combination TPC Cherenkov Detector with GEM Readout for Tracking and Particle Identification and its Potential Use at an Electron Ion Collider

Craig Woody[1,a], Babak Azmoun[1] , Richard Majka[2] , Michael Phipps[1,3] , Martin Purschke[1], Nikolai Smirnov[2]

[1]*Physics Department, Brookhaven National Laboratory, Upton, NY 11973 USA*
[2]*Physics Department, Yale University, New Haven, CT 06520 USA*
[3]*Physics Department, University of Illinois at Urbana-Champaign, Urbana, IL 61801 USA*

Abstract. A prototype detector is being developed which combines the functions of a Time Projection Chamber for charged particle tracking and a Cherenkov detector for particle identification. The TPC consists of a 10×10×10 cm^3 drift volume where the charge is drifted to a 10×10 cm^2 triple GEM detector. The charge is measured on a readout plane consisting of 2×10 mm^2 chevron pads which provide a spatial resolution ~ 100 μm per point in the chevron direction along with dE/dx information. The Cherenkov portion of the detector consists of a second 10×10 cm^2 triple GEM with a photosensitive CsI photocathode on the top layer. This detector measures Cherenkov light produced in the drift gas of the TPC by high velocity particles which are above threshold. CF_4 or CF_4 mixtures will be used as the drift gas which are highly transparent to UV light and can provide excellent efficiency for detecting Cherenkov photons. The drift gas is also used as the operating gas for both GEM detectors. The prototype detector has been constructed and is currently being tested in the lab with sources and cosmic rays, and additional tests are planned in the future to study the detector in a test beam.

1 Introduction

In many nuclear and particle physics experiments, one wants to both track and identify particles in various types of collisions. This is usually done using separate detectors for tracking and particle id. However, this can result in increasing the amount of material inside the sensitive volume of a large detector, and can also use up valuable space inside a magnetic spectrometer. It would therefore be desirable if these two features could be combined together in a single detector without compromising its performance. The detector described here combines the tracking features of a Time Projection Chamber (TPC) and a threshold Cherenkov detector for particle id.

This work is part of a Detector R&D Program for a future Electron Ion Collider that is being planned to be built at either Brookhaven National Lab (eRHIC) or Thomas Jefferson National Lab (MEIC) [1,2]. An EIC would collide beams of electrons with protons and heavy ions at high energies in order to study nucleon structure and QCD over a broad range of x and Q^2. A large multipurpose spectrometer would be used to measure deep inelastic electron scattering over a wide range of rapidity and solid angle, and the tracking system for the central detector would consist of a TPC and a precision vertex detector. The combined TPC-Cherenkov detector described here could be used to provide both tracking and particle id information for measuring the scattered electron and separating it from hadrons produced in the central region. An example of such a multipurpose spectrometer for EIC is described in the last section.

2 Prototype TPC Cherenkov Detector

We have constructed a prototype detector that combines the tracking features of a Time Projection Chamber (TPC) and a threshold Cherenkov detector for particle id. It utilizes two triple GEM detectors, one for measuring charged tracks in the TPC and another for detecting Cherenkov light produced by high momentum particles in the same gas volume. The charge from tracks passing through the detector is drifted to a triple GEM detector located at the bottom of the drift volume, and Cherenkov light is detected by a second triple GEM which has a photosensitive CsI photocathode deposited on its top layer. This technique for detecting Cherenkov light using a photosensitive GEM detector was also used in the PHENIX Hadron Blind Detector [3] for measuring low mass electron pairs in heavy ion collisions.

Figure 1 shows a 3D model of the detector. It consists of a three sided field cage made of a kapton foil with copper strips and a fourth side made of wires to allow light to pass through to the photosensitive GEM. There is no window between the drift volume and the

[a] Corresponding author: woody@bnl.gov

 https://doi.org/10.1051/epjconf/201817401001

photocathode which allows for maximum optical transparency. The photosensitive GEM is mounted on a movable stage to allow the distance between it and the TPC to be varied in order to study potential high voltage problems when the two detectors are brought into close proximity to each other. The drift volume is 10×10×10 cm^3 and the GEM detectors are 10×10 cm^2. The entire assembly is mounted inside a common enclosure and will be filled with a gas mixture that serves as the TPC gas and the operating gas for both GEMs, and also provides a highly UV transparent radiator for Cherenkov light.

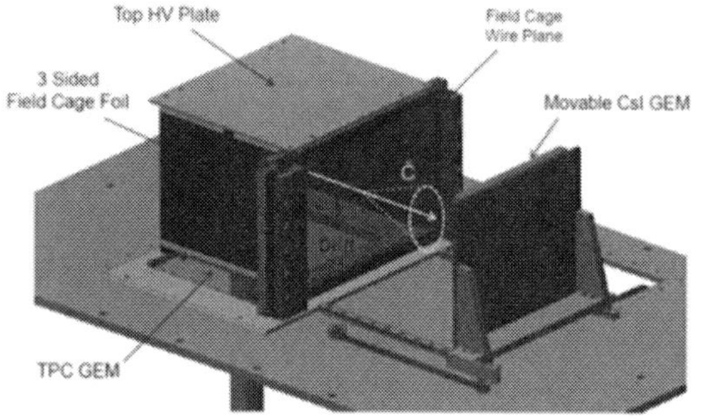

Figure 1. 3D model of the prototype TPC/Cherenkov detector.

Figure 2. Internal components of the actual TPC-Cherenkov prototype detector. The foil on the right is mounted on a movable track such that the distance between the photosensitive GEM and the TPC can be varied.

Figure 2 shows some of the components of the actual detector. The kapton foil field cage consists of 3.9 mm wide copper strips with 0.1 mm gaps in between. There are copper strips on both the front and back of the foil which are displaced by half a strip spacing (2 mm) to improve the field uniformity. For testing the TPC portion of the detector, a fourth side for the field cage made of a similar kapton foil is used as shown in Figure 3. The field cage has been tested up to 1 kV/cm, which is the maximum drift field we expect to use. The wire plane forming the fourth side consists of 75 μm wires spaced 1 mm apart that are connected in groups of four and held at the same potential in order to achieve the same field gradient as the copper strips on the kapton foil. Figure 4 shows the wire plane that is used for the fourth side of the field cage. The separate kapton foil and wire plane can be easily interchanged in order to study the detector as a conventional TPC or in combination with the Cherenkov detector.

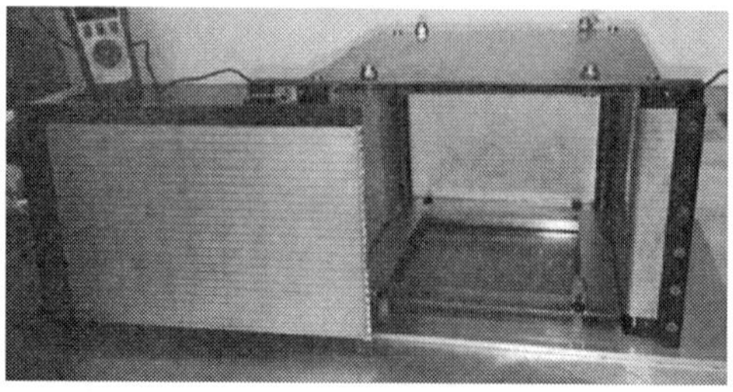

Figure 3. Three sided kapton field cage with a separate fourth kapton foil positioned to the side.

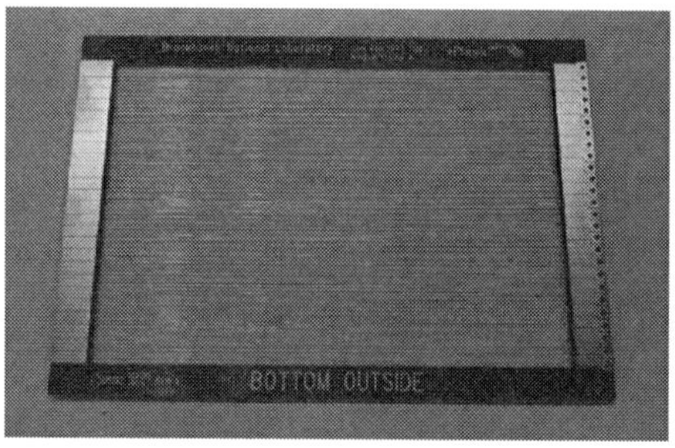

Figure 4. Wire plane used as the fourth side of the field cage when operating with the Cherenkov detector.

3 Electrostatic Field Simulations

The requirement of an optically transparent side of the field cage and the presence of the photosensitive GEM detector near the drift volume causes some distortion in the drift field of the TPC. This problem was studied using an electrostatic simulation program (ANSYS) in order to determine the magnitude of these distortions. Figure 5 shows the deviation of the nominal electric field vector in the drift volume as a function of distance along the drift direction and the distance perpendicular to the wire plane for the first mesh of the photosensitive GEM at a distance of x = -15 mm. The distortions caused by the wire plane and the presence of the photosensitive GEM are generally less than 1%.

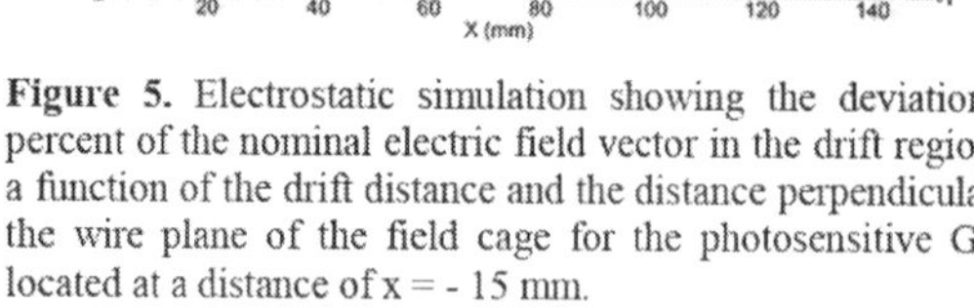

Figure 5. Electrostatic simulation showing the deviation in percent of the nominal electric field vector in the drift region as a function of the drift distance and the distance perpendicular to the wire plane of the field cage for the photosensitive GEM located at a distance of x = - 15 mm.

4 Initial Tests of the TPC

The TPC portion of the detector has been tested using radioactive sources and cosmic rays. The GEM detector for the TPC and its readout plane are shown in Fig. 6. The readout plane consists of an array of 512 chevron pads, where each pad is 2 mm wide in the chevron direction with a 0.5 mm chevron pitch and is 10 mm long.

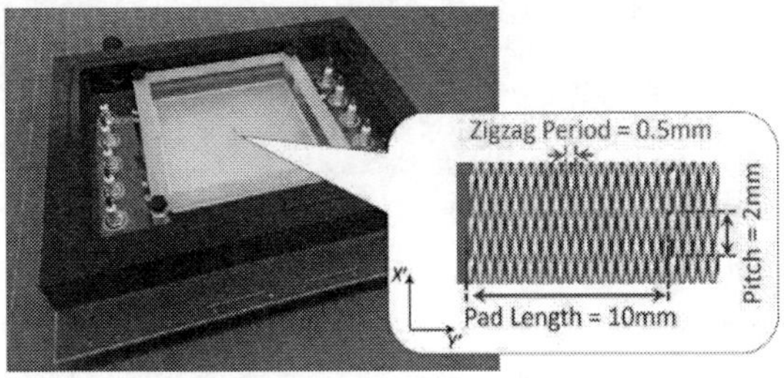

Figure 6. GEM detector with chevron readout board for the TPC portion of the prototype.

The chevron pads exhibit an inherent differential non-linearity due to the interleaving zig-zag pattern. However, this non-linearity can be corrected for by measuring the response of the pad in very fine steps. In order to measure this, the detector was configured as a "minidrift" detector with a 1.6 mm drift gap above the GEM [4,5] and scanned across the chevron direction using a highly collimated X-ray source. Figure 7 shows the measured response of the detector as a function of the source position and the differential non-linearity. Figure 8 shows the position resolution before and after the DNL correction. After correcting for the differential non-linearity, the position resolution decreased from 132 μm to 98 μm.

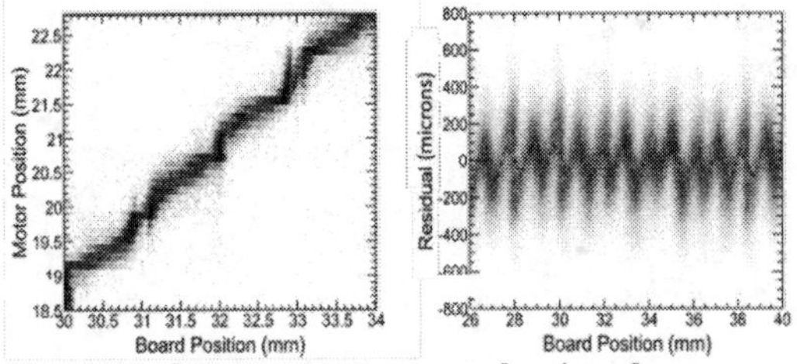

Figure 7. Left: Detector response as a function of source position; Right: Differential non-linearity

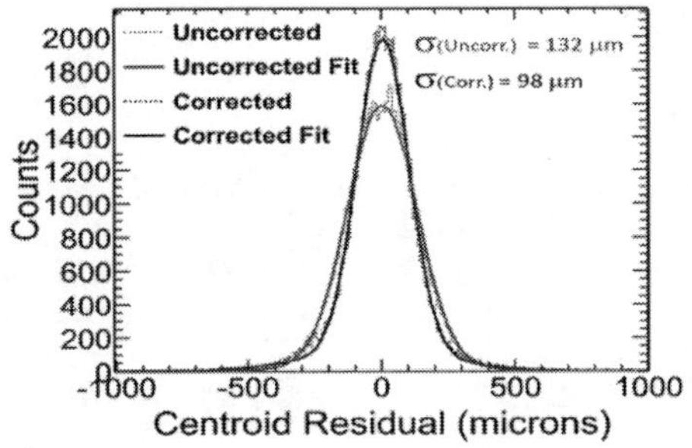

Figure 8. Position resolution across the chevron direction before and after applying the differential non-linearity correction.

The detector was also tested with the full field cage using cosmic rays. The readout used the CERN SRS system, which utilizes the APV25 readout chip and a 40 MHz clock which samples the preamp pulses every 25 ns. However, it has a buffer of only 27 samples, which limits the ability to measure the drift time over the entire drift region.

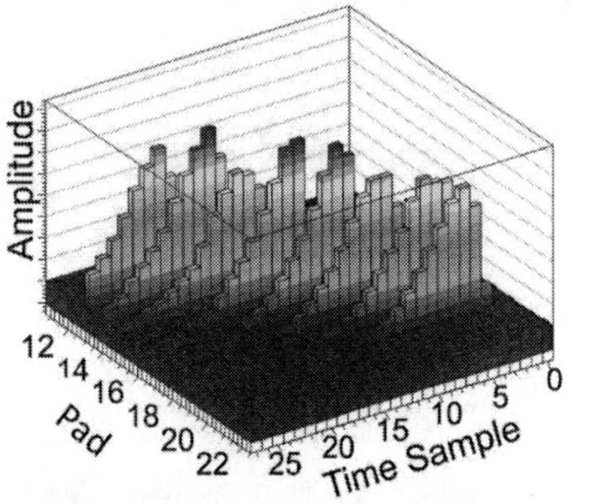

Figure 9. GEM detector with chevron readout board for the TPC portion of the prototype.

While we were not able to measure full length tracks in the drift volume, we were able to measure tracks over a range of a few cm in order to demonstrate that the TPC portion of the detector is working. Figure 9 shows the time sampled pulses from a number of pads hit from a cosmic ray traversing the detector at a moderate angle. The data was used to reconstruct the track using a "Time Slice Method", which computes a centroid for all pads hit above a given pulse height threshold in each time bin. This provides a series of points in transverse plane and drift direction that are then used to reconstruct the track. Figure 10 shows an example of a track found in the detector operating with an Ar/CO_2 (70/30) gas mixture at a drift field of 0.8 kV/cm, which has a drift velocity ~ 2.2 cm/μsec and corresponds to a distance ~ 14 mm along the drift direction.

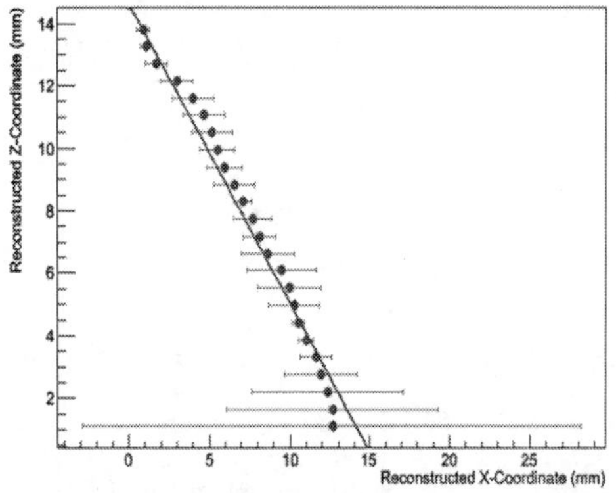

Figure 10. Cosmic ray track found in the TPC portion of the detector operating with Ar/CO_2 (70/30). The vertical path length corresponds to ~ 14 mm in this gas.

We have also studied the TPC using different gases such as $Ar/CO_2/CF_4$ (80/10/10) and Ar/CH_4 (80/20, P20). These gases give a somewhat higher drift velocity than Ar/CO_2, which is useful for testing the TPC, but are not suitable for operating with the Cherenkov detector. Figure 11 gives a compilation of drift velocities versus drift field for a number of gases we are considering using for the TPC alone and in the combined TPC/Cherenkov detector.

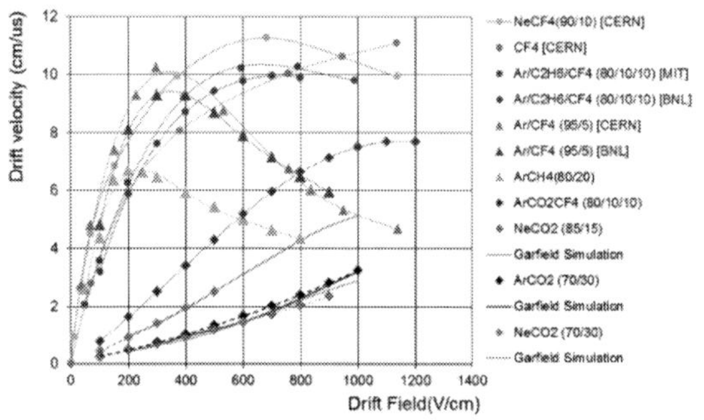

Figure 11. Compilation of drift velocities versus drift field for various gases that could be used for operating only the TPC or the combined TPC/Cherenkov detector.

For combined TPC/Cherenkov operation, a gas with high optical transparency into the deep VUV is required in order to have good photon detection efficiency for the photosensitive CsI coated GEM. The PHENIX HBD was operated with pure CF_4 and produced a yield of ~ 20 photoelectrons with a radiator length of 50 cm [3]. CF_4 is certainly a possibility for the TPC/Cherenkov detector, although a fairly high drift field is required to achieve a high drift velocity, as shown in Fig. 11. Other possibilities are mixtures of CF_4 and Ar, both of which have high transparency into the deep VUV as shown in Fig. 12, and can achieve a high drift velocity at a relatively low field as shown in Fig. 11.

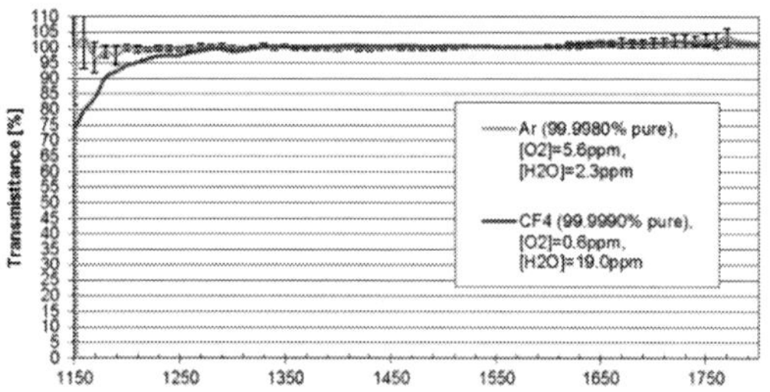

Figure 12. Transparency of Ar and CF_4 in the deep UV.

5 Application at an Electron Ion Collider

A combined TPC/Cherenkov detector would use ionization information for tracking and dE/dx and UV light from particles above the Cherenkov threshold in the same gas volume to provide both tracking and particle id. This would enhance the capabilities of a central TPC tracker for an Electron Ion Collider detector by helping to identify the scattered electron in ep and eA collisions.

Figure 13 shows an example of an EIC detector at eRHIC based on the upgrade of the proposed sPHENIX experiment [6,7]. It consists of a central spectrometer with a TPC tracker surrounded by calorimetry, a high resolution electromagnetic calorimeter in the electron going direction, and a hadronic spectrometer in the hadron going direction. Another example of an eRHIC detector with a central TPC is described in [2]. For the energy range at eRHIC (electrons up to 21.2 GeV/c, protons up to 250 GeV/c and ions up to 100 GeV/A), both electrons and hadrons are scattered into the central region. Therefore, good e/h separation is required over a large range of rapidity (~ $|\eta| < 3$), and a TPC tracker with enhanced electron id would be able to provide additional hadron rejection capability in the central region in a single detector.

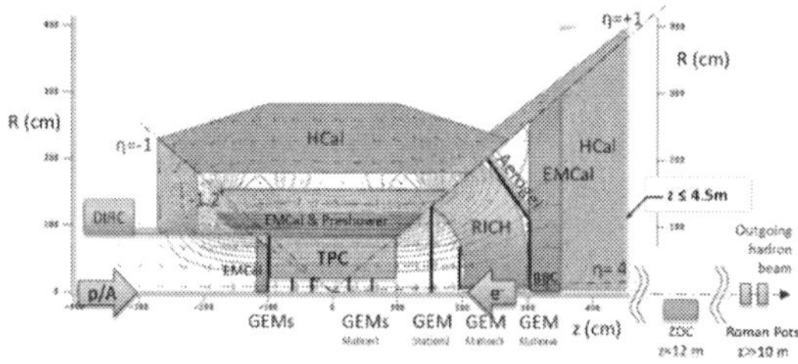

Figure 13. An EIC detector based on the sPHENIX experiment at RHIC.

6 Summary

A combined TPC/Cherenkov prototype detector is being developed that will provide both tracking and particle id. The detector has been designed and constructed and the TPC portion has been tested with radioactive sources and cosmic rays. The Cherenkov portion will be added next and the combined features of both detectors will be studied in a test beam at Fermilab or SLAC. A TPC/Cherenkov detector such as this would provide enhanced particle identification capabilities for a central tracker at a future Electron Ion Collider.

References

1. EIC White Paper, *Electron Ion Collider: The Next QCD Frontier,* arXiv:1212.1701 (May 2014)
2. eRHIC Design Study: *An Electron-Ion Collider at BNL*, arXiv:1409.1633 (Dec 2014)
3. W.Anderson et.al., Nucl. Inst. Meth. A646 (2011) 35-58.
4. M.L. Purschke et.al., Conf. Proc. of the 2013 IEEE Nuclear Science Symposium and Medical Imaging Conference, Seoul, Korea (Oct 2013)
5. B.Azmoun et.al., *A Study of a Mini-drift GEM Tracking Detector*, arXiv :1510.01747, submitted to the IEEE Trans. Nucl. Sci. (Sep 2015)
6. sPHENIX, *An Upgrade Proposal from the PHENIX Collaboration*, arXiv :1501.06197 (Feb 2015)
7. PHENIX Collaboration, *Concept for an Electron Ion Collider (EIC) detector built around the BaBar solenoid*, arXiv :1402 :1209 (Feb 2014)

Study of the d*E*/dx resolution of a GEM Readout Chamber prototype for the upgrade of the ALICE TPC

Andreas Mathis[1,2,a] on behalf of the ALICE collaboration

[1] *Physik Department E62, Technische Universität München, 85748 Garching, Germany*
[2] *Excellence Cluster 'Origin and Structure of the Universe', 85748 Garching, Germany*

Abstract. The ALICE Collaboration is planning a major upgrade of its central barrel detectors to be able to cope with the increased LHC luminosity beyond 2020. For the TPC, this implies a replacement of the currently used gated MWPCs (Multi-Wire Proportional Chamber) by GEM (Gas Electron Multiplier) based readout chambers. In order to prove, that the present particle identification capabilities via measurement of the specific energy loss are retained after the upgrade, a prototype of the ALICE IROC (Inner Readout Chamber) has been evaluated in a test beam campaign at the CERN PS. The dE/dx resolution of the prototype has been proven to be fully compatible with the current MWPCs.

1 Introduction

ALICE (A Large Ion Collider Experiment) [1] is a heavy-ion experiment designed to study the physics of the high-density, high-temperature phase of strongly interacting matter, and in particular the properties of the Quark–Gluon Plasma (QGP) in proton–proton, proton–nucleus and nucleus–nucleus collisions at the CERN Large Hadron Collider (LHC).

The second generation of LHC heavy-ion studies beyond 2020 (in LHC RUN3 after the so called Long Shutdown 2) will be enabled by a significant increase of the LHC instant luminosity up to 6×10^{27} cm^{-2}s^{-1}, thus allowing for the study of rare probes and their coupling to the medium [2].

In order to fully exploit this increased luminosity, a continuous readout of the ALICE detector at an interaction rate of 50 kHz is foreseen.

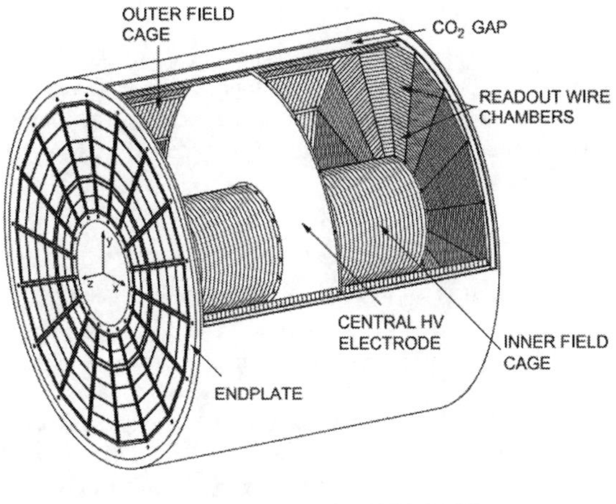

Figure 1. Schematic drawing of the ALICE TPC [3].

2 ALICE TPC - status and perspectives

The ALICE Time Projection Chamber [3] is the main device for charged-particle tracking, momentum measurement and particle identification (PID) in the central barrel of the ALICE experiment. It covers a transverse momentum range $0.1 < p_T < 100$ GeV/c within the pseudo rapidity range $|\eta| \leq 0.9$ and full azimuth.

The TPC consists of a hollow cylindrical barrel with the cathode in its center and two readout planes on its sides. The drift field of 400 V/cm is precisely defined by the cathode and a cylindrical field cage degrading the potential from 100 kV at the cathode to nearly ground potential close to the readout chambers. The active volume has an inner radius of 84.8 cm, an outer radius of 246.6 cm and an overall length along the beam axis of 499.4 cm. With an overall volume of about 90 m^3, it is the largest detector of its kind in the world.

A schematic layout of the TPC is shown in Fig. 1 with its most important components depicted. The two endplates are azimuthally segmented into 18 sectors, each covering 20°. A further division of each sector into Inner (IROC) and Outer Readout Chamber (OROC) is motivated by the different requirements for the readout chambers as a function of the radius due to the radial dependence of the track density. The design of the readout chambers is based on a gated MWPC with cathode pad readout operated at a gain of 7000-8000.

[a]e-mail: andreas.mathis@ph.tum.de

The necessity to operate the TPC in a gated mode arises from the imperative to minimize distortions of the drift field by ions released from the amplification region. However, this sets severe constraints on the maximally achievable readout rate, as it allows for no more than 3.5 kHz in pp. In Pb–Pb, the readout rate is further limited to ~300 Hz by the currently used readout electronics.

In RUN3 however, the interaction rate will be increased to 50 kHz, implying an average event pileup of 5 [2]. A triggered operation of the TPC with a gating grid will then no longer be possible, as it would cause unacceptable losses of data. Hence, a continuous operation of the TPC is mandatory, as is the minimization of the ion backflow with other means than a gating grid.

The baseline solution for the TPC upgrade consists of a stack of four large-size GEM [4] foils as amplification stage, containing both Standard (S, 140 μm) and Large Pitch (LP, 280 μm) GEM foils arranged in the order S-LP-LP-S. The high voltage configuration applied to the stack has been optimized in order to fulfill the design specifications in terms of ion backflow, energy resolution and stable operation under LHC conditions [5].

3 dE/dx resolution studies with an Inner Readout Chamber

In order to validate the performance of a large-size detector equipped with a stack of four GEMs in terms of dE/dx resolution, a prototype of the ALICE IROC has been built, commissioned and tested at the CERN Proton Synchrotron.

3.1 IROC prototype

The 4 GEM IROC prototype is assembled on a spare IROC chamber body [3]. It is a trapezoidal chamber with a size of $497 \times (292-467)$ mm^2. The mechanical structure of the chamber consists of four main components: an aluminum frame (alubody), a 3 mm Stesalit insulation plate, the pad plane (5504 pads of 4×7.5 mm^2) and the GEM stack with the cover electrode on top. Figure 2 shows an exploded view of the chamber. A set of four large-size GEM foils in the baseline configuration has been produced in the CERN MPGD workshop with single-mask technique and prepared according to the procedures as described in [6]. The cover electrode is a 1 mm thick PCB that surrounds the active area of the GEM foils and assures the homogeneity of the drift field at the borders of the chamber.

The chamber is mounted in a test box, which contains a drift cathode and a rectangular field cage with outer dimensions of 57×61 cm^2. This allows for the application of the nominal drift field of 400 V/cm over a drift length of 11.5 cm.

The detector is operated with a Ne-CO_2-N_2 (90-10-5) gas mixture.

3.2 Experimental setup

The dE/dx performance of the IROC prototype is evaluated in a test beam at the Proton Synchrotron (PS) at CERN. The prototype is installed in the PS East Area experimental hall (T10 beam line). The T10 beam line allows for the adjustment of the momentum of the secondary beam ranging from 1 to 7 GeV/c [7]. For the purposes of this performance study, a beam containing electrons and negative pions with a momentum of 1 GeV/c is chosen. For triggering and beam definition, a set of two scintillators is used. A threshold Cherenkov Counter provides a reference measurement for particle identification.

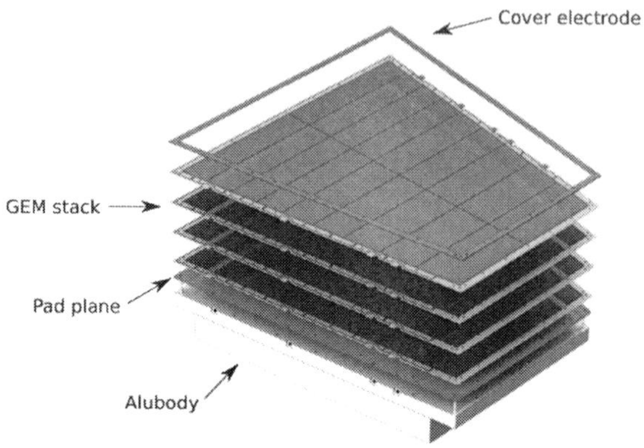

Figure 2. Exploded view of the IROC prototype with the individual components being depicted.

3.3 Readout

The IROC GEM prototype is equipped with 10 EUDET front-end cards (FEC), which were borrowed from the LCTPC (Linear Collider TPC) collaboration. This allows for the readout of about 1200 channels, covering a 6-7 cm wide corridor over the whole length of the chamber. The FECs are based on the PCA16 / ALTRO chips [8, 9]. The ALTRO chip was originally designed for the ALICE experiment. The RMS noise of the readout system has been determined to be about 600 electrons. A zero suppression threshold of 2 ADC counts was used, corresponding to around 2000 electrons (120 ns peaking time and 12 mV/fC conversion gain). The sampling frequency was set to 20 MHz.

The similarity of the FECs to the ALICE FECs allows for the usage of the current TPC readout system [3]: The data is read out via the backplane with two Readout Control Units (RCU) and transferred via optical links to a Local Data Concentrator PC, which contains the ReadOut Receiver Card (RORC) and runs the ALICE data acquisition system DATE [3]. The data from the two scintillators, the Cherenkov counter and a sensor monitoring the ambient conditions are read out via the CAMAC system. The two data streams are synchronized based on an event tag.

The average DAQ rate was ~ 300 events/spill with a spill length of 0.5 s, whereas the beam rate was of the order of 2000 particles/spill.

https://doi.org/10.1051/epjconf/201817401002

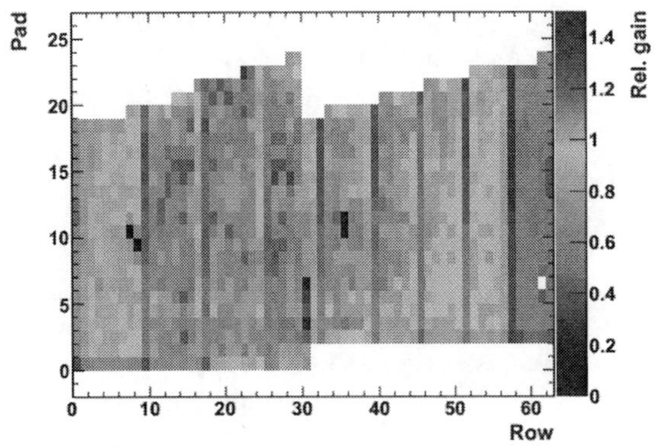

Figure 3. Normalized gain map. For further details see text.

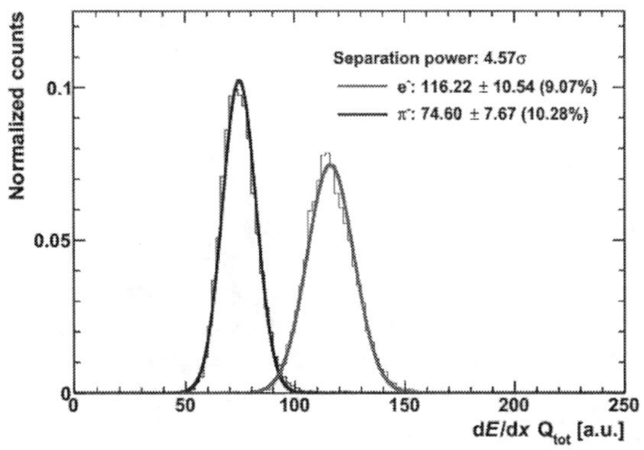

Figure 4. Truncated (0–70 %) energy loss distributions for electrons (*orange*) and pions (*blue*) with 1 GeV/c momentum measured with the IROC GEM prototype at a gain of 2000.

3.4 d*E*/d*x* measurements

The study of the particle identification capabilities of the detector prototype is conducted with AliRoot [10], the framework for reconstruction, simulation and analysis in ALICE.

Only events with single tracks are selected in order to make use of the particle identification provided by the Cherenkov counter. Additional cuts are applied on the number of clusters per track and the cluster drift time. Moreover, tracks with clusters at the acceptance edge are rejected.

Geometric imperfections of the GEM system as well as channel-by-channel variations of the front-end electronics may cause variations of the gain. In order to minimize the impact of such variations, the normalized gain is extracted for each pad individually with tracks from the test beam. Figure 3 shows the resulting gain map. Clearly visible are pad rows with lower gain, which are due to an overlap with supporting structures of the GEM stack. Also visible is a low-gain region for pad rows >57 due to a malfunctioning HV segment in one of the foils. For rows not affected by the mentioned effects, the observed gain spread is ~ 10.5 %. In order not to bias the $\mathrm{d}E/\mathrm{d}x$ measurement, the charge of individual clusters is corrected by the normalized gain factor on a pad-by-pad basis.

The specific energy loss $\mathrm{d}E/\mathrm{d}x$ is then given by the truncated mean of the 70 % lowest cluster charges Q_{tot} of the ionization clusters of a track, which are reconstructed on the 63 pad rows on the readout anode. The approach of the truncated mean is chosen in order to symmetrize the energy loss distribution, that can then be fitted with a Gaussian function to obtain its mean value $\langle \mathrm{d}E/\mathrm{d}x \rangle$ and width $\sigma(\mathrm{d}E/\mathrm{d}x)$. The relative $\mathrm{d}E/\mathrm{d}x$ resolution is then defined as $\sigma(\mathrm{d}E/\mathrm{d}x)/\langle \mathrm{d}E/\mathrm{d}x \rangle$. Figure 4 shows an exemplary $\mathrm{d}E/\mathrm{d}x$ spectrum of 1 GeV/c electrons and pions recorded at a gain of about 2000 and depicts the corresponding separation power, defined as

$$S_{AB} = \frac{2|\langle \mathrm{d}E/\mathrm{d}x \rangle_A - \langle \mathrm{d}E/\mathrm{d}x \rangle_B|}{\sigma(\mathrm{d}E/\mathrm{d}x)_A + \sigma(\mathrm{d}E/\mathrm{d}x)_B} \quad (1)$$

for two particle species A and B.

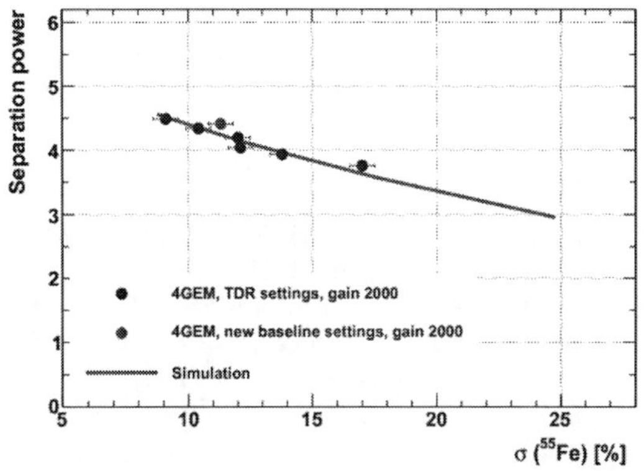

Figure 5. Separation power of pions and electrons with a momentum of 1 GeV/c as a function of the ^{55}Fe resolution at a gain of 2000. The green curve shows the result of a simulation. Reproduced from [5].

A set of different voltage settings was tested during the test beam. The respective voltage settings, taken from measurements carried out with small prototypes [6], are characterized in terms of the resolution of the ^{55}Fe peak and ion backflow at a gain of 2000.

The performance of the IROC prototype is shown as a function of the ^{55}Fe resolution in Fig. 5. As expected, the separation power decreases as the ^{55}Fe resolution is degraded. The measurement is compared to a microscopic simulation, in which the effective primary electron efficiency of the readout system is varied. This leads to a degradation of both the ^{55}Fe and the $\mathrm{d}E/\mathrm{d}x$ resolution, which reproduces the data very well.

The $\mathrm{d}E/\mathrm{d}x$ resolution measured with the GEM IROC under a voltage configuration optimized for the specific requirements of the ALICE TPC is fully compatible with that of the MWPC IROC.

Concluding, the current $\mathrm{d}E/\mathrm{d}x$ resolution of the ALICE TPC and thus its excellent PID capabilities will be preserved after the upgrade.

4 Summary and outlook

The ALICE TPC will be upgraded for the LHC RUN 3 and beyond to operate at a 50 kHz rate in Pb–Pb collisions, demanding for a non-gated and continuous readout. Employing a stack of four GEMs with different hole pitches, the requirements of the readout system in terms of ion back-flow, energy resolution and stability against discharges are met.

The $\mathrm{d}E/\mathrm{d}x$ performance of a large-sized prototype equipped with such a readout structure has been evaluated at the CERN PS and demonstrated, that the PID capabilities of the TPC will not be compromised after the upgrade.

References

[1] K. Aamodt *et al.* [ALICE Collaboration], "The ALICE experiment at the CERN LHC", JINST **3** (2008) S08002. http://dx.doi.org/10.1088/1748-0221/3/08/S08002

[2] B. Abelev *et al.* [ALICE Collaboration], "Upgrade of the ALICE Experiment: Letter of Intent", CERN-LHCC-2012-012 (2012). https://cds.cern.ch/record/1475243

[3] J. Alme *et al.*, "The ALICE TPC, a large 3-dimensional tracking device with fast readout for ultra-high multiplicity events", Nucl. Instrum. Meth. A **622** (2010) 316. http://dx.doi.org/10.1016/j.nima.2010.04.042

[4] F. Sauli, "GEM: A new concept for electron amplification in gas detectors" Nucl. Instrum. Meth. A **386** (1997) 531. http://dx.doi.org/10.1016/S0168-9002(96)01172-2

[5] J. Adam *et al.* [ALICE Collaboration], "Addendum to the Technical Design Report for the Upgrade of the ALICE Time Projection Chamber", CERN-LHCC-2015-002 (2015). https://cds.cern.ch/record/1984329

[6] B. Abelev *et al.* [ALICE Collaboration], "Upgrade of the ALICE Time Projection Chamber", CERN-LHCC-2013-020 (2013). https://cds.cern.ch/record/1622286

[7] D. J. Simon *et al.*, "Secondary beams for tests in the PS East experimental area", PS-PA-EP-Note-88-26 (1988). http://cds.cern.ch/record/1665434

[8] A. Kaukher, O. Schäfer, H. Schröder, and R. Wurth, "Status of TPC-electronics with Time-to-Digit Converters", EUDET-Memo-2009-08 (2009). http://www.eudet.org/e26/e28/e42441/e78196/EUDET-MEMO-2009-08.pdf

[9] R. Esteve Bosch, A. Jimenez de Parga, B. Mota and L. Musa, "The ALTRO Chip: A 16-Channel A/D Converter and Digital Processor for Gas Detectors", IEEE Transactions on Nuclear Science **50** (2003) 2460. http://dx.doi.org/10.1109/TNS.2003.820629

[10] AliRoot - ALICE Offline, http://aliweb.cern.ch/Offline/AliRoot/Manual.html.

EPJ Web of Conferences 174, 01003 (2018) https://doi.org/10.1051/epjconf/201817401003
MPGD 2015

Construction and Test of Full-Size Micromegas Modules for the ATLAS New Small Wheel Upgrade

Jonathan Bortfeldt[1,a] on behalf of the ATLAS Muon Collaboration

[1] *Ludwig-Maximilians-University Munich, Germany*

Abstract. In 2015 the first full size resistive-strip Micromegas modules for the ATLAS New Small Wheel upgrade will be realized. The goal is to provide precision muon tracking with spatial resolution below 100 µm on trapezoidal four-layer detector modules with areas between 2 and 3 m^2. This poses stringent limits on the overall accuracy of the modules with respect to strip positioning and planarity. Each module is built as a quadruplet of four resistive strip Micromegas layers, constructed from stiff sandwich panels. An assembly procedure has been developed to build the panels and modules with the required mechanical precision of 30 µm along the precision coordinate and 80 µm perpendicular to the chamber. The construction and quality assurance procedures ensure good module quality verified by checks done during the construction and the chamber assembly.

1 Introduction

ATLAS is a multi-purpose detector at the Large Hadron Collider (LHC) at CERN [1] with almost full solid angle coverage. The muon system represents the outermost layer of the detector. It consists of separate tracking and triggering chambers in toroidal magnetic fields, enabling a standalone muon momentum measurement [2]. Monitored Drift Tube Chambers (MDT) are used for tracking in the barrel and the endcap region. Only in the central region of the endcap station, in the so called Small Wheel, that is closest to the interaction point, Cathode Strip Chambers are used. Resistive Plate Chambers provide the trigger in the barrel and Thin Gap Chambers in the endcap region of the muon spectrometer.

The instantaneous luminosity of the LHC will be increased from its design value of $\mathcal{L} = 1 \times 10^{34}\,\mathrm{cm}^{-2}\mathrm{s}^{-1}$ up to $\mathcal{L} = 7 \times 10^{34}\,\mathrm{cm}^{-2}\mathrm{s}^{-1}$ in 2026. The background hit rates by photons and neutrons increase accordingly and can reach up to 15 kHz/cm^2 in the forward region of the muon spectrometer. In order to preserve the current muon triggering capabilities and momentum determination accuracy, the muon detectors in the Small Wheel need to be replaced.

Small strip Thin Gap Chambers (sTGC) have been chosen as main trigger and resistive strip Micromegas as main tracking detector [3], where each detector technology is capable of providing both functionalities. Each New Small Wheel consists of 8 large and 8 small sectors. Each sector consists of 2×4 layers of sTGCs, sandwiching 2×4 layers of Micromegas, mounted on a spacer frame. Each Micromegas part of a sector consists of upper and lower quadruplet chambers. The dimensions of these chambers are shown in figure 1. Four modules differing in dimension are constructed by the participating institutes in Italy (SM1), Germany (SM2), France (LM1) and Russia/Greece/CERN (LM2). Two layers of each Micromegas quadruplet are rotated with respect to the other two by ±1.5° such that these stereo layers provide a second coordinate hit measurement with millimeter accuracy.

In this paper, the construction and quality control procedures for the first full size Micromegas modules of the New Small Wheel are described and discussed.

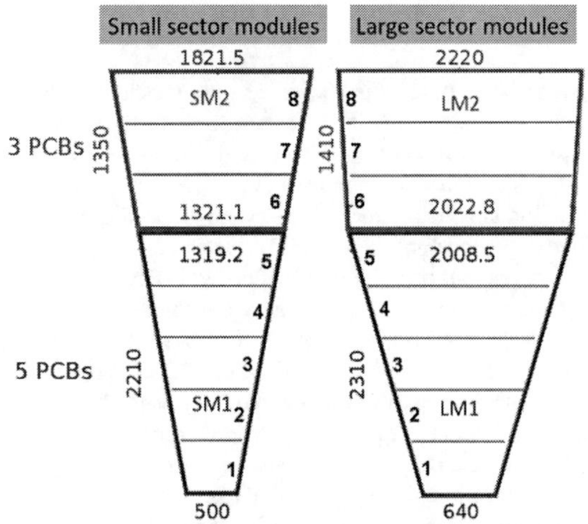

Figure 1. Dimensions of the small and large New Small Wheel Micromegas sectors in millimeters. Each Micromegas chamber consists of three or five printed circuit boards (PCB) per layer, numbered from 1 to 5 and 6 to 8.

[a] e-mail: jonathan.bortfeldt@cern.ch

2 Micromegas Chambers for the ATLAS New Small Wheel

2.1 Resistive Strip Micromegas

A resistive strip Micromegas is shown schematically in figure 2. The copper readout strips are covered by a congruent layer of screen-printed resistive strips. These anode strips are insulated with respect to the readout strips by a 50 μm thick Kapton foil and are supplied with high-voltage. This scheme suppresses the influence of discharges between anode and mesh on the detection efficiency [4], that are potentially created by strongly ionizing background particles. The micro-mesh is carried by a pre-

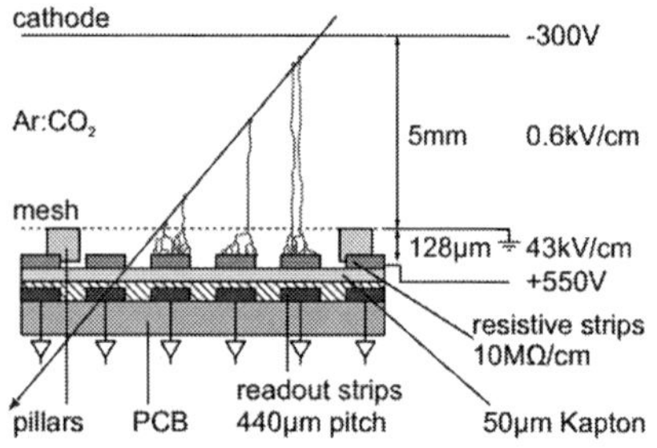

Figure 2. Schematic view of a resistive strip Micromegas.

cise aluminum frame, that is permanently attached to the cathode structure. This facilitates cleaning of the amplification region between mesh and anode during assembly and thus simplifies the commissioning. The readout strips are individually connected to the readout electronics using solderless connectors.

Each readout printed circuit board (PCB) carries several copper precision markers and chessboardlike coded alignment markers, so called RasNiK masks [5], a development used extensively for monitoring ATLAS muon detectors. They are produced in the same step as the readout strips and thus have a precise relative alignment with respect to the readout structure. Measuring the position of the circular or long-hole precision markers or the RasNiK masks with telecentric optics or contact-CCDs allows for a precise optical positioning of the readout structures and the final panels during assembly.

2.2 Module Design and Accuracy

A schematic view of a Micromegas quadruplet is shown in figure 3. The overall thickness of each module is about 70 mm and the total number of readout channels is on the order of $1.5 \cdot 10^4$ per module. It is constructed from two readout panels with readout anodes on both sides and three drift panels, that carry the cathode structures. The readout structure consists of strips with 0.43 mm pitch and up to 2 m length. Readout and cathode PCBs are used as faces of glued sandwich panels, consisting of aluminum honeycomb and aluminum reinforcement bars. The two outer drift panels are identical and carry cathode structure and micro-mesh on the inner side. The central double sided

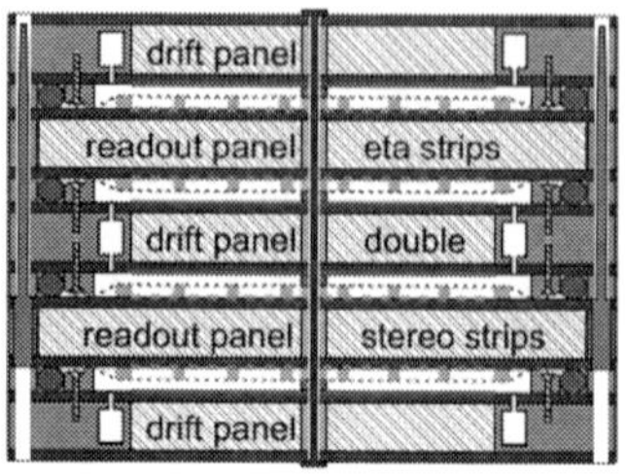

Figure 3. Schematic view of Micromegas quadruplet (not to scale). The cathode structure (magenta) is carried by three drift panels, the readout structure (green) is supported by two double readout panels. The micro-mesh (red) is attached to an aluminum frame, supported by the drift panels, its distance to the readout structure is defined by 128 μm high insulating pillars (yellow).

drift panel carries cathode and mesh on both sides. The two double sided readout panels carry the parallel eta- and the rotated stereo strips, respectively. The five panels are screwed together during assembly and sealed with O-rings, such that an opening of the chamber for maintenance is possible, if needed. Interconnections between the five panels minimize the deformation of the chamber, induced by a slight overpressure due to the continuous flushing of the chamber with detector gas.

The desired momentum resolution of 10% for muons with 1 TeV transverse momentum can be reached by chambers capable of measuring the overall muon track position with an accuracy better than 50 μm. This requirement can be translated into accuracy requirements (RMS) for the readout and cathode structure of a quadruplet for controlled temperature conditions. The absolute positioning of readout strips must be better than 35 μm, the relative alignment of the readout PCBs on one side of a panel and the relative alignment of the two sides of a readout panel must be better than 25 μm. Finally the relative alignment of the two readout panels should be better than 18 μm. The surface of readout and drift panel should be planar within 37 μm. During operation in the New Small Wheel the alignment of the chambers will be monitored with an optical system.

2.3 Concept of Panel Construction

The idea of panel construction is illustrated in figure 4. The PCBs are sucked upside-down against a granite table with a maximum deviation from planarity smaller than 10 μm. Thus the table planarity transfers to the PCB. 10 mm thick aluminum honeycomb and aluminum bars are

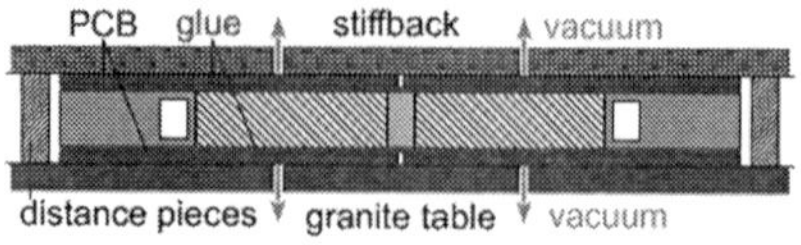

Figure 4. Panel construction on a planar granite table.

then glued by a 200 μm thick layer of Araldite 2011. The PCBs of the second face, with an equally thick layer of glue, are placed on top. During glue curing, the second side is sucked against a precise but light-weight sandwich-table, called stiffback, thus ensuring planarity of the second side. The parallelism of both sides is ensured by precision distance pieces between granite table and stiffback.

Two different gluing techniques are used at the contributing sites: The discussed stiffback method, when accurate absolute positioning of the two panel faces is required and alternatively a vacuum bag method, described in section 3. Also, two different drilling schemes are used: Drilling and milling of all holes and threads into the individual components before panel assembly or alternatively machining after gluing in a single step. Since the assembly techniques for SM1 and LM2 and for SM2 and LM1 are similar, the panel construction is discussed exemplarily for the SM1 and SM2 modules in sections 3 and 4.

3 Drift Panel Construction

The drift panels of the SM1 modules are assembled using the vacuum bag method, gluing both sides simultaneously. All machining is done prior to panel assembly. Each of the five drift PCBs features a pair of precision hole and slot. An aluminum bar with several removable precision dowel pins is mounted on the granite table. Glue is distributed on the PCBs before transferring them to the granite table. The upper PCB (number 5 in figure 1) is aligned with respect to the aluminum bar using two precision pins in the hole and slot of the PCB. The remaining four PCBs are then aligned relative to the upper using feeler gauges to define the distance between boards and a single precision pin in each slot. The aluminum honeycomb and the aluminum bars are placed into the glue and aligned relative to the PCBs using dowel pins in the mounting holes. This arrangement is shown in figure 5. The second layer of PCBs

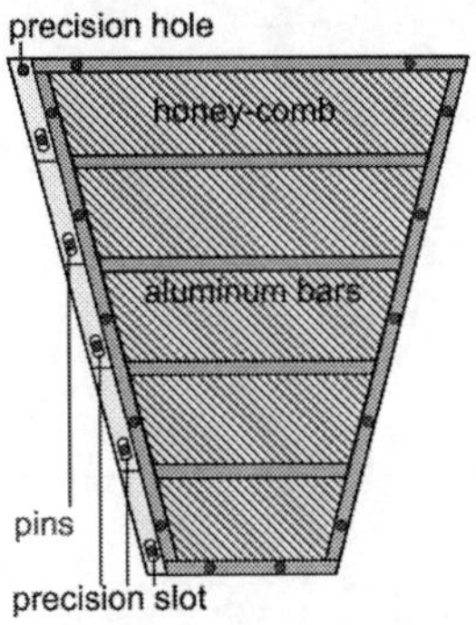

Figure 5. Open SM1 drift panel, before closing with the second layer of PCBs, which are pressed against the sandwich with a vacuum bag during glue curing.

is then placed on top, a vacuum bag covers the sandwich and the air is pumped out of the sandwich, pressing the components against the granite table. This method profits from the manufacturing precision of the thickness of PCBs, honeycomb and bars.

The SM2 drift panels are glued with the same method as the SM2 readout panels, c.f. section 4.2.

4 Readout Panel Construction

4.1 SM1 Module

While the lower side lies on a granite table, the upper side of the SM1 readout panels is supported by a stiffback during curing, both sides are glued simultaneously and all machining is done prior to gluing. The granite table and the stiffback permanently carry precision aluminum plates with milled vacuum suction channels and precision holes. Removable dowel pins fit into the precision holes. The readout PCBs and bars are aligned with respect to the aluminum plates using the dowel pins. The relative position of the two panel sides during glue curing is ensured by a tapered and a V-shaped interlock attached to the aluminum plates on the table and on the stiffback.

4.2 SM2 Module

The internal alignment of the SM2 module is performed relative to two steel cylinders with precise distance. Onto the reverse side of each readout PCB a brass slot and a brass hole washer are glued, concentric to the two precision markers, c.f. section 2.1. Alignment with an accuracy below 5 μm is achieved during adjustment when observing the washer and the precision marker with a telecentric camera. The three readout PCBs are then placed onto a granite table. A stiff aluminum alignment frame, carrying six precision dowel pins is placed on top of the three boards. The dowel pins fit into the hole and slot washers on the readout boards and thus ensure the precise relative alignment of the three boards. The alignment frame additionally carries two precision surfaces, straight and V-shaped. Pressing these against two cylinders, permanently mounted on the granite table, ensures the accurate absolute positioning of the readout PCBs, see figure 6. After sucking the three boards against the granite table,

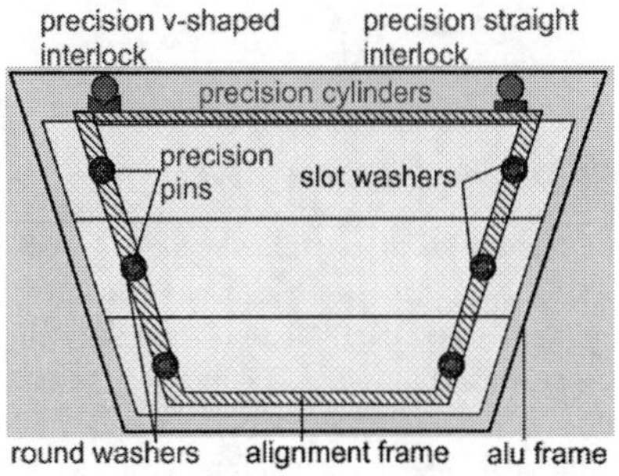

Figure 6. Readout PCBs (yellow) on the granite table, precisely aligned with respect to two cylinders (magenta) by a removable alignment frame (hatched).

the alignment frame can be removed. A V-shaped and a straight precision piece is glued onto the upper PCB, each precisely aligned with respect to the two cylinders. Aluminum honeycomb and bars are glued on the boards relative to a fixed aluminum frame using distance pieces. After glue curing, this semi-panel is removed and transferred to a stiffback. The alignment procedure on the granite table is repeated for the second PCB layer and the semi-panel, sucked against the stiffback, is glued onto it.

The readout panel is finally transferred to a large CNC machine, where all holes and threads are machined. All precision holes are manufactured in a single step without moving the panel, leading to an excellent relative alignment with machine precision.

4.3 Alignment Structure Gluing

The critical alignment between the two readout panels is achieved mechanically (figure 7). Two precision pins are glued into the stereo strip readout boards (c.f. figure 3) after glue curing for the SM1 panels and after milling for the SM2 panels, without moving the respective panel. Into

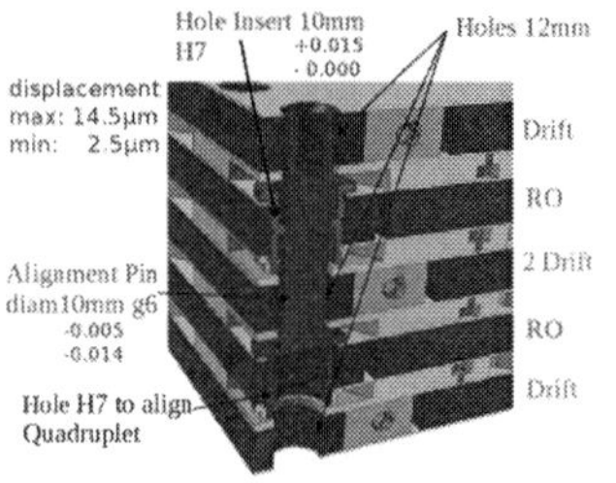

Figure 7. Two 10 mm alignment pins (green) with g6 precision in the stereo strip readout boards fit into matching slot and hole inserts (magenta) with H7 precision in the eta strip readout boards.

the eta strip readout boards two matching hole and slot inserts are glued consistently.

5 Quality Assurance

Many quality checks are performed on raw material and PCBs before panel assembly. Before and after panel gluing, the absolute positioning of the readout PCBs is confirmed optically. During module assembly, repeated high-voltage stability tests ensure the cleanliness of the detectors. In the following panel planarity and gas tightness measurements are discussed.

5.1 Planarity Measurements

Coordinate measurement machines equipped with tactile or with laser distance sensors are used to investigate the achieved planarity and parallelism of both readout panel sides after panel assembly. The accuracy of these systems is better than 10 μm. For three dummy panels the following planarities and min-max-deviations have been reached: 39 μm and 200 μm for a SM1 drift panel, 33 μm and 200 μm for a SM1 readout panel and 29 μm and 190 μm for a SM2 readout panel. The panels fulfill the ATLAS requirements.

5.2 Gas Tightness Measurements

The gas tightness of all panels is tested prior to module assembly with local and global methods. They rely on a measurement of the pressure drop as a function of time. Proper sealing of joints between adjacent PCBs and at interconnects is confirmed. The overall tightness of panels is tested for each side individually in a full-size vessel.

6 Mesh Stretching, Cleaning and Gluing

Woven stainless-steel meshes with 250 lines/inch and 30 μm wire diameter are stretched on pneumatic or mechanic clamps with a tension of 10 N/cm and glued to transfer frames. The meshes are cleaned with solvent and high-pressure deionized water and finally glued onto precision aluminum bars, screwed onto the drift panels. The residual mesh is cut off.

7 Quadruplet Assembly

Modules are assembled vertically in a clean room to avoid dust contamination. The relative alignment of the two readout panels is ensured by two pairs of precision pins and bushings, c.f. figure 7. For the SM2 modules an alternative alignment, based on the V-shaped and straight precision surfaces is foreseen. For the drift panels, an alignment with mounting hole accuracy is sufficient. After closing a final high-voltage test is performed, where each amplification structure is ramped up to almost 1 kV under normal air atmosphere. In case of frequent discharges, the respective layer is reopened and cleaned.

The completed quadruplets are tested with cosmic rays to validate the overall accuracy and confirm the desired detection efficiency and operational stability. Using drift tube chambers as cosmic muon track reference, the position and potential deformation of the readout strips can be measured with an accuracy of 20 μm [6].

8 Conclusions

The construction of quadruplet resistive strip Micromegas chambers with 30 μm spatial resolution in the precision direction for the forward ATLAS muon spectrometer is challenging. The assembly and quality assurance procedures are reviewed and described. The desired planarity is reached. During the next few months, the first four full-size quadruplet modules will be built with the desired precision.

References

[1] ATLAS Collaboration, JINST **3**, S08003 (2008)

[2] ATLAS Collaboration, CERN-LHCC-97-022, CERN (1997)

[3] ATLAS Collaboration, CERN-LHCC-2013-006, CERN (2013)

[4] T. Alexopoulos et al., Nucl. Instr. and Meth. A **640**, 110 (2011)

[5] H. van der Graaf et al., *RasNiK, an Alignment System for the ATLAS MDT Barrel Muon Chambers*, NIKHEF ETR94-10 (1994)

[6] Ph. Lösel et al., Nucl. Instr. and Meth. A (2015)

Status of COMPASS RICH-1 Upgrade with MPGD-based Photon Detectors

M. Alexeev[1,a], R Birsa[2], F. Bradamante[2,3], A. Bressan[2,3], M. Chiosso[1,4], P. Ciliberti[2,3], S. Dalla Torre[2], S. Dasgupta[2,3], O. Denisov[1], M. Finger[5,6], M. Finger Jr[5,6], H. Fishcher[7], B. Gobbo[2], M. Gregori[2], G. Hamar[2], F. Herrmann[7], K. Königsmann[7], S. Levorato[2], A. Maggiora[1], N. Makke[2,3], A. Martin[2,3], G. Menon[2], J. Novy[5,6], D. Panzieri[1,8], F. A. Pereira[9], C. A. Santos[2], G. Sbrizzai[2,3], P. Schiavon[2,3], S. Schopferer[7], M. Slunechka[5,6], K. Steiger[2,10,b], L. Steiger[2,10,b], M. Sulc[10], F. Tessarotto[2,c], and J. F. C. A. Veloso[9]

[1] *INFN Sezione di Torino, Torino, Italy*
[2] *INFN Sezione di Trieste, Trieste, Italy*
[3] *University of Trieste, Trieste, Italy*
[4] *University of Torino, Torino, Italy*
[5] *Charles University, Prague, Czech Republic*
[6] *JINR, Dubna, Russia*
[7] *Universität Freiburg, Physikalisches Institut, Freiburg, Germany*
[8] *University of East Piemonte, Alessandria, Italy*
[9] *Physics Department, University of Aveiro, Aveiro, Portugal*
[10] *Technical University of Liberec, Liberec, Czech Republic*

Abstract. A Set of new MPGD-based Photon Detectors is being built for the upgrade of COMPASS RICH-1. The detectors cover a total active area of 1.4 m^2 and are based on a hybrid architecture consisting of two THGEM layers and a Micromegas. A CsI film on one THGEM acts as a reflective photocathode. The characteristics of the detector, the production of the components and their validation tests are described in detail.

1 The COMPASS RICH-1 Upgrade

The COMPASS Experiment at CERN SPS has recently started a new set of measurements [1] imposing strict requirements in terms of rate capability, efficiency, and stability of the detector performance: several parts of the apparatus are being upgraded to meet these requirements, including the COMPASS RICH-1 detector [2], which provides π-K separation from 3 to 55 GeV/c over $\pm$200 mrad angular acceptance, at high rates.

COMPASS RICH-1 (Fig. 1, left) is a Ring Imaging Cherenkov counter with a 3 m long gaseous C_4F_{10} radiator, a 21 m^2 large focusing VUV mirror surface and Photon Detectors (PDs) covering a total active area of 5.5 m^2 with two technologies: Multi-Anode PMTs coupled to individual fused silica lens telescopes in the central region (25% of the surface) and MWPCs equipped with CsI-coated photocathodes in the remaining surface.

In spite of their good performance, MWPC-based PDs present intrinsic limitations: aging (decrease of quantum efficiency) after a few mC/cm^2 charge collection, feedback pulses with a rate increasing at large gain values, long recovery time after occasional discharge in the detector and long signal formation time. The MWPCs have to be operated at low gain and present a non-negligible detector memory and dead time. Among the eight MWPC's of COMPASS RICH-1 four, located above and below the centre of the detector show particularly critical performance [3]. The present upgrade aims at replacing them with novel MPGD-based PDs, developed in a seven year-long dedicated R&D programme [4]: the resulting detector architecture is a Hybrid MPGD including two THick GEM (THGEM) multiplication stages followed by a Micromegas.

The main technical characteristics and the production process of the MPGD-based PDs for COMPASS RICH-1 upgrade are described in this article.

2 The THGEM production and quality assessment

Each one of the four new chambers (Fig. 1, right) will host one of the existing 600 × 600 mm^2 panel of Multi-Anode PMTs and lens telescopes together with a new MPGD-based PD consisting of two identical hybrid modules covering about 600 × 300 mm^2, arranged side by side.

The basic structure of the hybrid module (Fig. 2) consists of two layers of THGEMs, one Micromegas, and two planes of wires. The top layer of the first THGEM (as seen

[a]On leave from JINR, Dubna, Russia.

[b]Their present address is Technical University of Liberec, Liberec, Czech Republic and Institute of Plasma Physics, Academy of Sciences of the Czech Republic, Turnov, Czech Republic.

[c]Corresponding author; email:fulvio.tessarotto@ts.infn.it

 https://doi.org/10.1051/epjconf/201817401004

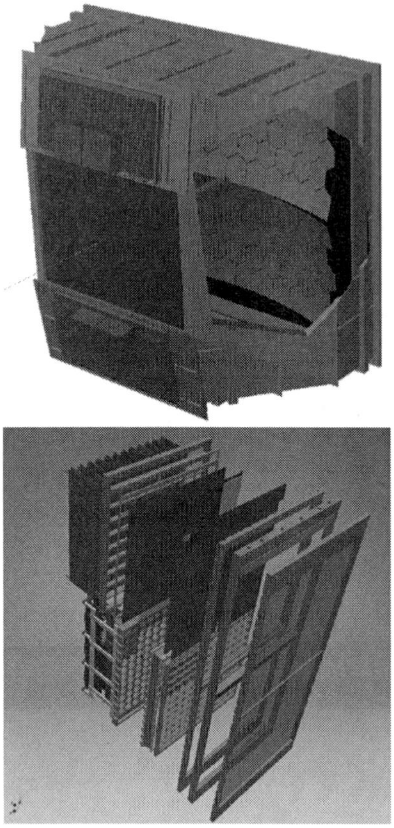

Figure 1. Artistic view of the COMPASS RICH-1 detector (left) and exploded view of the new PDs (right)

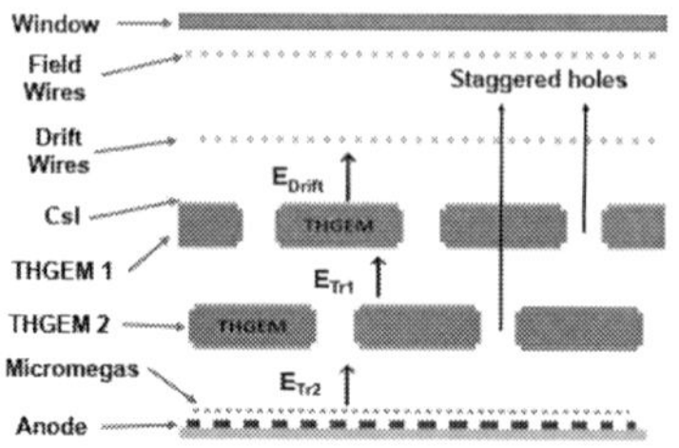

Figure 2. Schematic section of the hybrid module (not in scale).

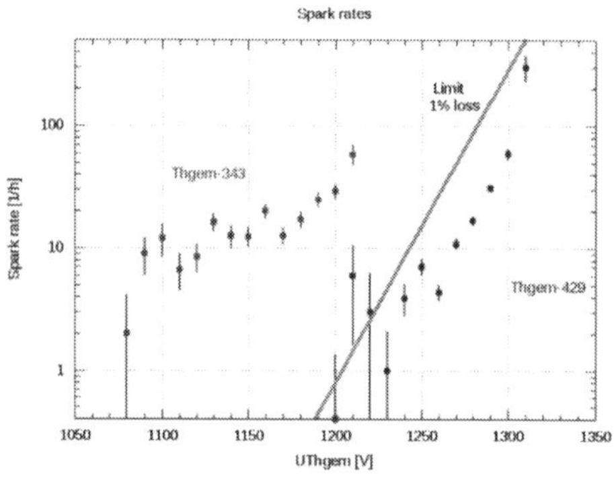

Figure 3. Rate of discharges in THGEMs as function of the bias voltage in Ar/CO_2 70/30 gas mixture.

from inside the RICH vessel) is coated with a CsI film and acts as a reflective photocathode for VUV photons.

The THGEM PCBs are all identical: they have an average thickness of 470 μm (400 μm of fiberglass and 2 × 35 μm Cu), a length of 581 mm and a with of 287 mm. They have a regular hexagonal array of holes of 400 μm diameter, with a pitch of 800 μm. The holes are produced by mechanical drilling and have no rim. The holes of the external border lines have 500 μm diameter, to avoid problems related to the asymmetric field configuration at the electrode edges. On both top and bottom face the active area is divided in 12 sectors which are 564 mm long and 23.3 mm wide (the two external ones are 17.9 mm wide), with a separation of 0.7 mm between neighboring sectors. Electrical connections are individually provided to each sector on one side of the THGEM. A set of 24 holes, used as fixation points, guarantees the correct planar positioning of the THGEM.

The PCB raw material (halogen-free EM 370-5 from Elite Material Co, Ltd.) has been carefully selected before the THGEM production phase: 50 foils (of 1245 mm × 1095 mm) have been purchased, reduced to 800 mm × 800 mm by cutting out the borders which usually present larger local thickness variations and mapped by means of a Mitutoyo EURO CA776 coordinate measuring machine hosted in a thermally stabilized room. The PCB foil was laying on the reference surface of the coordinate measuring machine, kept flat by under-pressure while its top surface was mapped in a square pattern of points at 20 mm steps; the measurement was than repeated after reversing the foil and good consistency between the two measurements was obtained. The reference surface was regularly re-measured and the point by point difference between reference surface and foil measurement provided the local foil thickness values. A typical thickness distribution for a good foil presents an average value of 472 μm and a standard deviation of 2 μm. Each PCB foil can host two raw THGEM PCBs, and a thickness uniformity quality factor δ_{th} is evaluated for each surface of the PCB corresponding to a raw THGEM PCB. The quality factor is defined as $\delta_{th} = (th_{max} - th_{min})/th_{min}$ where th_{max} and th_{min} are the maximum and minimum of the measured thickness values in the active area of the THGEM. The best raw THGEM PCBs have $\delta_{th} \leq 2\%$, and out of a total of 100 raw THGEM PCBs, 49 have $\delta_{th} < 3\%$, 38 have $3\% \leq \delta_{th} < 4\%$ and 13 have $\delta_{th} \geq 4\%$.

After the mapping and selection of the raw material 42 THGEMs have been produced. Individual foil labels and orientation marks allow to use the database of the local thickness values in the later stages of the THGEM production, test and operation. The mechanical drilling of the THGEM holes has been performed by ELTOS S.p.A. (Arezzo, Italy) after the image transfer, development, etching and dry film stripping operations have been completed.

Up to 6 THGEMs can be drilled in parallel by using a multi-spindle machine (Posalux 6000-LZ). To guarantee a good quality of the hole surface no more than 1000 holes are drilled by a single tool. No electroplating process is applied and electrical continuity between top and bottom paths for the voltage bias is provided by copper rivets where needed.

A specific smoothing and cleaning procedure has been developed [5] and is applied to the 42 THGEMs in the Trieste COMPASS Laboratory: it consists in a long, careful polishing operation using fine-grain pumice powder and a manually guided vibration-rotation machine, followed by cleaning using high pressure water and ultrasonic bath in a basic solution (Sonica PCB detergent with PH=11), distilled water rinsing and drying by blowing nitrogen and storing in a vacuum tank or nitrogen flushed box.

For the quality assessment tests the THGEM is mounted in a box equipped with a pad-segmented anode and a wire plane biased to provide a proper drift field; it is flushed with an Ar/CO_2 70/30 gas mixture at 10 l/h flow and connected to a remotely controlled power supply programmed to provide from each electrode current monitoring and discharge counting. Discharges are defined as monitored current values larger than 50 nA. The electrical stability test program increases the bias voltage in steps of 10 V until a few discharges are registered; the bias voltage is thus lowered and the test cycle restarts. Sparks are sometimes associated to discharges and allow to visually identify the location of the discharge. Simultaneous discharges often appear in neighboring sectors and also in distant sectors and even in different test boxes, suggesting cosmic shower triggered events. A discharge rate below 1 event per hour is considered acceptable and the maximum stable voltage is defined accordingly. If all sectors of the THGEM have a maximum stable voltage larger than 1200 V (gain $\approx 3\times$ the operational gain) then the THGEM is validated for electrical stability. In Fig. 3, the rate of discharges for a typical validated THGEM (n. 429), and for a rejected one (n. 343) are plotted as function of the bias voltage. The line corresponding to the typical discharge rate for THGEMs at the validation threshold limit is shown in gray. The failure rate for this validation stage is about 30% but reapplying the polishing and cleaning procedure allows to recover some of the THGEMs which fail the first electrical stability test.

The gain uniformity of each THGEM has been measured for the first produced pieces by using a ^{55}Fe X-ray source sequentially positioned in 72 points and collecting the signal amplitude spectra from the the set of anodic pads (covering about 20 cm^2) in front of the source position via a CREMAT CR-110 preamplifier, a CANBERRA AFT Research Amplifier model 2025 and an AMPTEK 8000A MCA. Typical standard deviation values range from 7% to 10% thanks to the strict thickness selection criteria applied for the raw THGEM PCBs.

For all THGEMs except the first ones a new procedure has been implemented for the gain uniformity measurement: the X-ray source is an AMPTEK Mini-X with Au target and a Cu filter, providing uniform illumination over

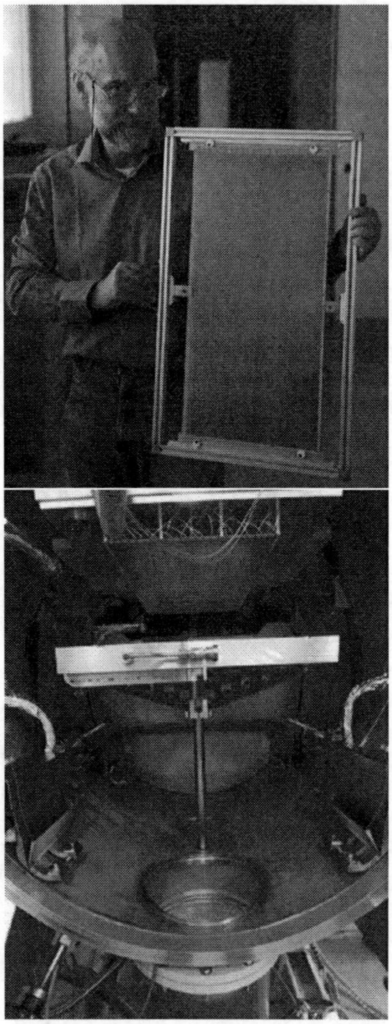

Figure 4. A THGEM in its handling frame (top) and inside the CsI deposition plant (bottom).

the entire THGEM area at a rate of 5 kHz cm^{-2}, the readout being performed by the RD51 Scalable Readout System (SRS) based on the APV-25 analogue readout chip. This procedure is about 10 times faster and provides a detailed characterization of the local gain variations.

The validated and characterized THGEMs are transported to CERN for the Ni-Au coating process. A layer of about 5 μm thickness of chemically deposited nickel and about 0.5 μm chemically deposited gold are used as protection against copper oxidation and as substrate for the photoconverting CsI layer.

The procedure developed by the RD26 Collaboration for the photocathode production [6] is followed, whenever possible, for the CsI coating of the THGEMs faces to be used as reflective photocathodes. After cleaning and preparation the THGEM to be coated is mounted in a dedicated gas tight box with a two-pins electrical feed-through connection: the electrodes of the top layer of the THGEM are connected together via one pin and the electrodes of the bottom layer are connected via the other pin. The box is flushed with clean gas and used for THGEM storage but thanks to the electrical connections the THGEM can be biased and its electrical strength checked before and after the coating. The 300 nm thick CsI layer is deposited by evaporation using the dedicated CERN plant, with the typical conditions previously used for COMPASS and ALICE photocathode production (vavuum level around 10^{-6} mbar, T $\approx$ 60 °C). After the CsI evaporation a measurement of photocurrent is performed by illuminating with VUV light from a D_2 lamp a spot of about 1 cm^2 and recording the ratio between the THGEM photocurrent and the current from a reference PMT sensitive in the UV; the measurement is performed at 60 different points to map the response uniformity and it is repeated after lowering the temperature to $\approx$ 25 °C, before extracting the coated THGEM from the evaporation plant.

3 The Micromegas and the other components

COMPASS Micromegas are produced at CERN using the bulk technology (standard photolithographic lamination technique) to fix a stainless steel woven square micromesh (18 μm diameter wire, 63 μm pitch, tensioned at 15 N) onto an anode PCB specifically designed for COMPASS RICH-1 Micromegas. The anode PCB is 3.2 mm thick and has four copper layers: the anode layer is segmented in square pads of 8.0 mm pitch (with 0.5 mm inter-pad distance and pads of 7.5×7.5 mm^2 Cu with 0.3 mm radius at the corners); a second layer, with the same segmentation, sits inside the PCB at 70 μm from the anode, providing a capacitive coupling of $\approx$ 40 pF between each anode pad an its facing (readout) pad, a value about 10 times larger than the capacitance between an anode pad and the micro-mesh; two layers are needed to route the paths from each readout pad to the corresponding pin of the connector for the front-end electronics and from each anode pad (trough a hole at the center of the readout pad) to the pin of the connector distributing the positive high voltage bias. Individual resistors of 470 M Ohm are separating each anode pad from the high voltage power supply, to reduce the effects of a local discharge and let the non discharging pads unaffected by the discharge and recharge processes. The uniformity of the Micromegas amplification gap (128 μm nominal distance) is guaranteed by a square array of cylindrical micro-pillars with 300 μm diameter and 2 mm pitch (16 per pad) obtained from the photo-imageable polyimide film. A 5.5 mm wide coverlay border surrounding the detector's active area helps holding the micromesh and 24 circular (6 mm diameter) coverlay spots are prepared as bases for the THGEM fixation pillars. At the end of the production process the Micormegas is cut at its final 283×586 mm^2 dimension, readout connectors and bias resistor connectors are mounted and two identical Micromegas are glued onto the detector holder, side by side. The detector holder consists of an aluminum frame machined (by electro-erosion) from a 30 mm thick plate to provide stiffness, planarity and specific grooves for the insertion of the front-end boards, glued to a frame made of isolating material to avoid parasitic capacitance and machined to host the Micormegas. Strong Micromegas grounding is provided by embedding in conductive glue a large surface (5×260 mm^2) of each micromesh, on the side which is not in the active area of the final detector.

The gain uniformity of the Micromegas has been studied using a prototype detector, an Ar/CO_2 70/30 gas mixture and a ^{55}Fe source: for an average gain of about 3000 the the local gain variations typically show a standard deviation of 5%.

The micromesh, being the most extended unique electrode in the photon detector, is always kept at ground potential and in case of a discharge in the Micromegas the potential of the anode pad where the discharge takes place completely drops. To study the influence of such events on the neighboring pads, discharges have been systematically generated on a specific pad (by biasing it with an anomalously large potential) and the adjacent pads have been biased using for each pad a different high voltage channel from the power supply. The probability of correlated (within ± 100 ns) discharges turned out to be negligible (no discharge cross-talk) when the adjacent pad was operated at normal voltage values, while raising the bias voltage of a neighboring pad led to more and more frequent correlated discharges, with a probability up to 0.7 when the same anomalous bias was applied, with a large majority of correlated events having a time difference $\leq$ 10 ns.

After gluing the 48 THGEM pillars (24 for each Micromegas, M3 threaded, made of PEEK,) on the prepared coverlay bases and the high voltage distribution PCBs on the detector frame the hybrid PD is assembled: the gap between the Micromesh and the closer THGEM is 5 mm wide while the distance between the THGEMs is 3 mm. The THGEMs are mounted in a staggered holes configuration (by a $\approx$ 462 μm displacement with respect to the aligned holes configuration) in order to favor the spread of the avalanche charges.

The electrostatic field above the photocathode is defined by two planes of wires (100 μm diameter, Cu-Be alloy with a flash of Ni and Au), 60 cm long, with a pitch of 4 mm and guard wires of 200 μm diameter at the edges. The wires are manually soldered on half of a large (645 × 1400 mm^2) mechanical frame which is also used, together with a similar size spacing frame, as holder of the existing Multi-Anode PMT based photon detectors. Special electrodes, embedded in the isolating material protections of the chamber frames are used to shape the field at the edges of the active volume, preventing large field values on any of the electrodes in the gas volume: the optimization of the final geometry required detailed numerical calculations of the fields taking into account all electrodes and their real shapes.

The preparation and mounting of the wire and spacing frames too requires several different steps since the overall mechanical structure of the detector is rather complex (see Fig. 1, right) and the procedures for assembling and installation of the new PDs are quite elaborated.

The high voltage distribution system provides independent bias values to each half of a THGEM face (6 sectors are connected to one HV channel via a resistive splitter with protection diodes to prevent HV drop cross-talk between sectors). Specific switch-on and safe-mode procedures prevent anomalous bias values to be applied to the THGEMs. The effective values of the applied bias will be continuously corrected to counteract the effects of environmental variables (p,T) variation: a gain stability in the order of $\pm10\%$ is expected, thanks to these corrections.

The signal from the readout pads are collected by front-end electronic cards [7] designed for COMPASS RICH-1: they host four APV25-S1 chips, each reading 108 pads. Three front-end cards are connected to a 10-bit flash ADC digitizer board equipped with a FPGA performing on-line zero suppression. A cooling system using under-pressure water flow assures efficient removal of the heat produced by the readout.

The new hybrid PDs will be tested and installed on COMPASS RICH-1 before the 2016 run.

4 Conclusions

COMPASS RICH-1 will operate during the physics run of 2016 in an upgraded version with four new Photon Detectors based on hybrid THGEM + Micromegas technology. The expected high performance of the new PDs will allow to meet the strict requirements of the new COMPASS physics measurements. The production of all elements of the new PDs is almost complete and the procedures for test and validation of the components have been defined and applied. The recent developments of the field confirm the great potentialities of MPGD-based PDs and offer promising perspectives for their future applications.

References

[1] The COMPASS Collaboration, CERN/SPSC/2010-014, SPSC-P-340, May 17, 2010; CERN/SPSC/2010-022, SPSC-M-772, September 3, 2010.

[2] E.Albrecht et al., Nucl. Instr. and Meth. A 502 (2003) 112; E.Albrecht et al., Nucl. Instr. and Meth. A 502 (2003) 236; E.Albrecht et al., Nucl. Instr. and Meth. A 502 (2003) 266; E.Albrecht et al., Nucl. Instr. and Meth. A 510 (2003) 262; E.Albrecht et al., Nucl. Instr. and Meth. A 518 (2004) 586; M.Alekseev et al., Nucl. Instr. and Meth. A 553 (2005) 53; S.Costa et al., Nucl. Instr. and Meth. A 553 (2005) 135; E.Albrecht et al., Nucl. Instr. and Meth. A 553 (2005) 215; P.Abbon et al., Czech. J. Phys. 56, Suppl. F (2006) 307; P.Abbon et al., Czech. J. Phys. 56, Suppl. F (2006) 315; P.Abbon et al., Czech. J. Phys. 56 Suppl. F (2006) 323; P.Abbon et al., Nucl. Instr. and Meth. A 567 (2007) 114; P.Abbon et al., Nucl. Instr. and Meth. A 572 (2007) 419; P.Abbon et al., Nucl. Instr. and Meth. A 580 (2007) 906; P.Abbon et al., Nucl. Instr. and Meth. A 581 (2007) 419; P.Abbon et al., Nucl. Instr. and Meth. A 587 (2008) 371; P.Abbon et al., Nucl. Instr. and Meth. A 595 (2008) 23; P.Abbon et al., Nucl. Instr. and Meth. A 595 (2008) 177; M.Alexeev et al., Nucl. Instr. and Meth. A 595 (2008) 194; P.Abbon et al., Nucl. Instr. and Meth. A 595 (2008) 204; S.Dalla Torre et al., Nucl. Instr. and Meth. A 595 (2008) 220; P.Abbon et al., Nucl. Instr. and Meth. A 595 (2008) 233; P.Abbon et al., Nucl. Instr. and Meth. A 616 (2010) 21; P.Abbon et al., Nucl. Instr. and Meth. A 623 (2010) 330; P.Abbon et al., Nucl. Instr. and Meth. A 631 (2011) 26; P.Abbon et al., Nucl. Instr. and Meth. A 639 (2011) 15; M.Alexeev et al., Nucl. Instr. and Meth. A 639 (2011) 219; S.Dalla Torre et al., Nucl. Instr. and Meth. A 639 (2011) 271; F.Tessarotto et al., JINST 9 (2014) C09011; M.Alexeev et al., Nucl. Instr. and Meth. A 766 (2014) 208.

[3] M.Alexeev et al., JINST 9 (2014) P01006; M.Alexeev et al., Nucl. Instr. and Meth. A 766 (2014) 199.

[4] M.Alexeev et al., Nucl. Instr. and Meth. A 610 (2009) 174; M.Alexeev et al., Nucl. Instr. and Meth. A 617 (2010) 396; M.Alexeev et al., Nucl. Instr. and Meth. A 623 (2010) 129; M.Alexeev et al., JINST 5 (2010) P03009; M.Alexeev et al., Nucl. Instr. and Meth. A 639 (2011) 130; M.Alexeev et al., JINST 7 (2012) C02014; M.Alexeev et al., Nucl. Instr. and Meth. A 695 (2012) 159; M.Alexeev et al., Physics Procedia 37 (2012) 781; M.Alexeev et al., Nucl. Instr. and Meth. A 732 (2013) 264; M.Alexeev et al., JINST 9 (2014) C03046; M.Alexeev et al., JINST 9 (2014) C09017; M.Alexeev et al., JINST 10 (2014) P03026;

[5] M.Alexeev et al., Nucl. Instr. and Meth. A 766 (2014) 133.

[6] A.Braem et al., Nucl. Instr. and Meth. A 502 (2003) 205.

[7] P.Abbon et al., Nucl. Instr. and Meth. A 567 (2006) 104.

Construction and Performance Studies of Large Resistive Micromegas Quadruplets

E. Farina[1,2,a], P. Iengo[1], M. Bianco[1], O. Sidiropoulou[1,3], F. Kuger[1,3], G. Sekhniaidze[4], M. Vergain[1], J. Wotschack[1], H. Danielsson[1], J. Degrange[1], R. De Oliveira[1], M. Schott[5], Tai-Hua Lin[5], C. Valderanis[5], and A. Düdder[5]
on behalf of the ATLAS Muon Collaboration

[1]*CERN*
[2]*Universita e INFN, Pavia (IT)*
[3]*Bayerische Julius Max. Universitaet Wuerzburg (DE)*
[4]*Universita e INFN, Napoli (IT)*
[5]*Johannes-Gutenberg-Universitaet Mainz (DE)*

Abstract. In view of the use of Micromegas detectors for the upgrade of the ATLAS muon system, two detector quadruplets with an area of 0.3 m^2 per plane serving as prototypes for future ATLAS chambers have been constructed. They are based on the resistive-strip technology and thus spark tolerant. The detectors were built in a modular way. The quadruplets consist of two double-sided readout panels and three support (or drift) panels equipped with the micromesh and the drift electrode. The panels are bolted together such that the detector can be opened and cleaned, if required. Two of the readout planes are equipped with readout strips inclined by 1.5 degree. In this talk, we present the results of detailed performance studies based on X-Ray and cosmic ray measurements as well as measurements with 855 MeV electrons at the MAMI accelerator. In particular, results on reconstruction efficiencies, track resolution and gain homogeneity is presented.

1 Introduction

Micromegas detectors are parallel plate gaseous detectors composed by two distinct electric field regions, the drift and amplification gap, separated by a metallic mesh. The first large-area (0.3m ^{2}) Micromegas quadruplet has been realized following the design foreseen for the next ATLAS [1] Muon Upgrade, the so-called *New Small Wheel Upgrade* consisting in the replacement of the innermost muon station [2]. In the layout adopted for the construction, the readout strips are covered by a Kapton® foil carrying a pattern of carbon resistive strips[1], in order to reduce the effects of sparks. The mesh is integrated directly in the drift panel (it is not embedded in the pillars) allowing for the opening and cleaning of the readout plane. In this work the construction and the characterization of one of the two quadruplets, including efficiency, gain uniformity, alignment and resolution studies, will be presented.

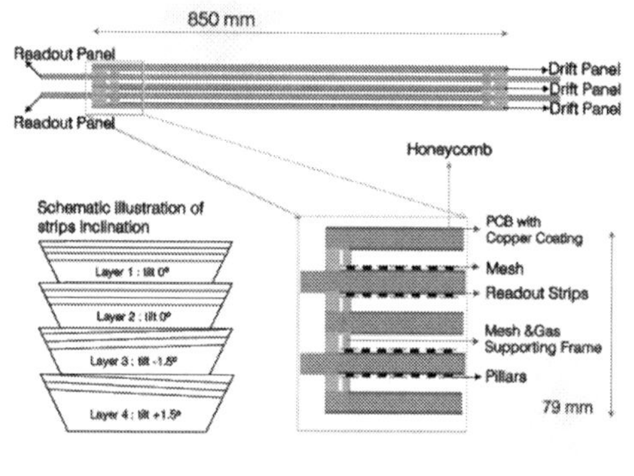

Figure 1: Internal structure of MMSW quadruplet

2 Detector design

The internal structure of the quadruplet is shown in Fig. 1. The detector is composed of three drift and two double sided readout panels. Each trapezoidal panel is formed by two FR4 printed circuit boards (PCBs) and an Al honeycomb stiffener in between, surrounded by an aluminium profile. Each readout PCB hosts 1024 readout strips with a pitch of 415 μm. On two readout panels the readout strips are parallel to the base of the trapezoid, to measure the precision coordinate (called η coordinate), while on the other two they are rotated by ±1.5° with respect to the base of the detector (stereo layers). The information provided by the stereo layers allows for the reconstruction of both coordinates (η and ϕ). A resolution of about 100 μm for the precise coordinate and 2.5 mm for the second one are expected.

[a]e-mail: edoardo.maria.farina@cern.ch
[1]Thanks to A. Ochi (Kobe University) for the resistive Kapton foils.

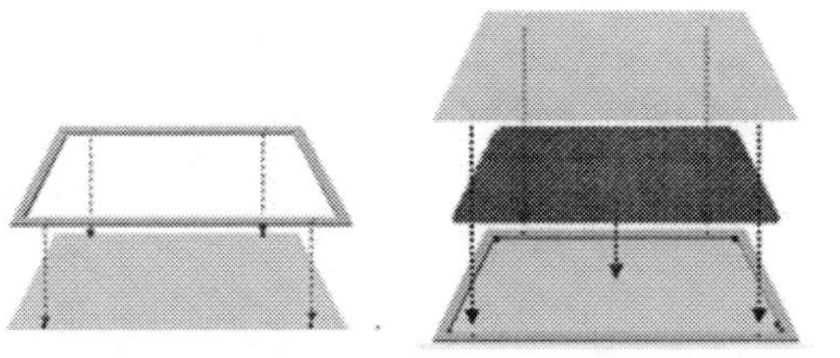

(a) Step 1 (b) Step 2

Figure 2: Schematic of the drift panel construction

3 Detector Construction

In order to reach the required panel planarity, two dedicated stiff-backs, composed of an aluminium honeycomb core between two skins of carbon fibre and coated by a 0.5 mm thick gel-coat layer, have been realized. The panel construction is almost identical for drift and readout panels. In a first step a PCB is sucked face down on one stiff-back, in order to transfer the planarity of the stiff-back to the PCB, then the aluminium frame is glued on the PCB. In a second step the honeycomb is put on the PCB and glued. Finally the second PCB, previously prepared and sucked face down on the second stiff-back, is glued on the honeycomb in order to close the panel. After the glue curing, an aluminium frame, called *mesh frame*, is then assembled on the drift panels, in order to host the pre-stretched metallic mesh. More details about the construction and assembly of the quadruplet can be found in [3], [4].

4 Characterization of the detector

The characterization of the quadruplet has been realized in the RD51 laboratory at CERN by using both X-rays and cosmic rays. A dedicated test stand has been built to host large area detectors up to 2.5 m^2. The characterization of the detector includes efficiency, gain uniformity and alignment studies. Resolution studies have been moreover performed at the MAMI accelerator in Mainz.

4.1 Efficiency studies

The efficiency of each layer has been evaluated as a function of the amplification voltage (drift voltage set at 300 V) by using cosmic rays. A specific tracking algorithm has been developed in order to track muons without requiring any external tracking system, but only using the information provided by the four layers of the detector. In Fig. 3 the efficiency curves for the four layers as a function of the HV are shown. All layers reach the same efficiency, higher than 98%, at 580 V.
The difference in the efficiencies are related to the different gains for the four planes. These in turn are coming

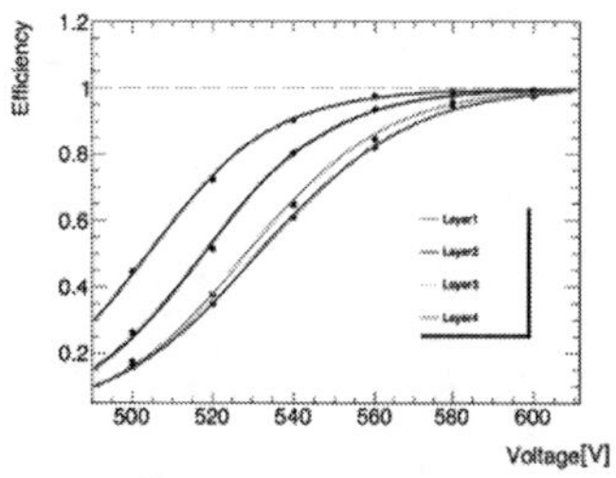

Figure 3: Efficiency curves for the four layers as a function of the voltage

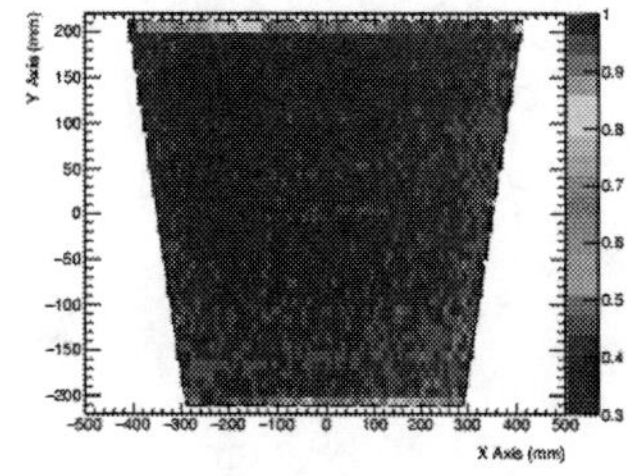

Figure 4: 2D distribution of the efficiency for a stereo layer. Note the inefficient corners due to the absence of the readout strips

from variations of the pillar height and the mesh tension and are consistent with the measured values. It should be noted that most of the observed variations can and will be avoided in the future. However, even variations as large as observed in the MMSW gains are acceptable for the NSW detectors. Bi-dimensional distributions of the efficiency have been realized in order to spot local inefficiencies of the panels and to study the efficiency homogeneity within a layer at lower voltage with respect to the nominal value. As an example in Fig. 4 the 2D map of the efficiency at 580 V for a stereo layer is shown. In particular it is possible to notice two small low efficient areas at the corners: in these areas there are no readout strips due to the rotation of the strips in the stereo layers. For all the layers an excellent homogeneity of the efficiency is reached at 580 V (differences within 5% with respect to the average value).

4.2 Gain uniformity

The uniformity of the gain has been measured both with X-rays and cosmic rays. In the first case an Ag X-rays gun equipped with a 2 mm collimator (5 degrees of opening angle) has been used. The X-rays gun has been operated at 50 kV and a current of 50 μA. The X-ray spectrum can be found on the Amptek website (http://amptek.com/product/mini-x-ray-tube/#5). It extends up to 50 keV, with peaks at 22.5 and 25 keV. However it should be noted that almost all of the lower-energy photons are absorbed in the first detector panel such that the average photon energy detected lies somewhere between 30 keV and 50 keV. A scan of the

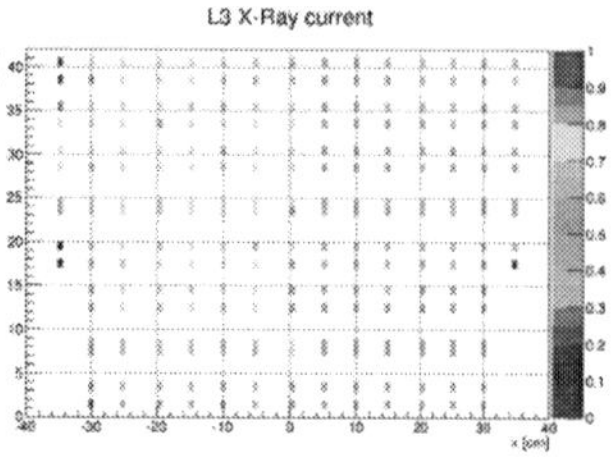

Figure 5: 2D map of the current for 228 points irradiated with the X-ray gun normalized to the maximum value

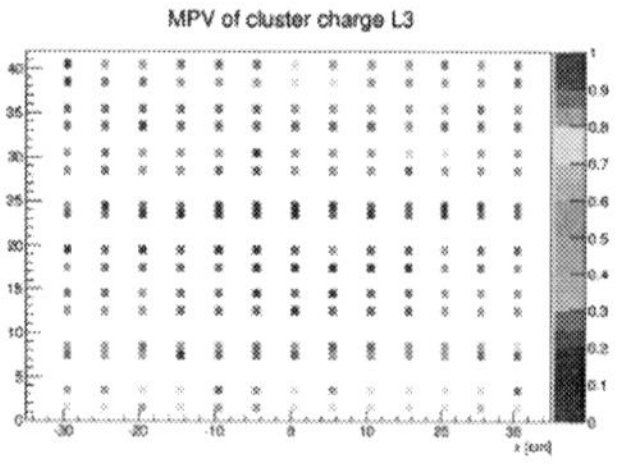

Figure 6: 2D distribution of the mean cluster charge normalized to the maximum value

whole surface of the detector has been realized by moving the gun in 228 different points on the detector. The relative gain, as a function of the position of the gun, has been measured by recording the current on the resistive strips for each point. A 2D map of the current, normalized to the maximum value, is reported in Fig. 5. Despite the large area of the layers (about 0.3 m^2) this study shows an excellent homogeneity of the amplification within the layers.

An equivalent study has been realized by using cosmic rays. This time the detector has been equipped with the readout electronics (APV 25) and the data obtained from the DAQ system have been analysed. In order to study the relative gain, the cluster charge distribution as a function of the position on the chamber has been studied. In more details, the cluster charge distribution of hits reconstructed around each of the same 228 points already used for the X-rays analysis has been fitted with a Landau distribution. The most-probable-value (MVP) of the fit has been then plotted in the corresponding position in a 2D map of the layer. An example of a 2D map of the cluster charge related to a stereo layer and normalized to the maximum value is reported in Fig. 6. Also in this case a very good gain homogeneity has been found. The results from the two methods, X-rays and cosmic rays, have been compared showing a good agreement for all the four layers.

4.3 Alignment studies

The alignment of the PCB skins on the two sides of the readout panel has been measured with a laser interferometer. Some of the strips have been routed up to the edge of the PCBs. The relative position of these strips with respect to a reference pin has been evaluated on both sides of the

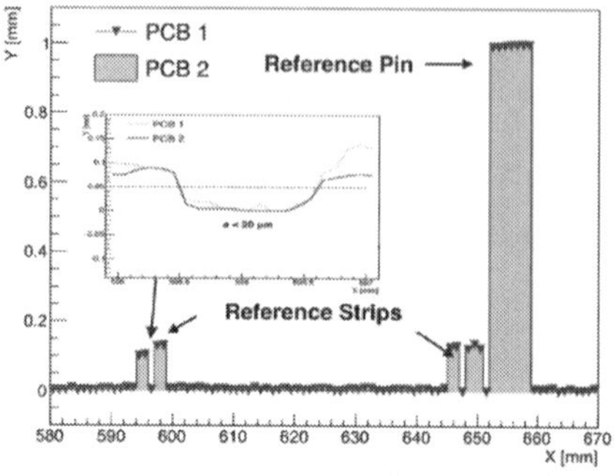

Figure 7: Results of the measurement of the distance between strips and the reference pin on both sides of the R.O. panel. In the insert a zoom of a group of strips

readout panel and compared. In Fig. 7 the measurements obtained on one layer (blue) and on the other (yellow) are reported. An upper limit on the misalignment of the two PCBs of 20 μm has been extracted.

The alignment of the two PCBs has been studied also after the assembly of the quadruplet by using the X-rays gun. The detector has been equipped with readout electronics. In order to give an estimation of the alignment of the planes, the profiles of the strips irradiated by the X-rays on the two sides of the panel have been compared. For each point, two different measurements have been taken, rotating the position of the box containing the gun by 180 degrees, in order to get rid of a possible effect due to the inclination of the gun inside the box. For each measurement the strip profile has been fitted with a Gaussian. The mean value between the two measurements (same position, different orientation of the gun) for each point has been calculated and compared with the other layer. A misalignment between the two layers of about 19 μm has been found, compatible with the result obtained using the laser interferometer.

5 Performance study at MAMI accelerator

The spatial resolution of the detector has been studied in a test beam campaign at the MAMI accelerator in Mainz. The accelerator delivers an almost continuous electron beam at 855 MeV. The setup was composed of the MMSW detector followed at 30 cm distance by a 10×10 cm^2 Micromegas chamber with 70 μm spatial resolution. The distribution of the residuals, consisting of the difference between the cluster position on the two η layers of the MMSW has been fitted as reported in Fig. 8. The contribution of the multiple scattering and the beam divergence has been accounted for and a comprehensive resolution of 90 μm for the coordinate perpendicular to the base of the trapezoid has been found. The contribution of multiple scattering and beam divergence has been determined by comparing the position measurement in the η coordinate between MMSW and the reference chamber. The second coordinate has been measured combining the information

https://doi.org/10.1051/epjconf/201817401005

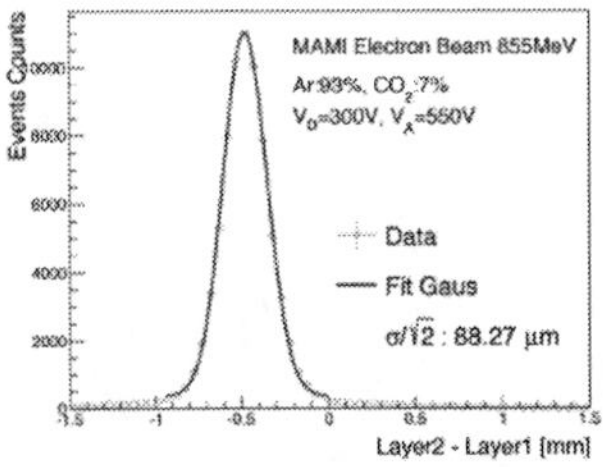

Figure 8: Difference between cluster position measured on the two sides of a readout panel

of the two MMSW stereo layers and comparing it with the coordinate measurement provided by the small tracking chamber. The residuals and the fit are shown in Fig. 9. Different from the η case, given the large distance between MMSW and the reference detector, multiple scattering and beam divergence contribute for about 1.8 mm, leading to a second coordinate resolution of MMSW of 2.3 mm. These

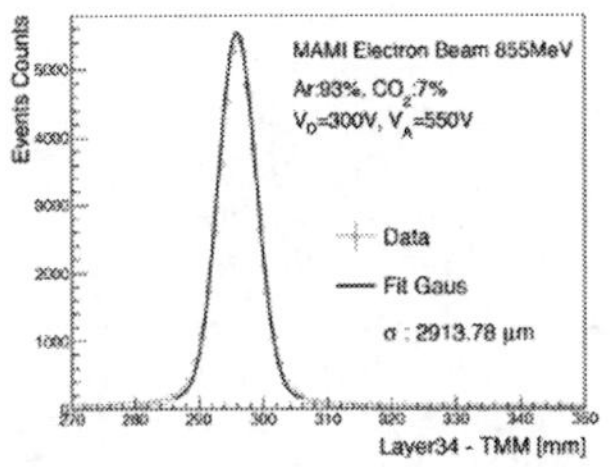

Figure 9: Difference between the cluster position calculated combining the stereo layer information and the measurement provided by the tracking chamber

results are in line with the expectations [5] and meet the ATLAS requirement for the NSW upgrade.

6 Conclusion

The construction of the first Micromegas quadruplet, realized according to the general design foreseen for the ATLAS Muon upgrade has been described. A detailed characterization of the prototype has been performed by using cosmic rays and X-rays. All the layers show efficiency higher than 98% at 580 V. The 2D efficiency maps has not shown any local inefficiency of the panels. The alignment of the readout layers has been measured during the construction using the laser interferometer and after with X-rays showing an alignment better than 20 µm.

The spatial resolution of the chamber has been studied at the MAMI accelerator in Mainz providing a resolution of 90 µm for the most accurate coordinate (η) and 2.3 mm for the second coordinate (ϕ). These results meet the ATLAS requirements.

References

[1] ATLAS Collaboration, *The ATLAS Experiment at the CERN Large Hadron Collider*, JINST 3 S08003 (2008)

[2] ATLAS Collaboration, *New Small Wheel Technical Design Report*, ATLAS-TDR-020, June 2013, http://cdsweb.cern.ch/record/1552862

[3] P. Iengo, et al., *Construction of two large-size four plane micromegas detectors* [arXiv:1511.03884]

[4] P. Iengo, et al., *Construction of a large-size four plane micromegas detector* PoS 058 TIPP2014 2014

[5] Ntekas, Konstantinos, *Study of the performance of the Micromegas chambers for the ATLAS muon spectrometer upgrade.* ATL-MUON-PROC-2014-011, http://cdsweb.cern.ch/record/1971986

EPJ Web of Conferences **174**, 01006 (2018) https://doi.org/10.1051/epjconf/201817401006
MPGD 2015

Status of the R&D activities for the upgrade of the ALICE TPC

Alexander Deisting[1,a] for the ALICE collaboration

[1] *GSI Helmholtzzentrum für Schwerionenforschung GmbH, Planckstraße 1, 64291 Darmstadt*

Abstract.
After the Long Shutdown 2 (LS2) the LHC will provide lead–lead collisions at interaction rates as high as 50 kHz. In order to cope with such conditions the ALICE Time Projection Chamber (TPC) needs to be upgraded.
After the upgrade the TPC will run in a continuous mode, without any degradation of the momentum and $\mathrm{d}E/\mathrm{d}x$ resolution compared to the performance of the present TPC. Since readout by multi-wire proportional chambers is no longer feasible with these requirements, new technologies have to be employed. In the new readout chambers the electron amplification is provided by a stack of four Gas Electron Multiplier (GEM) foils. Here foils with a standard hole pitch of 140 µm as well as large pitch foils (280 µm) are used. Their high voltage settings and orientation have been optimised to provide an energy resolution of $\sigma_E/E \leq 12\%$ at the photopeak of ^{55}Fe. At the same settings the Ion BackFlow into the drift volume is less than 1% of the effective number of ions produced during gas amplification and the primary ionisations. This is necessary to prevent the accumulation of space charge, which eventually will distort the field in the drift volume. To ensure stable operation at the high loads during LHC run 3 the chambers have to be robust against discharges, too. With the selected configuration in a quadruple GEM stack the discharge probability is kept at the level of 10^{-12} discharges per incoming hadron.
An overview of the ALICE TPC upgrade activities will be given in these proceedings and the optimised settings foreseen for the GEM stacks of the future readout chambers are introduced. Furthermore the outcome of two beam time campaigns at SPS and PS (at CERN) in the end of 2014 is shown. At this campaigns the stability against discharges and the $\mathrm{d}E/\mathrm{d}x$ performance of a full size readout chamber prototype was tested. In addition it is reported on charging-up studies of 4GEM stacks and on tests of electromagnetic sagging of large GEM foils.

1 The ALICE TPC

ALICE (A Large Ion Collider Experiment) is located at the Large Hadron Collider (LHC) at CERN and was specially designed to examine heavy-ion (lead–lead) collisions. In its central barrel it features a Time Projection Chamber (TPC), which is currently the largest of its kind. The TPC is a cylindrical vessel with a length of 5 m and an outer (inner) diameter of roughly 5.5 m (1.1 m) (see figure 1). A central drift cathode divides the TPC into two drift volumes with a maximal drift length of 2.5 m each. Opposite to the drift cathode, on the endcaps, the readout chambers are mounted. These chambers are multi-wire proportional chambers, employing a gating grid to prevent ions from escaping the

[a] e-mail: alexander.deisting@cern.ch

gas amplification region. This is necessary to avoid the build-up of space charges in the drift volume, which would distort the field lines and thereby the tracks of particles measured in the TPC. Since the gating grid has to be closed for some time to collect the ions efficiently, it limits the maximal readout rate to about 3.3 kHz.

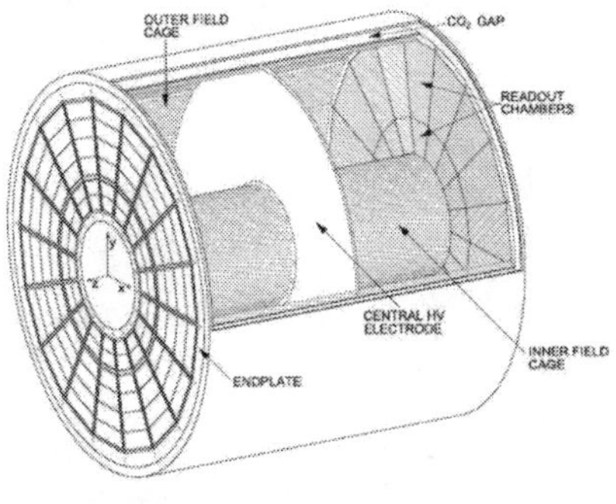

Figure 1. Schematic of the ALICE TPC [1].

2 ALICE TPC upgrade

2.1 Requirements on new readout chambers

During the long shutdown 2, the LHC will be upgraded to provide rates of up to 50 kHz in Pb–Pb collisions, which will lead to (on average) 5 events piled up in the TPC. Hence a gated readout is no longer feasible and new readout chambers are needed, which allow a continuous readout. In addition to this, the new chambers should preserve the momentum and dE/dx resolution of the current chambers. This translates into an energy resolution of $\sigma_E/E_{^{55}Fe} \leq 12\%$ at the ^{55}Fe photopeak. Furthermore the new chambers must have an Ion BackFlow (IBF) of less than 1% in order to keep the space charges in the TPC at a tolerable level and they must operate stably at LHC Run 3 conditions.

2.2 Baseline settings

A system using Gas Electron Multiplier (GEM) foils was found promising to fulfil all these requirements. Extensive R&D was performed in order to find the optimal settings for the future readout chambers of the ALICE TPC. Eventually the so-called baseline settings were fixed. In the following sections (section 2.3 to 2.5) the different requirements will be revisited and the corresponding results of our R&D are given. In addition the conclusions of charging up studies with a quadruple GEM stack are shown in section 2.6 and the outcome of measurements of sagging due to the electrostatic attraction of large GEMs is given in section 2.7.

2.3 Ion back flow and energy resolution

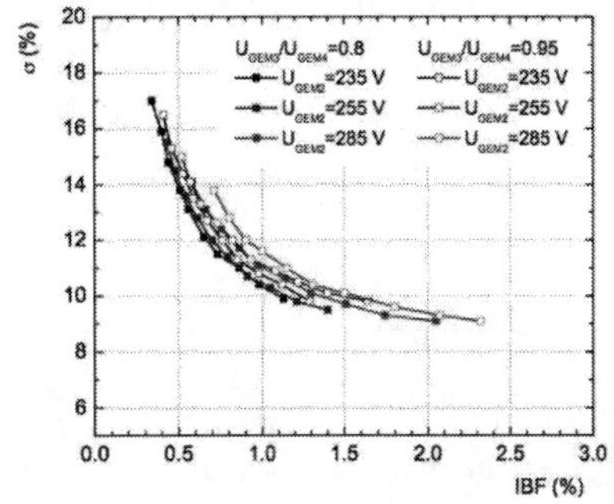

Figure 2. Energy resolution at the ^{55}Fe photopeak versus the IBF. The data points were measured with small detectors equipped with quadruple GEM stacks (GEM configuration: S-LP-LP-S; Gas: Ne-CO_2-N_2 (90-10-5)) at a gas gain of 2000 – the scan through different settings was done by varying the voltage on GEM1 from 225 V to 315 V, while adjusting the gain by changing the voltage on GEM3 and GEM4. Furthermore every second foil is rotated by 90° relative to the other foils in order to avoid interference effects of the hole pattern [1].

In Ne-CO_2-N_2 (90-10-5) different configurations of quadruple GEM stacks were tested. The parameters varied were the High Voltage (HV) settings of the GEMs and the kind of GEM used at the different positions in the GEM stack. For example GEMs with standard hole pitch of 140 μm and large pitch foils with 280 μm pitch were tested as well as foils with other hole pitches. Using high ionising sources (like X-ray guns) for the IBF measurements and ^{55}Fe for the energy resolution measurements it was found that optimising the IBF and the energy

resolution are competing effects. This is illustrated in figure 2.
By these measurements a configuration using two Standard (S) and two Large Pitch (LP) foils in the order S-LP-LP-S with 2 mm induction gap and the same gap between the GEMs could be identified, which fulfills all the requirements. This setting provides an IBF of 0.6 %-0.7 % as well as an energy resolution of 11 %-12 %. The exact HV settings are listed in table 1. Here a low transfer field 3 (E_{T3}) ensures that most of ions produced in GEM4 aren't extracted from the GEM holes they were produced in.

2.4 Stability tests

The stability against discharges of the future readout chambers was tested with a full size prototype of an Inner ReadOut Chamber (IROC) equipped with a quadruple GEM stack. Particle showers were produced with a pion beam from the CERN SPS pounding on 40 cm of iron. After the Fe bricks and in the particle shower the detector was put with the readout plane perpendicular to the beam axis thus facing the particle shower. With the particle rates at the SPS it was possible to improve the upper limit on the discharge probability measured previously with small detectors. For the tested HV setting a discharge probability of $6(4) \times 10^{-12}$ discharges per incoming hadron was measured [2]. For comparison: The expected number of particles traversing a GEM stack in the future ALICE TPC during one month of 50 kHz Pb–Pb data taking is 7×10^{11} particles. Hence reasonable few, non damaging discharges are expected at the rates in the LHC Run 3.

2.5 d*E*/dx measurement

To study the $\mathrm{d}E/\mathrm{d}x$ performance the same IROC as used for the stability tests was tested at the CERN PS with electrons and pions. It was found that the separation power between these particle species – based on the energy loss inside the detector – is similar to the separation power of the current readout chambers of the ALICE TPC. In another paper in these proceedings this topic is discussed in detail [3].

2.6 Charging up studies

As operated at a collider experiment the new readout chambers will have to change state from a "safe" state without gas amplification to a "ready" state, where the chamber HV is fully applied. As soon as the HV is on, charging-up processes of the GEM material take place which affect the gain until it eventually stabilises.

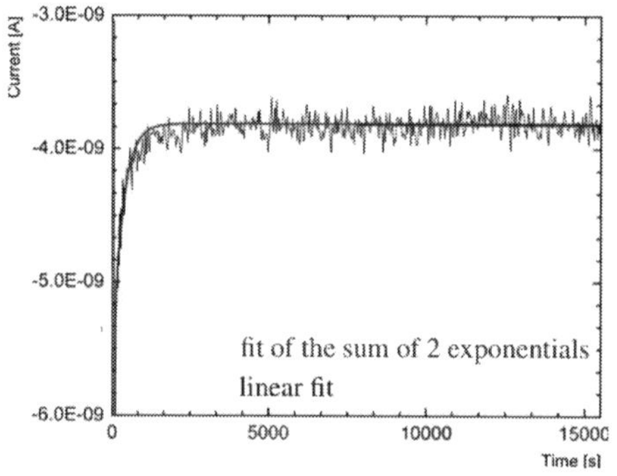

Figure 3. Example plot showing the time dependence of the current. The decreasing of the absolut current value after the ramping up procedure can be described by the fit of the sum of two exponential functions with a constant offset [4].

To measure the time scales of this charging-up effects a small prototype with a quadruple GEM stack was irradiated with an ionising source while the current was measured on the pad plane. Figure 3 shows the result of such a measurement. The current while and after ramping can be described by:

$$I(\mathrm{t}) = I_0 - A_1 \exp(-\frac{\mathrm{t} - t_0}{t_1}) - A_2 \exp(-\frac{\mathrm{t} - t_0}{t_2})$$

where I_0 is the equilibrium value and t_1 and t_2 are the time constants for charging-up the HV elements (cables, etc) and for charging-up the foils, respectively. In the studies described in [4] it was found that $t_1 \sim 35$ s and $t_2 \sim 12$ min. Furthermore it could be shown, that these time constants do not depend on the time the detector stayed in the safe state before the ramping up of the HV. This – as well as the value of the time constants – shows that GEM based readout chambers can be operated at ALICE without delaying the start of data taking of the whole experiment. The charging-up is a slow enough process to follow and correct for it by means of online calibration.

2.7 Tests of electrostatic GEM foil sagging

With the upcoming mass production of GEM foils and GEM frames for the new readout chambers the following question came up: A how extensive support structure ("spacer grid") to the GEM frame is needed in order to prevent the foils from moving towards each other as soon as HV is applied to the GEM stack? To answer this, the foil sagging due to electrostatic attraction was studied.
For the sagging tests a framed, trapezoidal GEM foil (a = 87 cm, b = 73 cm, h = 40 cm – the largest foil foreseen for the TPC) was placed on the pad plane of a readout chamber prototype. Then a potential difference between the pad plane and the foil was applied and the sagging of the foil towards the pad plane was observed. This was repeated with support structures being added to the framed foil.
It was found that the foil sags the full 2 mm distance between the pad plane and the GEM in case of no supporting structure. If a bar was added in parallel to the parallel sides of the trapezoidal foil, then sagging of the full 2 mm was already prevented. With two crossed bars however no visible sagging was observed, which leads to the assumption that a moderate "spacer grid" will be necessary in the frames of the future GEM foils.

3 Conclusions

To allow for continuous operation the readout chambers of the ALICE TPC will be upgraded. The necessary R&D activities are almost finished – they led to GEM based readout chambers. A system consisting of a quadruple GEM stack employing S and LP foils (S-LP-LP-S) fulfills all requirements in Ne-CO_2-N_2 (90-10-5) while operated at the voltage settings given in table 1. The new chambers provide an IBF in the range of 0.6 %–0.7 % and a concurrent energy resolution $\sigma_E/E_{^{55}\mathrm{Fe}}$ of 11 %-12 %. Furthermore stability tests yield a discharge probability of $6(4) \times 10^{-12}$ discharges per incoming charged hadron, hence stable operation during LHC Run 3 is expected. The chamber's $\mathrm{d}E/\mathrm{d}x$ resolution is compatible with the separation power of the existing TPC and tests of the charging-up of 4GEM stacks yielded that this process is slow enough to follow it by means of online calibration. For the upcoming mass production of GEM foils and the corresponding GEM frames it has been shown that a suitable spacer grid is needed to prevent the foils from sagging due to electrostatic attraction.

Table 1. Baseline HV settings for the GEM based readout chambers of the upgraded ALICE TPC – the transfer fields are denoted by E_{Ti} and the induction field by E_{Ind}. The drift field in the TPC will be 400 V cm^{-1}.

ΔV_{GEM1}	ΔV_{GEM2}	ΔV_{GEM3}	ΔV_{GEM4}	E_{T1}	E_{T2}	E_{T3}	E_{Ind}
270 V	250 V	270 V	340 V	$\frac{4\,kV}{cm}$	$\frac{2\,kV}{cm}$	$\frac{0.1\,kV}{cm}$	$\frac{4\,kV}{cm}$

References

[1] ALICE Collaboration, *Technical Design report for the Upgrade of the ALICE Time Projection Chamber* (CERN-LHCC-2013-020)

[2] ALICE Collaboration, *Addendum to the Technical Design Report for the Upgrade of the ALICE Time Projection Chamber* (CERN-LHCC-2015-002)

[3] A. Mathis, *Study of the dE/dx resolution of a GEM Readout Chamber prototype for the upgrade of the ALICE TPC* (These proccedings)

[4] Y. Vetter, *Studies on charge-up effects and gain stability for the ALICE TPC upgrade with GEMs* (Bachelor thesis 2015, Universität Heidelberg)

Recent advances with a hybrid micro-pattern gas detector operated in low pressure H_2 and He, for AT-TPC applications

Marco Cortesi[1,a], Wolfgang Mittig[1], Daniel Bazin[1], Yassid Ayyad Limonge[1], Saul Beceiro-Novo[1], Rim Soussi Tanani[1], Michael Wolff[1], John Yurkon[1], and Andreas Stolz[1]

[1] *National Superconducting Cyclotron Laboratory, Michigan State University, East Lansing (MI) 48824, U.S.A*

Abstract. In view of a possible application as a charge-particle track readout for an Active-Target Time Projection Chamber (AT-TPC), the operational properties and performances of a hybrid Micro-Pattern Gaseous Detector (MPGD) were investigated in pure low-pressure Hydrogen (H_2) and Helium (He). The detector consists of a MICROMEsh GAseous Structure (MICROMEGAS) coupled to a two-cascade THick Gaseous Electron Multiplier (THGEM) as a pre-amplification stage. This study reports the effective gain dependence of the hybrid-MPGD at relevant pressure (in the range of 200-760 torr) for different detector arrangements. The results of this work are relevant in the field of avalanche mechanism in low-pressure, low-mass noble gases, in particularly for applications of MPGD end-cap readout for active-target Time Projection Chambers (TPC) in the field of nuclear physics and nuclear astrophysics.

1 Introduction

In the field of low-energy nuclear physics, Active-Target Time Projection Chambers (AT-TPCs) have been developed into unique devices to study reactions induced by short-live radioactive isotopes [1], and to break the limits imposed by conventional methods of investigation based on high-resolution spectrometer [2]. AT-TPCs provide several significant features, including 4π acceptance of the reaction products; high effective thickness; full detection efficiency and high sensitivity to very low energy events; and an event-by-event reconstruction in three dimensions.

In its basic design, the AT-TPC consists of a relative large, cylindrical volume of gas, surrounded by a field-shaping cage and terminated at one end by a position-sensitive end-cap chamber. The type of gas that fills the detector volume and acts simultaneously as target and tracking medium, depends upon the requirements of the particular physics case of interest. Most relevant gases are hydrogen (H_2) as proton target, deuterium (D_2) as deuteron target, helium-4 (^4He) as an alpha target, or helium-3 (^3He) for studies involving elastic and inelastic scattering, as well as transfer and charge exchange reactions. For investigation of inverse kinematic reactions with TPCs in active target mode, two operative conditions are important:

-) a pure target, specifically no quencher added to the noble gas in order to suppress background reactions.

-) a tuneable pressure of the filling gas (down to a few tens of torr), so that the thickness of the target is adjusted according to the kinematics of the reaction under study.

However, the operation of the position-sensitive end-cap detector in low-pressure, pure noble gases is rather difficult: in these conditions, the charge amplification is strongly limited by photo-mediated secondary effects (gas photoionization, feedback loops, etc.) that cause early transition from the proportional avalanche to the streamer mode, and thus leads to spark breakdown [3, 4]. This results in a general degradation of the performance of the detector in terms of energy and spatial resolution, counting rate capability, and gain stability.

In view of a possible application as a high-gain, charged-particle track readout for AT-TPC applications, the present work reports recent results of the development program carried out at the National Superconducting Cyclotron Laboratory (NSCL), Michigan State University, on the development of a Hybrid Micro-Pattern Gas Detector (MPGD) operated in low pressure low-mass noble gas.

2 AT-TPCs at NSCL

At NSCL, two AT-TPCs of new generation have been developed for accurate studying of low-energy nuclear reactions with rare isotopes beams:

-) a half-scale prototype (pAT-TPC) [5], with a volume of 25 litres (50 cm long, 25 cm diameter), equipped with a pad-plane segmented into more than 250 pads. The geometry of anode pad readout can be designed according the type of experiment. The pAT-TPC, being a compact, versatile, and transportable device, has been used to validate the detector concept in first feasibility studies.

-) a full-scale AT-TPC of 200 litres (100 cm long, 50 cm diameter), with an anode plane of more than 10,000 pads read out by the General Electronics for TPC (GET [6]);

[a] e-mail: cortesi@nscl.msu.edu

this device is stationed inside a large bore solenoid magnet which applies a longitudinal magnetic field, up to 2T [1].

Both detectors are equipped with an endcap based on Micromegas [7], which serves as electron avalanche multiplier and tracking device of the reaction products released in the AT-TPC effective volume. Micromegas is unquestionably one of the most successful MPGD structures, providing an excellent position resolution (down to a few tens of μm [8]) and an excellent charge resolution [8, 9]. A series of successful experiments have been conducted with the pAT-TPC, including studies of sub-barrier fusion reaction in ^{10}Be (with Ar/10%CH_4 as the gas target) and resonant scattering to investigate α-cluster states in ^{10}Be [10] and ^{14}C [11] (with He/20%CO_2 as gas target). However, Micromegas is a single-stage, open-geometry multiplier structure and suffers from photon-induced secondary feedback effects, which limit the gas gain when operated without sufficient quench gas. Figure 1 shows examples of Micromegas gain curves measured in He at different pressures (760, 500 and 300 torr) performed on a standard Micromegas detector (see section 3). A maximum achievable gain of around 30 was measured at 760 torr, while for lower pressure the value decreases to less than 10 for pressure below 500 torr.

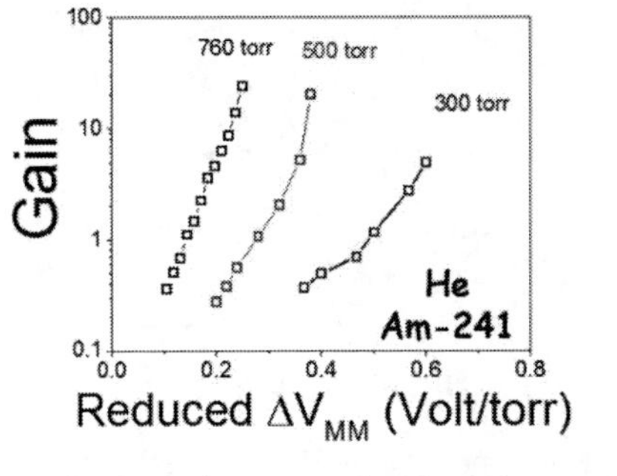

Figure 1: Effective gain measured with a Micromegas detector in He at different pressures (760, 500 and 300 torr), using a 241-Am source.

In order to achieve a suitable gas gain during operation with pure low-mass noble gas, the Micromegas readout was combined with pre-amplification stages based on the hole-type THGEM multiplier [12]. This hybrid configuration allow to cascade many multiplier elements and to obtain an overall higher effective gain by manifold electron avalanche processes. Furthermore, THGEM-based detectors permit stable operation at high gain in pure noble gas at low pressure due to (see [13–15]):

-) the extended dimension of the THGEM holes, typically several times larger than the electron mean-free path even at low pressure;

-) the confinement of the avalanche within the holes, resulting in smaller photon-mediated secondary effects;

-) the quenching effect of small amounts of impurities from natural outgassing of detector components - e.g. N_2 acts as wavelength shifter suppressing UV-photons emitted during the avalanche.

Maximum effective gains of 10^4–10^7 have been recently reported [14] with multi-THGEMs structures in pure He at pressure ranging from 100 torr up to 760 torr. In the case of H_2 and D_2, maximum achievable gains of above 10^4, for pressures of 200 torr and above, were obtained with a double-THGEM detector from single-photoelectrons avalanche [15]. A recent work [16] also reported that the hybrid Micromegas/THGEM detector reduces ion backflow, which in turn results in a substantial increase of the detector gain stability as well as an improved charge resolution.

3 Gain in He and H_2

In this work, basic performance evaluation studies have been carried out using a hybrid (Micromegas/two-cascade THGEM) detector prototype of an effective area of 10x10 cm^2 (Figure 2). The Micromegas structure consisted on a micromesh (18 μm thick stainless steel wire woven at a pitch of 63 μm) held at a distance of 128 μm from the anode plane. The THGEM electrodes are economically produced by mechanically drilling of 0.5 mm diameter holes, spaced by 1 mm (pitch) in a 0.6 mm thick printed circuit board (PCB), followed by Cu etching of the hole's rim (0.1 mm).

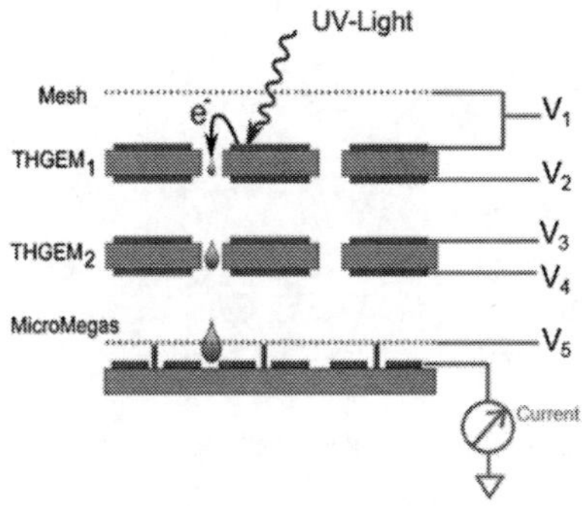

Figure 2: Schematic drawing of the detector prototype, consisting of a Micromegas structure coupled to two-cascade THGEM.

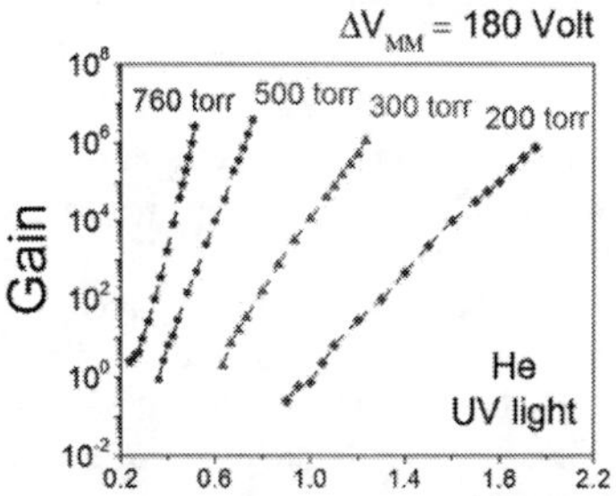

Figure 3: Gain curves in He as a function of the THGEM reduced bias.

Figure 3 depicts the effective gain as a function of the reduced bias applied symmetrically to the two THGEMs, while the voltage difference across the Micromegas was kept at a constant value of 180 Volt. The measurements were performed in pure He at different pressures (from 760 to 200 torr), and carried out by illuminating the bare Cu-cladded top surface of the first-cascade THGEM with UV-light. The maximum achievable gains, defined by the onset of the discharges, was roughly 10^6, regardless of the pressure of the filling gas. Notice that for pressures below 200 torr the frequency of discharge increases considerably and no stable operation was possible at high gain. Following the encouraging results obtained in He with the small-area, test-bench detector [14, 15], two large THGEM electrodes (270 mm diameter) have been commissioned and assembled on the pAT-TPC endcap plane.

The performance evaluation and characterization of the new hybrid endcap detector have been performed in H_2, at a pressure of 200 and 300 torr (figure 4 and 5, respectively). The detector was irradiated with a collimated fission source (252-Cf), mounted on the cathode plane of the pAT-TPC. At these pressures, all of the heavy fragments from the 252-Cf spontaneous fission were stopped inside of the gas volume; the total kinetic energy of the fragments is around 180 MeV.

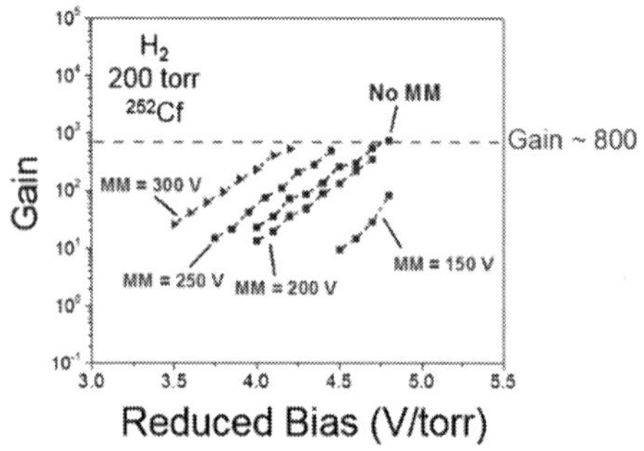

Figure 4: Gain curves in pure H_2 at 200 torr.

In Figure 4 and 5, all the gain curves were obtained by varying the voltage bias applied symmetrically to the two THGEMs (in the graphs expressed as reduced bias – Volt/torr), while the voltage across the Micromegas structure was kept at a constant value (indicated as MM in the figures). The transfer field between the various multiplier stages was kept at a value of 250 Volt/cm. The gain curves of the hybrid detector (green graphs) are compared to the effective gain measured with only THGEM elements as amplification stage (purple graph); in this case the pulse-heights were recorded on the Micromegas mesh. Independently from the voltage bias applied to the Micromegas, upper limits for the maximum achievable gain was achieved at values of 800 and 300, for pressure of 200 and 300 torr respectively. Notice that, for low biases applied to the Micromegas, the significant loss of gain is the result of a low electron transparency of the Micromegas mesh – a substantial number of electrons coming from the THGEM pre-amplification stage are collected on the mesh, instead of being transferred to the anode readout.

As a typical diatomic molecule, H_2 (as well as D_2) is characterized by many rotational and vibration modes that

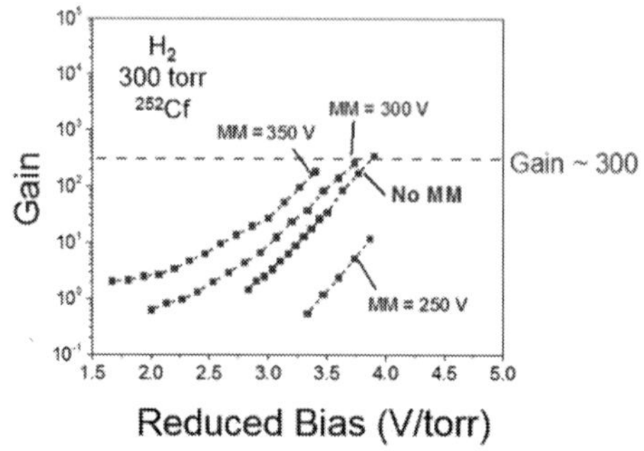

Figure 5: Gain curves in pure H_2 at 300 torr.

can be excited by electron impact [15]. These radiationless processes have high cross sections that extend beyond the energy threshold of the ionization process, with the result that an extremely high electric field strength is necessary for a substantial gas avalanche multiplication. Nevertheless, higher voltage biases applied to the various detector elements induced a higher probability of sporadic discharges, resulting in a reduced maximum achievable gain compared to operation in other noble gases.

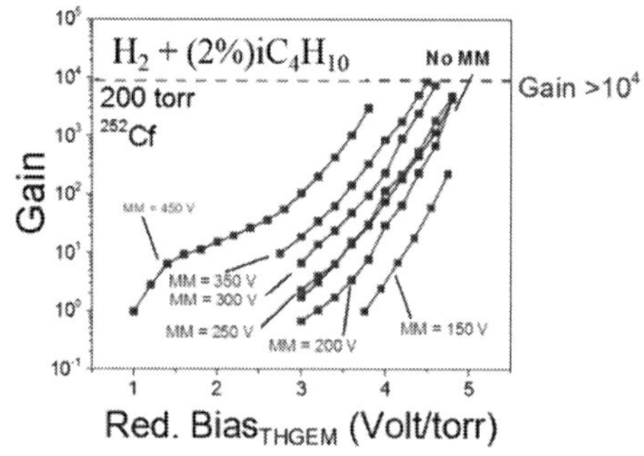

Figure 6: Gain curves in $H_2/(2\%)iC_4H_{10}$ at 200 torr

If higher gain operation is necessary, a small percentage of a quench gas (isobutane or other complex gas molecule, e.g. CH_4, CO_2, etc.) can be added to the primary noble gas component - H_2 (or He). Spark-free operation with an order of magnitude higher effective gain can be achieved, though at the expense of a lower reaction (detection) efficiency and a more complex data analysis - reactions induced with Carbon (or other elements, e.g. Oxygen) need to be identified and rejected. Figure 6 illustrates examples of gain curves measured in H_2 with 2% Isobutane admixture at 200 torr.

4 AT-TPC calibration

The tracking calibration of the AT-TPC, performed by online monitoring of operating and ambient parameters, is among the various required procedures for keeping a strict tolerance on all the possible sources of inaccuracy. The main goal is to provide the precise information needed for the reconstruction of the particle tracks with sufficient precision, so that the design performance can be achieved.

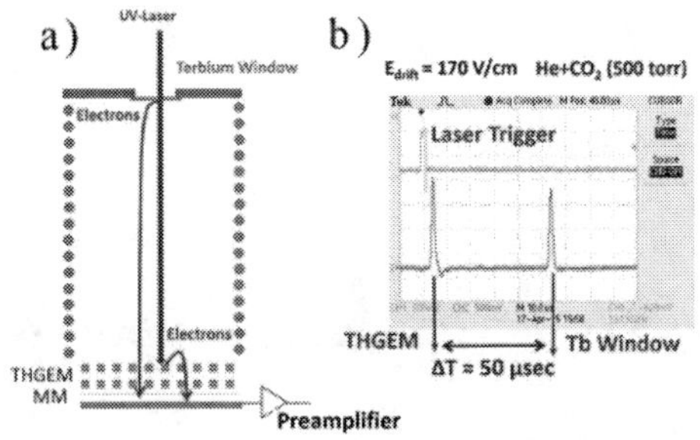

Figure 7: Part a): operational principle of the pAT-TPC drift-velocity monitor system. Part b): example of recorded photoelectrons pulse-heights signals.

The calibration of the drift velocity, which in conjunction with the drift time provides the z-position of the traversing particles, is essential. Variations in drift velocity during operation can be caused by changes of the filling gas composition (due to outgassing of the detector components and/or the gas handling system), as well as temperature, pressure, electric (and magnetic) field fluctuations. The simple approach for determining the drift velocity is by measuring the difference of the arrival time of electrons emitted from two known distances (so-called pick-up electrodes), inside the AT-TPC volume.

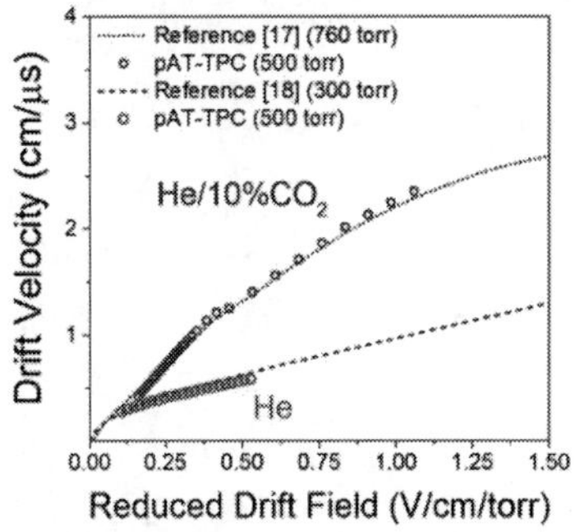

Figure 8: Drift velocity in He and He/CO_2 as function of the reduced drift field. Data from this work are compared to data available in literature.

The monitor system, which will be implemented on the pAT-TPC, consists of a terbium-coated quartz windows mounted on the cathode plane and illuminated by an intense, pulsed nitrogen laser (337 nm wavelength). Photoelectrons are extracted from the 150 Å thick terbium layer and drifted towards the anode. Part of the UV-light is transmitted through the quartz windows and illuminates the Cu-cladded top surface of the first-cascade THGEM (Figure 7a). A copious number of photoelectrons are extracted from the copper; part of them are focused inside the THGEM holes and then multiplied. Thus each UV-pulse generates two signals, one due to photoelectrons emitted at the cathode and one due to photo-electrons emitted at the anode plane. The time difference between the two signals (Figure 7b) corresponds to the drift-time of electrons across the entire AT-TPC length (50 cm).

Figure 8 shows the drift velocity as function of the reduced drift field, measured in He and He/CO_2 with pAT-TPC monitor system prototype described above. Experimental data from this work are in good agreement with values reported in the literature [17, 18].

5 Summary

AT-TPCs are a powerful tool for exploiting recent major progress in accelerator technology to perform accurate investigation of non-perturbative quantum systems, by testing weakly bound nuclear species using simple reaction mechanisms. In this work we report recent progress on the development of a hybrid MPGD for AT-TPC applications, consisting of Micromegas coupled to double-cascade THGEMs as pre-amplification stage. The hybrid readout plane detector provides relative high gas gain in pure noble gas (with no gas quencher) at low pressure. This condition is particularly important for beams composed of rare isotopes far from stability, whose intensities are inherently low but their investigations are of prime scientific interest. It is also crucial for reaction studies in the astrophysical domain, where the cross sections are small and an unambiguous event-by-event kinematic reconstruction is essential for identifying particular reaction channels with high sensitivity.

References

[1] S. Beceiro-Novo, et al., Prog. Part. Nucl. Phys. **84**, 124–165 (2015).
[2] W. Mittig, et al., NIM A **784**, 494–498 (2015).
[3] V. Peskov and P. Fonte, ArXiv09110463 (2009).
[4] J. Miyamoto et al., JINST **5**, P05008 (2010).
[5] D. Suzuki, et al., NIM A **691**, pp. 39–54, Nov. 2012.
[6] E. Pollacco, et al., Phys. Procedia **37**, 1799–1804 (2012).
[7] Y. Giomataris, et al., NIM A **376**, 29–35 (1996).
[8] S. Andriamonje, et al., JINST **5**, P02001 (2010).
[9] P. Bhattacharya, et al. NIM A **732**, 208–212 (2013).
[10] D. Suzuki, et al., Phys. Rev. C **87**, 054301 (2013).
[11] A. L. Fritsch, PhD Thesis: The search for cluster structure in 14C with the prototype AT-TPC, 182 p. (2014), ISBN: 9781321124811 .
[12] A. Breskin, et al., NIM A **598**, 107–111 (2009).
[13] C. K. Shalem et al., NIM A **558**, 468–474 (2006).
[14] M. Cortesi, et al., JINST **10**, P02012 (2015).
[15] M. Cortesi, et al., JINST **10**, P09020 (2015).
[16] S. Duval et al., NIM A **695**, 163–167 (2012).
[17] Assran Y. et al., arXiv:1110.6761 (2011).
[18] A. Phelps et al., Phy. Rev. **117**, 470 (1960).

Micromegas for dark matter searches: CAST/IAXO & TREX-DM experiments

J. G. Garza[1,a], S. Aune[2,b], F. Aznar[1], J. F. Castel[1], S. Cebrián[1], T. Dafni[1], E. Ferrer-Ribas[2], J. Galán[1], J. A. García[1], I. Giomataris[2], F.J. Iguaz[1], I.G. Irastorza[1], G. Luzón[1], H. Mirallas[1], T. Papaevangelou[2], A. Peiró[1], A. Tomás[3], and T. Vafeiadis[4]

[1]*Grupo de Física Nuclear y Astropartículas, Universidad of Zaragoza, Zaragoza, Spain.*
[2]*Centre d'Études de Saclay, CEA, Gif-sur-Yvette, France.*
[3]*Brackett Laboratory, Imperial College, London, UK.*
[4]*CERN, European Organization for Particle Physics and Nuclear Research, Geneva, Switzerland.*

Abstract. The most compelling candidates for Dark Matter to day are WIMPs and axions. The applicability of gasesous Time Projection Chambers (TPCs) with Micromesh Gas Structures (Micromegas) to the search of these particles is explored within this work. Both particles would produce an extremely low rate at very low energies in particle detectors. Micromegas detectors can provide both low background rates and low energy threshold, due to the high granularity, radiopurity and uniformity of the readout. Small (few cm wide) Micromegas detectors are used to image the axion-induced x-ray signal expected in the CERN Axion Solar Telescope (CAST) experiment. We show the background levels obtained in CAST and the prospects to further reduce them to the values required by the Internation Axion Observatory (IAXO). We also present TREX-DM, a scaled-up version of the Micromegas used in axion research, but this time dedicated to the low-mass WIMP detection. TREX-DM is a high-pressure Micromegas-based TPC designed to host a few hundreds of grams of light nuclei (argon or neon) with energy thresholds potentially at the level of 100 eV. The detector is described in detail, as well as the results of the commissioning and characterization phase on surface. Besides, the background model of TREX-DM is presented, along with the anticipated sensitivity of this search, which could go beyond current experimental limits.

1 Micromegas in the search of Dark Matter: axions and WIMPs

The development of Micromegas-based gaseous Time Projection Chambers (TPCs) for the search of rare events has been the goal of the ERC-funded T-REX project [1]. In this document we present the application of this concept to the search of two of the most compelling dark matter candidates: axions and WIMPs, which arise in independent extensions of the Standard Model.

Although the detection of axions and WIMPs is conceptually very different, both searches share the extreme low rates and low energy of the expected signal events. The experimental challenge is therefore to achieve background rates and energy thresholds as low as possible.

The mainstream of WIMP experiments has impressively increased their sensitivity thanks to the use of very large target masses and the capability to distinguish between electron and neutron-induced recoils. However, the efficiency of this discrimination technique is reduced below a relatively large energy, which implies a lost of sensitivity to low-mass WIMPs. Due to the intrinsic charge amplification happening in gaseous detectors, Micromegas are a promising technology to achieve very low energy thresholds. Although electron/nuclear discrimination in atmospheric gaseous TPCs is modest, high sensitivity to light WIMPs could be reached. TREX-DM is a prototype that pretends to test this detection concept. The apparatus, commissioning and first results are presented in section 3.

On the other hand, if axions exist, they would be copiously produced in the Sun's core with ~keV energies and they could be converted back to detectable x-rays in a laboratory magnetic field aligned with the Sun. The expected signal is an excess of x-rays at the exit of the magnet over the background measured at non-alignment periods. Thus, ultra-low background x-ray detectors are a technological pillar of axion helioscopes. Here, we present the small Micromegas TPCs used in the CERN Axion Solar Telescope (CAST) since 2003, where they pioneered the application of these readouts to rare event searches. The background reduction techniques and the rates achieved are reviewed in section 2, as well as the prospects for using this detection system for the future International Axion Observatory (IAXO) [2], the next generation axion helioscope, now under proposal.

[a]e-mail: jgraciag@unizar.es
[b]Present address: Centro Universitario de la Defensa, Universidad de Zaragoza, Zaragoza, Spain

2 Low background x-ray detection in CAST

One of the experimental challenges of axion helioscopes is the detection of a very low x-ray flux in the keV energy range. For this, the use of low background x-ray detection techniques is needed, potentially coupled to x-ray optics to focalize the parallel beam of photons into a small spot, in order to further increase the signal-to-noise ratio.

The concept of detecting low energy x-rays with Micromegas detectors has been extensively developed and used in CAST since an early stage of the experiment [3–5]. The detector is a small TPC with a Micromegas readout at the anode, and whose cathode faces the magnet bore from where x-rays enter the detector. The conversion volume of the chamber is set to efficiently absorb the signal photons, while minimizing background, and typically has 3 cm height and is filled with 1.4 bar argon in addition to a small quantity of quencher. The x-rays coming from the magnet enter the conversion volume via a gas-tight window made of 5 μm aluminized mylar foil. This foil is also the cathode of the TPC, and is supported by a metallic strong-back, in order to withstand the pressure difference with respect to the magnet's vacuum system.

The ionization produced by the interactions in the conversion volume drifts and is projected onto the Micromegas readout plane, where signal amplification takes place. The Micromegas active area is 36 cm^2 that comfortably covers the 14.55 cm^2 projection of the magnet's bore, which fixes the analysis fiducial area. The readout plane is finely pixelized with a typical pitch of ~0.5 mm. The pixels are read out in X and Y columns using state-of-the-art TPC data acquisition (DAQ) electronics based on the AFTER chip [6]. This provides relative time-of-flight information of the ionization charge arriving to the readout plane and hence 3D information of the primary ionization cloud.

The sources of background for the Micromegas detectors are: γ-rays, cosmic rays, the presence of radon around the detector, intrinsic radioactivity of the detector components or shielding, neutrons and cosmogenic activation of the detector materials. All these radiations can produce secondary or fluorescence emissions that can reach the detector active volume. The strategies to reduce these backgrounds are: the use of high-Z material (lead, copper) passive shielding to block the pass of γ-rays; active shielding to tag muons, such as plastic scintillators; the continuous flux of vaporized LN_2 into the detector environment to avoid the presence of air-borne radon; neutron moderator and absorber; and the use of radiopure components.

Muon-induced events are an important part of the background level of the Micromegas detectors when operated at surface level. In order to suppress these events, an active detector can be installed over the Micromegas detectors to act as a muon veto. The strategy is to use the time difference between the signal in the muon veto and the delayed Micromegas trigger to tag muon-induced events. A plastic scintillator acting as a muon veto was installed over the detectors for the data taking of 2012, and

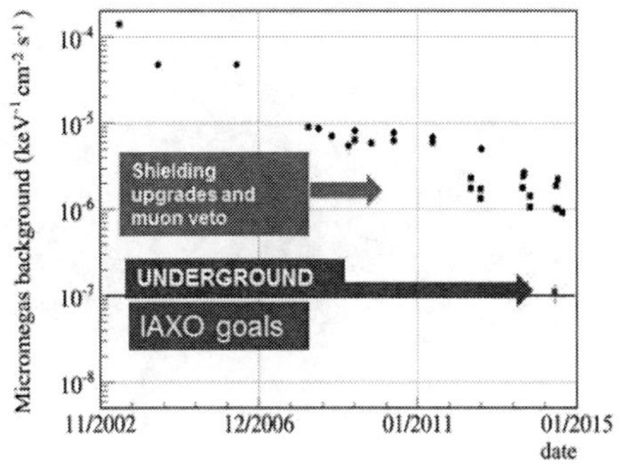

Figure 1. Evolution of the background level in CAST, the limit obtained in the LSC and IAXO goals in the 2-7 keV energy range.

it was subsequently upgraded in 2013 due to its promising results.

The application of these techniques along with the reliability of the microbulk technology, the upgrade of the readout electronics, the high granularity of the readout – which offers topological information of the event, a powerful tool for signal/background discrimination– and the tunning of the rejection algorithms led to a background level reduction of a factor 10^2 in CAST-MM detectors over the last ten years. Figure 1 shows a compilation of background levels achieved in CAST in the keV energy range, at surface and underground test benches. The best level achieved at surface is already below 10^{-6} keV^{-1}cm^{-2}s^{-1} [5]. In particular, the best detector of CAST has demonstrated a background level as low as (0.8 ± 0.1) keV^{-1}cm^{-2}s^{-1} during the last data-taking campaigns. The background energy spectrum is characterized by a fluorescence peak at 8 keV (from copper K_α emission), its escape peak at 5 keV and by the argon K_α line at 3 keV.

The operation of a replica of the detector in the Canfranc Underground Laboratory (LSC) sets a level as low as $\sim 10^{-7}$ keV^{-1}cm^{-2}s^{-1}, almost at the required IAXO levels. According to our background model, based on in-situ measurements at CAST, tests underground and at surface as well as on Geant4 simulations, the contributions from γ-rays, radon and internal radioactivity have been reduced to negligible levels. The dominant contribution is due to cosmic muons and their secondary products, generated after their interaction in the setup components. On the other hand, the origin of the events limiting the LSC performance is uncertain. Some hypotheses are the β-decay of ^{39}Ar present in the detection medium, neutrons or the cosmic activation of the materials.

2.1 Prospects for IAXO

The developments described in previous sections have a strong motivation as part of the preparatory activities for IAXO, the future next-generation axion helioscope now

Figure 2. Different views of the TREX-DM setup showing the field cage and cathode, Micromegas readout and electronics.

under proposal [2]. As part of this effort, we plan to build a prototype setup, IAXO-D0, with the aim of demonstrating the levels of background needed by IAXO.

This prototype is conceived as a generic test-bench for testing other technologies. The main improvement with respect to the CAST-MM is that a thicker and more compact shielding (20–30 cm) and a 4π enlarged muon veto system will be implemented. Pushing the lowest underground limit requires a replacement of the target gas by xenon or depleted argon from underground sources, or the installation of a neutron shielding. These activities are being developed in the context of the R&D phase for the IAXO technical design report.

3 TREX-DM prototype and characterization

In some respects, TREX-DM [7] is a scaled-up version of the CAST-MM, but with a 10^3 times larger active mass. Figure 2 shows different views of the apparatus. The vessel is comprised of a copper sleeve of 0.5 m diameter and 0.5 m length and two 6 cm thick flat end caps that can hold up to 12 bar. The inner volume is divided in two active volumes, separated by a central cathode made of 4 μm aluminized mylar foil. Around each active volume there is a 19 cm long and 25 cm wide square section field shaper that guarantees the homogeneity of the drift field, independently of the applied voltages in the cathode and mesh. The field shaper is composed of copper strips printed on single-layer of kapton, each strip separated by a 10 MΩ resistor.

The Micromegas detectors are built using the bulk technology, with an amplification field of 128 μm. The active surface is 25.2×25.2 cm^2, divided in square pads of 312 μm length with a pitch of 582 μm. The pads are alternatively interconnected in 432 strips per axis, which are rooted to four connector prints at the PCB sides. A flat cable links each detector footprint to the electronics by means of a commercial 300-pin connector. The mesh signal is extracted from the vessel by a coaxial low-noise cable and a feedthrough, and it is subsequently amplified and the spectrum recorded. The mesh also triggers the acquisition of the strip pulses, which come out from the vessel through flat cables and feed the AFTER-based front-end cards.

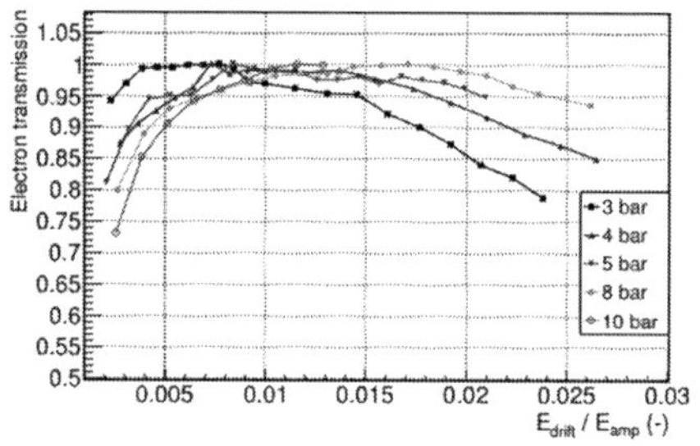

Figure 3. Dependence of the electron transmission with the ratio of drift and amplification fields in Ar+2%iC_4H_{10}.

A ^{109}Cd source is used to calibrate the detector at 22.1 (K_α) and 24.9 keV (K_β). The Micromegas detectors are characterized in argon- and neon-based mixtures at high pressure with this source in order to find the optimum point of operation for a physics run.

A bulk Micromegas detectors was installed at each end of the symmetric TPC, and were simultaneously from 1.2 to 10 bar in steps of 1 bar in terms of electron transmission, detector gain, energy threshold and gain uniformity, in Ar + 2%iC_4H_{10}. This is the first systematic characterization of bulk Micromegas detectors above atmospheric pressure to the authors' knowledge. A gas flow of of 3–5 l/h is kept during the measurements.

The spectra are characterized by a peak at around 22 keV and by a fluorescence emission at around 6.4 and 8 keV from the iron and copper components. The spectral parameters are defined through an iterative multi-Gaussian fit corresponding to the K_α and K_β emission lines of the source and their escape peaks.

First, the drift voltage is varied for a fixed mesh voltage to obtain the electron transmission curve at each pressure (see figure 3). The detectors show a plateau of maximum electron transmission for a wide range of ratios of drift and amplification fields at all pressures. The electron transmission drops at very low drift fields (~50 V/cm/bar) due to electron attachment and recombination of the primary electrons. For high drift fields, the mesh stops being transparent for primary electrons, and the energy resolution also degrades.

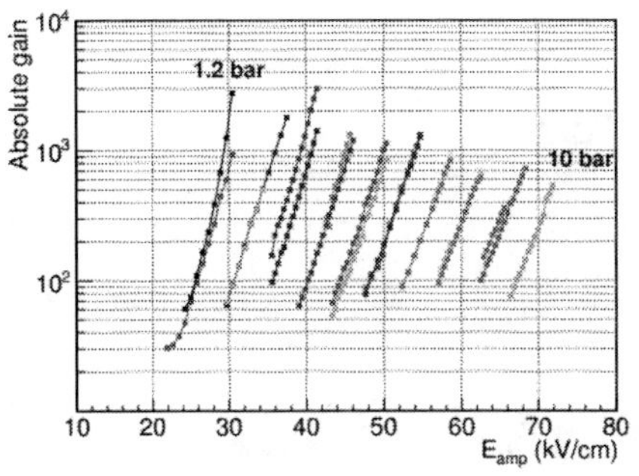

Figure 4. Dependence of the absolute gain with the amplification field between 1 and 10 bar, in steps of 1 bar. At each pressure the two Micromegas detectors of TREX-DM were characterized simultaneously.

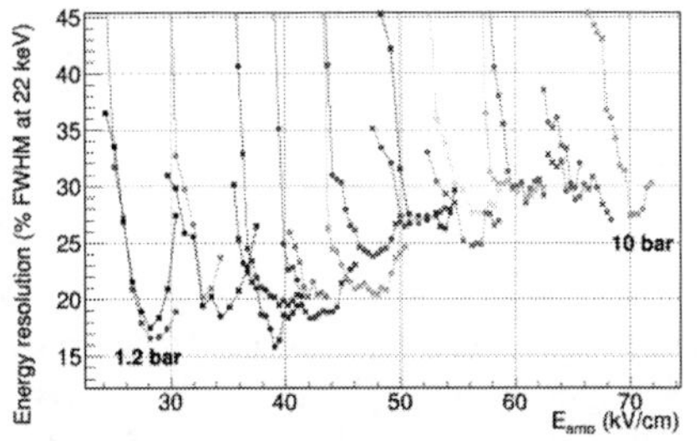

Figure 5. Dependence of the energy resolution at 22 keV with the amplification field for the bulk Micromegas of TREX-DM in $Ar+2\%iC_4H_{10}$ between 1.2 and 10 bar. At each pressure the two Micromegas detectors of TREX-DM were characterized simultaneously.

The absolute gain of the detector as a function of the amplification field is shown in figure 4 for all the pressure settings between 1 and 10 bar. The maximum gain decreases with pressure from 3×10^3 at 1.2 bar to 600 at 10 bar.

The dependence of the energy resolution with the amplification field for all the pressures settings is shown in figure 5. At each pressure there is a range of amplification fields for which the energy resolution is optimized. At low gains, the energy resolution degrades because the signal becomes comparable with noise. At high fields, the resolution degrades due to increase in the gain fluctuations by the UV photons generated in the avalanche. As can be noticed, the best energy resolution degrades with the pressure, being around 16% FWHM at 22.0 keV at 1.2 bar and above 25% FWHM at 10 bar; modest values for bulk detectors. The gain homogeneity over the detector surface has been also studied, showing gain fluctuations below 10%.

3.1 Prospects for light WIMP searches in TREX-DM

The role of the quencher fraction will be studied in near term. The modest gains achieved may be explained by the low quantity of quencher (2%) in the gas. A 5-10% isobutane mixture will be tested in the near future, while neon-based mixtures will also be studied as they are expected to show hihger gains and consequently lower thresholds.

A first background model of the experiment in argon and neon-based mixtures in an underground environment has been developed in order to study the sensitivity of TREX-DM. Two target materials at 10 bar with an active mass of 0.3 and 0.16 kg have been considered, respectively for argon and neon. The simulation process is described in detail in [7].

The expected background level for the argon-based gas is around 2×10^2 keV^{-1}kg^{-1}day^{-1}, dominated by the ^{39}Ar isotope. If this contribution could be eliminated by using depleted argon from underground sources, the background level would be reduced to $\sim$ 1 keV^{-1}kg^{-1}day^{-1} in both argon and neon, limited by the copper vessel and connectors.

Assuming that the TREX-DM detector reaches an effective energy threshold of 0.4 keV and a conservative background level of 10^2 keV^{-1}kg^{-1}day^{-1}, the experiment could be sensitive to a relevant fraction of the low-mass WIMP parameter space with a reasonable exposure. The experiment can set upper limits below to the low-mass "region of interest" invoked by some positive interpretations of some dark matter experiments.

Acknowledgements

We thank our colleagues of CAST for many years of collaborative work in the experiment. We thank R. de Oliveira and his team at CERN for the manufacturing of the microbulk readouts. We also thank the LSC staff for their help in the support of the Micromegas setup at the LSC. Authors would like to acknowledge the use of Servicio General de Apoyo a la Investigación-SAI, Universidad de Zaragoza. We acknowledge support from the European Commission under the European Research Council T-REX Starting Grant ref. ERC-2009-StG-240054 of the IDEAS program of the 7th EU Framework Program. We also acknowledge support from the Spanish Ministry of Economy and Competitiveness (MINECO) under contract ref. FPA2013-41085, and under the CPAN project ref. CSD2007-00042 from the Consolider-Ingenio 2010 program. Part of these grants are funded by the European Regional Development Fund (ERDF/FEDER).

References

[1] I. G. Irastorza *et al.*, EAS Publications Series **53**, 147–154 (2012)
[2] I. G. Irastorza *et al.*, CERN-SPSC-2013-022 (2013).
[3] P. Abbon *et al.*, New J. Phys. **9**, 170 (2007)
[4] S. Aune *et al.*, JINST **9**, P01001 (2014)
[5] S. Aune *et al.*, JINST **8**, C12042 (2013)
[6] P. Baron *et al.*, IEEE Trans. Nucl. Sci. **55**, 1744 (2008)
[7] F. J. Iguaz *et al.*, Submitted to Eur. Jour. Phys. C.

WA105: A Large LAr Double Phase TPC Prototype

Thorsten Lux[1,a] on behalf of the WA105 collaboration

[1]Institut de Física d´Altes Energies (IFAE) - The Barcelona Institute of Science and Technology, Campus UAB, Bellaterra (Barcelona), Spain

Abstract. For the next generation of long baseline neutrino oscillation experiments detectors with several tens of ktons of detection medium will be needed. Liquid argon (LAr) is ideal for this application since it allows the construction of fully sensitive time projection chambers (TPCs). A mandatory step between the current setups and the final detector for the long baseline experiment is the construction of a large prototype to prove the technical feasibility of the scaling. WA105 with a size of 6x6x6 m^3 is this intermediate step for the detector concept of a LAr double phase TPC and will be presented in these proceedings.

1 Introduction

The next generation of long baseline neutrino oscillation experiment, the Deep Underground Neutrino Experiment (DUNE), foresees as detector technology for the far detector 4 liquid argon (LAr) time projection chambers (TPC) of about 10 kton sensitive mass each. LAr has the advantage to provide the required large detector masses at reasonable costs and to allow the construction of a fully active tracking calorimeter. The basic principle of a LAr TPC is the following: A charged particle traversing the LAr ionizes the argon atoms creating a large amount of electron-ion pairs on its way through the gas. For example minimal ionizing particles like cosmic muons create in 1 cm about 50.000 electron-ion pairs. By applying an external electric field, the drift field, the electrons and ions are separated and drift to the the anode and cathode respectively. The field strength of the drift field is normally chosen to be between 0.5 and 1 kV/cm. While the ions are simply neutralized at the cathode, the electrons measured at the anode are used to reconstruct the event. Two different detector concepts exist for the anode side. In the *single phase* detector a set of wires is immersed in the liquid on which the electrons create a signal, either due to induction or direct absorption. In the *double phase* detector the electrons are extracted from the liquid to a gaseous volume within a strong electric field of 2 to 3 kV/cm and undergo afterwards a charge amplification before the signal is read out at the segmented anode. While the first detector concept is simpler, the double phase concept has several advantages: LAr is critically sensitive to oxygen impurities which causes the loss of electrons while drifting to the anode as illustrated in Fig. 1 (left). Due to the amplification in the gas phase these losses can be partly compensated allowing lower energy detection thresholds and simpler electronics. The signals recorded at the segmented anode allow to reconstruct the trajectory of the charged particle in 3D. However, the signal measured at the anode does not provide information about the absolute z coordinate, the position between anode and cathode, of the event. This information can be obtained from the primary scintillation light which is emitted isotropically during the ionization process. Measuring the time difference between this light pulse and the arrival of the electrons at the anode provides the z coordinate assuming the knowledge of the electron drift velocity. The concept of a LAr double phase TPC is shown in Fig. 1 (right).

The aim of the WA105 project [1] is the construction of a large double phase LAr TPC prototype to test all technological aspects necessary for the scaling to kton detectors. In addition the detector will be characterized using a well defined testbeam. In the following the WA105 detector, the testbeam characteristics and the time schedule will be presented in more detail.

2 The WA105 Detector

The design of the WA105 detector is shown in Fig. 2. The cryostat has walls 1 m thick which limit the heat transfer to maximal 5 W/m^2. Its inner volume has dimensions of 8.3x8.3x8.1 m^3 (WxLxH) containing about 700 ton of LAr. The sensitive volume, defined by the field cage, anode and cathode, has dimensions of 6x6x6 m^3 and contains about 300 ton of active mass. The large depths of the detector lead to voltages of up to 600 kV at the cathode which consists of wires to achieve a high transparency for the photons of the primary scintillation light. A set of 36 PMTs with a diameter of 8" (Hamamatsu R5912-02mod) is installed about 1 m below the cathode. Since the PMTs are not directly sensitive to the 128 nm of the primary scintillation light, acrylic plates coated with tetraphenyl butadine (TPB) will be installed directly in front of the PMTs. TPB shifts the light from 128 nm to about 430 nm,

[a]e-mail: Thorsten.Lux@ifae.es

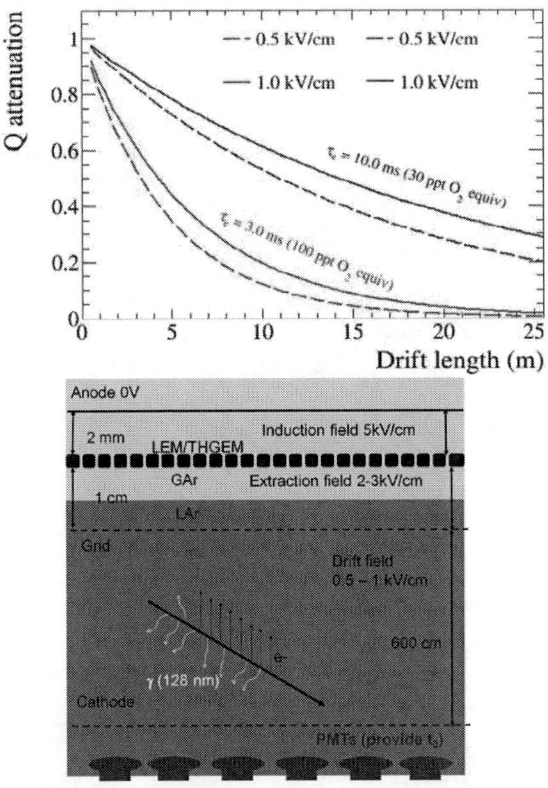

Figure 1. The effect of oxygen contamination on the charge loss (left). The principle of a double phase LAr (right).

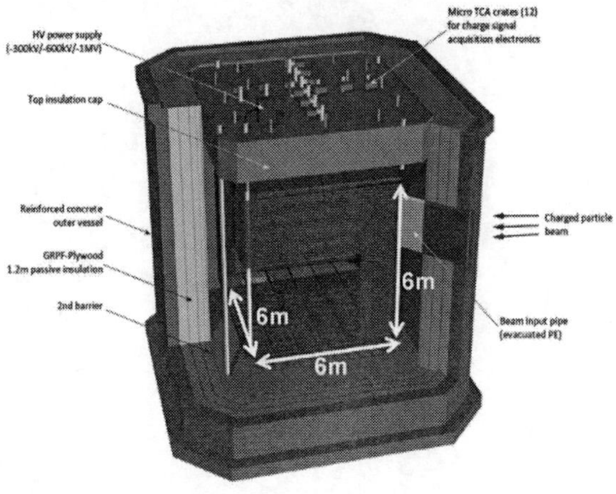

Figure 2. Schematic view of the WA105 detector. The sensitive volume has dimensions of 6x6x6 m^3 and contains about 300 ton of LAr.

a wavelength at which the PMTs highly sensitive. On the anode side the electrons are extracted from the liquid to the gaseous phase applying an electric field of 3 kV/cm at the phase transition zone. For this a steel mesh with wire diameter of 100 μm and wire pitch of 3.125 mm is immersed 5 mm into the liquid phase. In the gaseous phase, 5 mm above the liquid surface, Large Electron Multipliers (LEMs) are installed. A LEM is a printed circuit board (PCB), 1 mm thick, coated on both sides with copper and highly perforated by drilled holes. The diameter of the holes in the PCB is 500 μm and the holes are distributed in a hexagonal pattern with 800 μm pitch. The holes in the copper have a 40 μm larger radius to improve HV stability. By applying a voltage between the two copper sides, the electrons extracted from the liquid are guided into the holes where the electric field is so high that charge amplification occurs. The performance of LEMs was intensively studied in pure Ar at cryogenic temperatures [2]. It was found that at a voltage difference of about 3500 V/cm, an effective gain of 20 can be achieved. The amplified electrons are drifting afterwards towards the anode. The gap between LEM and anode is 2 mm. The anode is segmented in strips in x and and y, providing two independent views. The pitch of the strips is 3.125 mm and each strip has a length of 3 m. Their geometry is optimized to achieve equal charge share between the two views and to minimize parasitic capacitance between adjacent strips. In total there will be 7680 strips, 3840 for each view. The signals are extracted on the top of the cryostat by a system of feedthrough chimneys which reach into the gaseous phase to minimize the cable length between anode and analog preamplifier which are installed at the bottom of the chimneys, outside of the cryostat inner volume. This arrangement is possible thanks to the pre-amplification in the gaseous argon and is a clear advantage of the double phase concept since it allows the maintenance of the electronics without the need to open the cryostat. The analog preamplifiers are based on CMOS technology and are optimized in respect of low power consumption, 18 mW per channel, and low noise. For latter also the ambient temperature of 110 K is an advantage. The digital part of the electronics is installed on top of the chimneys in μTCA crates. The principle design is shown in Fig. 3. The signals are digitized with a sampling rate of 2.5 MHz over a time period of 4 ms per event given my the maximal drift time of the electrons. Two trigger modes are implemented in the electronics: for the testbeam data taking an external trigger synchronized with the arrival of the particles is used, while data taking with cosmics is performed with a self-triggering mode based on the PMT signals.

3 Location and Testbeam

The WA105 detector will be installed in the extension of the EHN1 testbeam facility, currently under construction, at the CERN Prevessin site. This location allows to expose the detector to a beam consisting of muons, electrons and pions of both polarities. The energy of the beam is between 0.4 and 10 GeV/c. The beam allows studying the detector performance as hadronic and electromagnetic calorimeter, the particle identification capability of the detector and especially the $e/\pi^{\pm}$ separation. In addition it enables the possibility to develop and optimize reconstruction software and its validation.

4 Conclusions and Outlook

For the next generation of neutrino oscillation experiment detectors with about 10 kton active mass will be needed. The WA105 detector is a necessary intermediate step for these detectors to prove their technical feasibility and will provide also a deeper understanding of the detector performance due to its exposing to a well defined beam of particles at CERN. The WA105 collaboration consisting of about 120 physicists from more than 20 institutes and 10 countries is working towards the construction of a large double phase LAr TPC with an active mass of about 300 ton. The start of the data taking is foreseen for 2018.

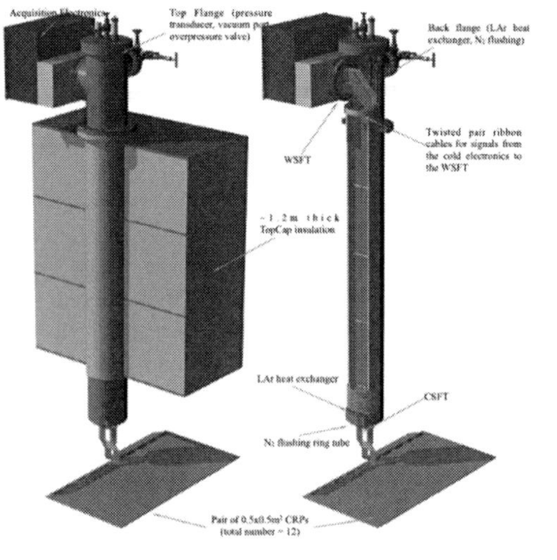

Figure 3. Conceptual drawing of the chimney approach. The analog preamplifier are installed at the bottom of the chimneys, outside of the inner cryostat volume, while the digitial electronics is mounted on top of the chimneys.

Acknowledgment

The author acknowledge the support received from the Ministerio de Ciencia e Innovacion under grants FPA2014-59855-P, TEC2012-39150-C02-02 and Centro de Excelencia Severo Ochoa SEV-2012-0234, some of which include ERDF funds from the European Union.

References

[1] L. Agostino *et al.*, SPSC-TDR-004, arXiv: 1409.4405 (2014)
[2] C. Cantini et al, JINST 9 P03017, arXiv: 1312.648 (2014)

MICROMEGAS calibration for ACTAR TPC

B. Mauss[1,a], T. Roger[1], J. Pancin[1], S. Damoy[1], and G. F. Grinyer[1]

[1] *Grand Accélérateur National d'Ions Lourds (GANIL), CEA/DSM-CNRS/IN2P3, Bvd Henri Becquerel, 14076 Caen, France*

Abstract. Active targets, such as the ACtive TARget and Time Projection Chamber (ACTAR TPC) being developed at GANIL, are detection systems that operate on the basis of a time projection chamber but where the filling gas also serves as a thick target for nuclear reactions. In nuclear physics experiments, the energy resolution is of primary importance to identify the reaction products and to precisely reconstruct level schemes of nuclei. These measurements are based on the energy deposited on a pixelated pad plane. A MICROMEGAS detector is used in ACTAR TPC for the ionization electron collection and amplification, and it is a major contributor to the energy dispersion through, for example, inhomogeneities of the amplification gap. A variation of one percent in the gap can lead to an amplitude variation of more than two percent which is of the same order as the resolution obtained with an energy deposition of 5 MeV. One way to calibrate the pad plane is through the use of a two dimensional source scanning table. It is used to calibrate the gain inhomogeneities and, using MAGBOLTZ calculations, deduce the corresponding gap variations. The inverse of this method would allow the relative gain variations to be calculated for the different gas mixtures and pressures used in experiments with ACTAR TPC.

1 Introduction

Nuclear matter under extreme conditions of excitation energy or extreme isospin exhibits properties which strongly differ from those of nuclei near stability. The appearance of nuclear matter haloes near the drip lines or the formation of clusters at high excitation energies are examples of such exotic phenomena. Access to these extreme conditions requires the study of nuclei far from stability that are produced with low beam intensities or through reactions with small cross sections. The use of very efficient detection systems and thick targets is needed to cope with these difficulties. In an active target, detection of the reaction products is performed inside the target, the detection efficiency and effective target thickness are increased without losing resolution since the position of the reaction inside the target is measured.

Active targets and Time Projection Chambers (TPC) were introduced in nuclear physics about 20 years ago with the detector IKAR [1, 2] in GSI. It was followed by the MAYA active target at GANIL [3–5] and the CENBG TPC [6, 7] used for decay studies. In order to gain in sensitivity and precision on the reconstruction of trajectories, the next generation active target ACTAR TPC is now being developed at GANIL. ACTAR TPC is foreseen to be used for direct and resonant reactions, and decay studies. A demonstrator of this future active target has been built. Its working principle is the following: ionization electrons produced along charged particle tracks drift under the influence of an electric field to a readout plane composed of 2048 pads of 2×2 mm^2 each. The drift field

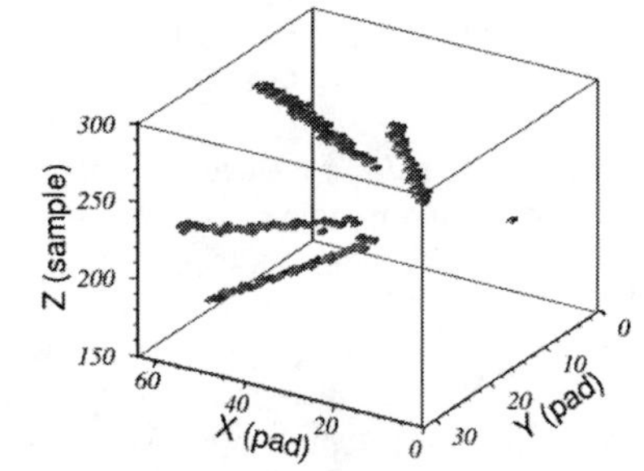

Figure 1. Example of a $^{12}C(^{4}He, ^{4}He)\,^{12}C^{*} \rightarrow 3\alpha$ event recorded with the ACTAR TPC demonstrator. The color scale corresponds to the value of the charge deposited per pad. The 6.6 MeV/u ^{12}C beam entered the active volume from the right, along the X axis. The chamber was filled with a He:iC_4H_{10} 90:10 gas mixture with a pressure of 1 bar.

cage is 17 cm high with 168 evenly-spaced metallic wires on the four sides. By degrading the wires potential from the cathode potential to the MICROMEGAS potential, this field cage permits to have an homogeneous vertical electric field. The amplification of the primary ionization electrons is achieved through the use of a 220 μm amplification gap obtained with a bulked MICROMEGAS [8, 9], which matches the requirements of the TPC in terms of timing, gain, types of gases used and operating pressure [10]. The readout of the pads is performed using GET digital elec-

[a] e-mail: mauss@ganil.fr

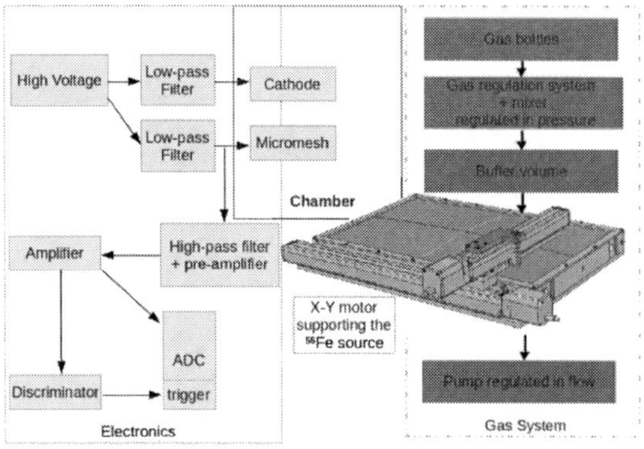

Figure 2. Test bench layout, with the electronics for the supply of the cathode and mesh (upper part) and the reading of the signal on the mesh (lower part). In green the pumping system for the regulation in pressure of the gas inside the chamber.

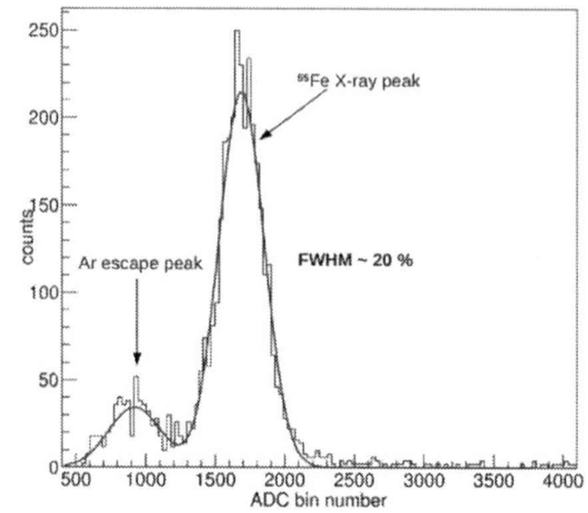

Figure 3. Typical ^{55}Fe X-ray source spectrum in Ar:iC_4H_{10} 97.15:2.85 at P = 1021 mbar and V_{mesh} = -420 V.

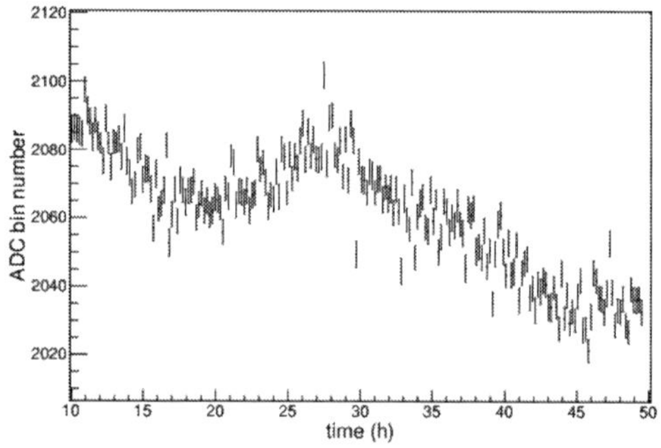

Figure 4. Variations of the gain at a reference position over a two-day period. The variation in gain is ≃ 2.5%.

tronics [11], thus enabling a full 3D reconstruction of the tracks, as illustrated in Figure 1.

The identification of the reaction products can be achieved using the correlation between their average energy loss measured along their path and their total energy. The latter can be estimated either by reconstructing the range of the particles if they stop in the gas or using auxiliary detectors (such as silicon detectors) if the particles escape the active volume. The identification capabilities of the active target therefore depend on the local gain variations which arise from inhomogeneities of the gap over the whole pad plane.

In this article, we propose a method to correct gain inhomogeneities for each gas/target and pressure. This method is based on the measurement of the response of the MICROMEGAS to an ^{55}Fe X-ray source as a function of the position of the source over the pad plane. This will be described in Section 2. The reconstruction of the relative gap for each pad of the pad plane using MAGBOLTZ calculations to obtain the associated gain corrections for any gas/target is developed in Section 3.

2 Experimental set-up

The MICROMEGAS was installed in a chamber where the gas pressure was regulated to be stable at better than ±0.1 mbar. A drift gap of 2 cm was achieved through the placement inside the chamber of a meshed cathode above the MICROMEGAS. An X-Y motorized holder was used to position the collimated ^{55}Fe X-ray source 7 cm above each pad with a precision of ±0.1 mm, the X-ray beam irradiating a zone of 3×3 pads. The irradiated zone has the shape of a 2D Gaussian function centered on the central pad, which collects more than 60% of the events. A similar method was used to calibrate the gain of the MICROMEGAS detectors for the T2K experiment [12]. The recorded events correspond to the photoionization of the gas. The photoelectrons drift towards the pad plane and are amplified in the MICROMEGAS gap, between the micromesh and the pad plane. The avalanche signal from the micromesh is sent to a pre-amplifier and an amplifier. If the signal is large enough, a discriminator sends a trigger to the ADC that records the charge. The blue part of Figure 2 shows a schematic of the acquisition chain.

During the calibration, 1500 events were acquired for each pad. A typical spectrum from a single measurement, which takes approximately 90 s, is shown in Figure 3. The centroid position of the 5.9 keV photoionization peak was obtained through a double Gaussian fit to this peak and the Argon escape peak observed on the figure. The centroid position was associated to the corresponding (x,y) position of the pad.

Over a few hours of data taking, the gas characteristics may change slightly because of temperature variations. To take these effects into account, a systematic measurement every 30 minutes was performed on a reference position in order to realign the amplitude response. An example of such variations are shown in Figure 4 where large day/night variations were observed. The gain variation of (2.5±0.8)% is coherent with a temperature variation on the order of 1 K but a more precise evaluation is being investigated.

The MICROMEGAS was bulked [9] using 4 layers of

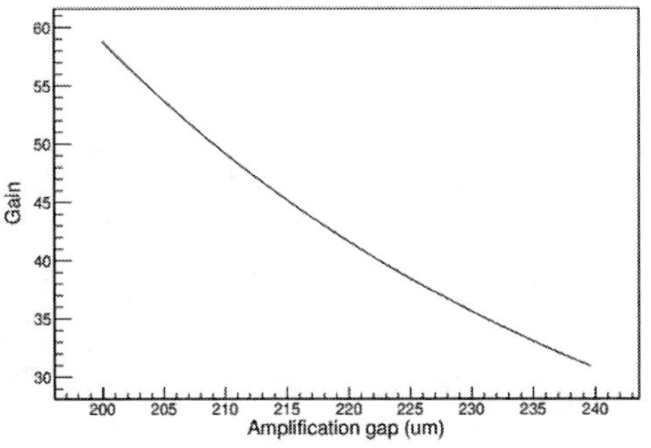

Figure 5. Gain as a function of the MICROMEGAS amplification gap as obtained from MAGBOLTZ calculations for an $Ar{:}iC_4H_{10}$ 97.15:2.85 mixture at P = 1020 mbar and T = 24°C

vacrel below the mesh (provided by the CERN PCB Workshop). A pillar of 300μm diameter is placed at the center of each pad, which should not have any effect on the measurements. This unusual gap size of over 200 μm was chosen because of the wider range of operational pressure [10]. For the calibration, the detector was operated at a voltage V_{mesh} = -420 V with a constant drift field of about 300 V·cm^{-1}. The chamber contained a mixture of $Ar{:}iC_4H_{10}$ 97.15:2.85 supplied by a premixed bottle at a few mbars above atmospheric pressure. In this context, the absolute gain of the MICROMEGAS was found to be of the order of $8 \cdot 10^2$. Measurements at each position had a resolution of about 20% FWHM.

3 Calibration of the MICROMEGAS

3.1 Analysis and results

From the centroid position obtained for each pad position, it was then possible to obtain a relative correction in gain of the pad plane. It was referred in Section 1 that, depending on the target required, the gas inside the chamber will change. The software MAGBOLTZ [13] can be used to calculate the gas characteristics such as the electron drift velocity or the Townsend coefficient as a function of the electric field and for most of the gas mixtures used in detection, at fixed temperature and pressure [14]. Thus, it is possible to adapt the correction factor to the particular target/gas used inside the chamber.

The Townsend coefficient α was calculated for several values of the electric field and a cubic spline was used for the interpolation. Taking the working voltage V_{mesh} of the detector, the Townsend coefficient was evaluated as a function of the amplification gap d. The gain G was then calculated from the approximation $G = e^{\alpha d}$. The size of the amplification region d changes the value of the amplification through the active length and through the Townsend coefficient that depends on the electric field. In figure 5, the gain is plotted as a function of the gap in our region of interest, around 220 μm. This gap value was obtained through optical measurements of the gap size performed previously. Using a similar curve for any gas and pressure

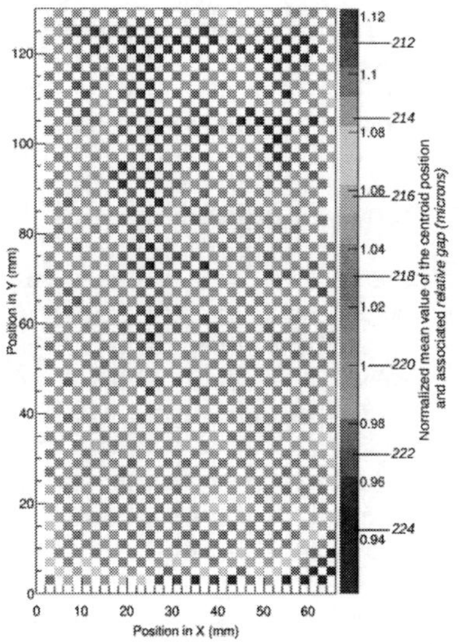

Figure 6. Result obtained for the $Ar{:}iC_4H_{10}$ 97.15:2.85 gas mixture at 1021 mbar. The gap, as calculated from MAGBOLTZ using 220 μm as reference, is also indicated on the color scale.

used, it is possible to convert from gap to gain and *vice versa*.

Some effects like the Penning effect are not completely taken into account for certain gas mixtures in MAGBOLTZ calculations. These effects are assumed to be constant within the range of temperature/gap considered in this study. Therefore, the average gain that was deduced was then scaled to the one calculated by MAGBOLTZ for a gap of 220 μm. Gap variations were then obtained using the curve of Figure 5. It confirms that the gain calculated by MAGBOLTZ is very different from the measured gain. This discrepancy can easily be explained by variations of the Townsend coefficient smaller than a factor of two and due for example to the Penning effect [15]. The result of a scan for every other pad is shown in Figure 6 with the relative normalized gain and the corresponding gaps ranging from 212 to 224 μm. The uncertainties for the gap values were extracted from the uncertainty on the measured amplitude response. The uncertainty on the centroid determination was about 0.3% and propagation to the uncertainty on the gap led to uncertainties below 0.4 μm. To study the sensitivity of the method on the chosen mean gap, the calculation was also performed using mean gaps of 200 and 240 μm. The subsequent gap variations were only lowered and increased by 1 μm, respectively.

3.2 Application to other gases and mixtures

Once a map of the amplification gap over the pad plane is obtained with this calibration, it is possible to calculate gain variations for other gases. Once again, there are unaccounted for effects in MAGBOLTZ and it is only possible to calculate relative gain variations. However, considering that the contributions of these effects is constant with small gap variations, they don't influence the relative variations of the gain. This will be verified using another mixture like Ne+iC_4H_{10}. The new relative gain variations calculated using MAGBOLTZ should be the same as the new relative gain variations measured in this mixture.

4 Conclusion and outlooks

The technique presented here can be used on any MPGD detector to efficiently verify the homogeneity of the amplification stage. Relative corrections can be performed for any gas/target described by MAGBOLTZ which can be useful for any kind of active target or TPC.

The corrections calculated with MAGBOLTZ will be compared with the mapping of the same pad plane in a different gas mixture to validate the method. If successful, the gap/gain corrections will be used in the data analysis of the experiment performed with the ACTAR TPC demonstrator to realign the pad plane response and improve the energy resolution. The scanning table is adapted to the final ACTAR TPC pad planes of 16384 channels with sizes 25.6×25.6 cm^2 and 12.5×51.2 cm^2. Acceptance tests will be performed using this scanning table and the gaps will be calculated to be able to correct the local gains.

Acknowledgments

The research leading to these results has received funding from the European Research Council under the European Union's Seventh Framework Program (FP7/2007-2013)/ERC grant agreement n° 335593.

References

[1] A.V. Dobrovolsky *et al.*, Nucl. Phys. B **214**, 1 (1983)
[2] S. Ilieva *et al.*, Nucl. Phys. A **875**, 8 (2012)
[3] C.-E. Demonchy *et al.*, Nucl. Instr. and Meth. A **431**, 341 (2007)
[4] T. Roger *et al.*, Phys. Rev C **79**, 031603(R) (2009)
[5] M. Vandebrouck *et al.*, Phys. Rev. Lett. **113**, 032504 (2014)
[6] B. Blank *et al.*, Nucl. Instr. and Meth. A **613**, 65 (2010)
[7] P. Ascher *et al.*, Phys. Rev. Lett. **107**, 102502 (2011)
[8] I. Giomataris *et al.*, Nucl. Instr. and Meth. A **376**, 29 (1996)
[9] I. Giomataris *et al.*, Nucl. Instr. and Meth. A **560**, 405 (2006)
[10] J. Pancin *et al.*, Nucl. Instr. and Meth. A **735**, 532 (2014)
[11] E.C. Pollaco *et al.*, Physics Procedia 37, 1799 (2012)
[12] A. Delbart, Nucl. Instr. and Meth. A **623**, 105 (2010)
[13] magboltz.web.cern.ch/magboltz/
[14] garfield.web.cern.ch/garfield/help/garfield_40.html
[15] M.A. Chefdeville, PhD thesis, Univ. of Twente, http://doc.utwente.nl/62815/ (2009)

A beam monitor based on MPGD detectors for hadron therapy

P. R. Altieri[1,2,a], D. Di Benedetto[1,2], G. Galetta[1,2,b], R. A. Intonti[1,2,c], A. Mercadante[1,2], S. Nuzzo[1,2,d], and P. Verwilligen[1,2,e]

[1] *Dipartimento Interateneo di Fisica, Università degli Studi di Bari, Bari, Italy*
[2] *INFN Sezione di Bari, Bari, Italy*

Abstract. Remarkable scientific and technological progress during the last years has led to the construction of accelerator based facilities dedicated to hadron therapy. This kind of technology requires precise and continuous control of position, intensity and shape of the ions or protons used to irradiate cancers. Patient safety, accelerator operation and dose delivery should be optimized by a real time monitoring of beam intensity and profile during the treatment, by using non-destructive, high spatial resolution detectors. In the framework of AMIDERHA (AMIDERHA - *Enhanced Radiotherapy with HAdron*) project funded by the Ministero dell'Istruzione, dell'Università e della Ricerca (Italian Ministry of Education and Research) the authors are studying and developing an innovative beam monitor based on Micro Pattern Gaseous Detectors (MPDGs) characterized by a high spatial resolution and rate capability. The Monte Carlo simulation of the beam monitor prototype was carried out to optimize the geometrical set up and to predict the behavior of the detector. A first prototype has been constructed and successfully tested using ^{55}Fe, ^{90}Sr and also an X-ray tube. Preliminary results on both simulations and tests will be presented.

1 Introduction

Hadron therapy is nowadays the most precise technique of external radiation therapy [1]. Unlike particles used in conventional radiation therapy (photons or electrons), hadrons (mainly protons and light ions) can penetrate through the body and deposit most of their kinetic energy at the end of their path, giving rise to the Bragg Peak. Since hadron therapy is very sensitive to uncertainties during treatment planning and dose delivery, it is imperative to know the exact location of the dose deposition to ensure the tumor target is being irradiated, and not the surrounding healthy tissues. The hadron beams used in particle therapy are characterized by high currents (the beam flux is of the order of 10^9 particles cm^{-2}s^{-1}), the increasing use of dose delivery technique based on the Active Scanning System [2] and, as a consequence, the need to synchronize the therapeutic beam with the movements of organs such as lungs, require precise monitoring of the beam parameters such as beam intensity, beam profile, beam position and transverse beam direction, in order to ensure that the prescribed dose is correctly delivered [3].
Since GEM (GEM - *Gas Electrons Multiplier*) [4] detectors provide excellent spatial (less than 1 mm) and time resolution (1 ns) and linearity of response over a wide dynamic range, in the framework of AMIDERHA project [5] the authors are studying and developing an innovative beam monitor based on GEM detectors. Two main devices for beam monitors are under study: a Triple-GEM based detector and a Time Projection Chamber (TPC) prototype. In the present work we report on some preliminary results on both Monte Carlo simulations and first laboratory tests of GEM based beam monitors for hadron therapy application.

[a] e-mail: palma.altieri@ba.infn.it
[b] e-mail: giannicola.galetta@ba.infn.it
[c] e-mail: rosaria.intonti@ba.infn.it
[d] e-mail: salvatore.nuzzo@ba.infn.it
[e] e-mail: piet.verwilligen@ba.infn.it

2 Triple GEM detector prototype

The triple GEM beam monitor consists of three GEM foils, the drift volumes is $3 \times 100 \times 100$ mm^3, the two transfer gap are 1 mm and 2 mm wide, while the induction gap is 1 mm wide. The read-out electrode is segmented in strips (strips are 300 μm wide and spaced 400 μm apart), the device is filled with a gas mixture of Ar and CO_2. Fig. 1 shows a scheme of such a triple GEM beam monitor. Fig. 2 shows the unit cell of the GEM stack simulated with the ANSYS software [7]. The geometrical parameters of the GEM foils, also used for the Monte Carlo simulation, are the following: hole outer radius = 35μm, hole middle radius = 25 μm, kapton thickness = 50 μm, hole pitch = 140 μm, copper thickness = 5.0 μm.

2.1 Monte Carlo simulation

The detector simulation study was based on the Monte Carlo simulation code Garfield++ [6] which as been developed by the RD51 collaboration for the detailed simulations of drift chambers with gaseous media. Garfield++

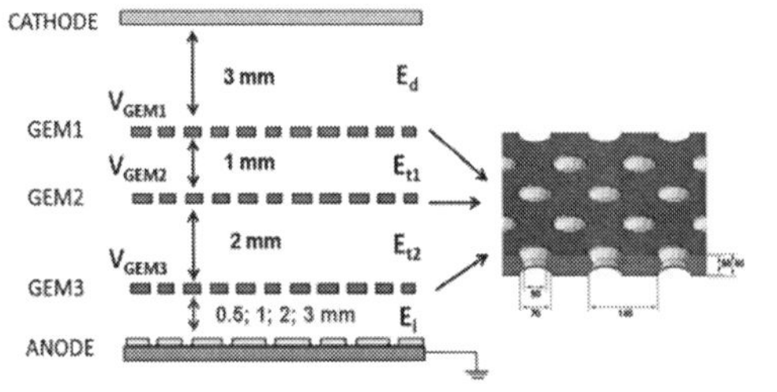

Figure 1. Triple GEM beam monitor set up.

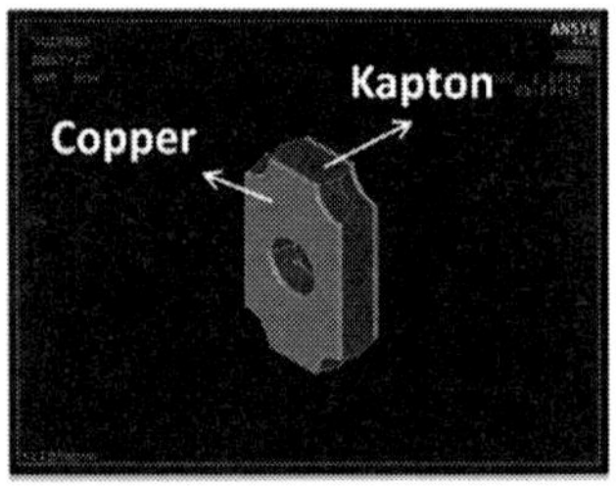

Figure 2. Unit cell of GEM foil simulated with ANSYS showing copper part (light blue) and kapton part (violet).

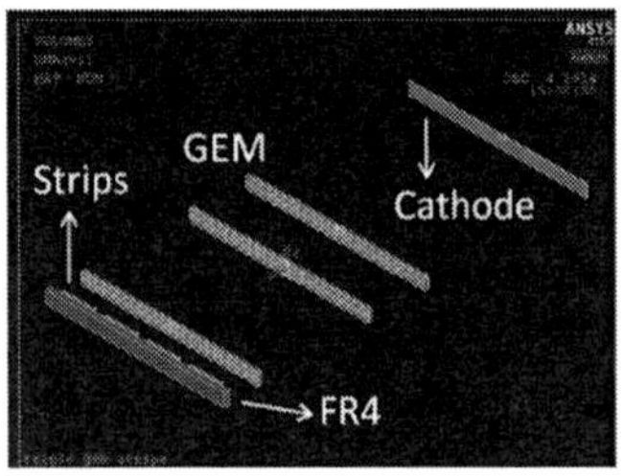

Figure 3. Triple GEM beam monitor ANSYS CAD model showing the cathode at the top, the three GEM foils and the readout strips at the bottom.

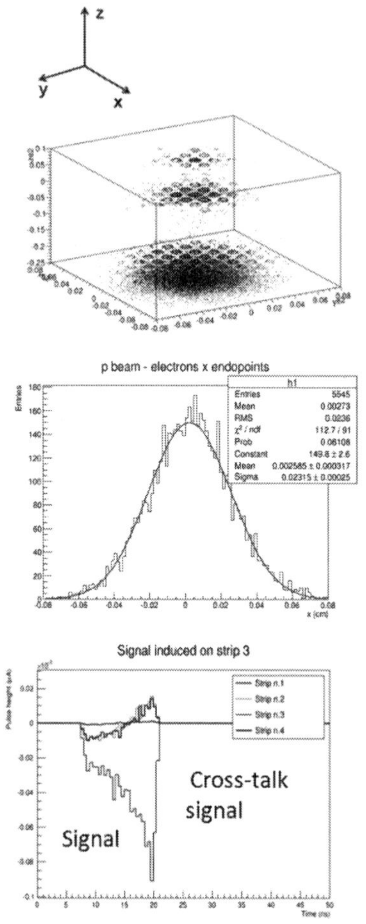

Figure 4. Avalanche production (top), spatial resolution (center) and the raw signal (bottom) produced on the anode strips simulated with Garfield++ code.

code can provide the simulation of primary and secondary tracks inside the gas and it allows the simulation of simple electric fields, for example produced by electrodes with thin-wire or plate geometry. MPGDs, like GEMs, require the simulation of electric field in complex geometries, for this reason it is necessary to calculate the electric field using finite element methods (FEM) software. The triple GEM chamber CAD model implemented in ANSYS software is shown in Fig. 3.

With the electric field map produced with ANSYS the planar triple GEM detector response to a 60 MeV proton projectile was simulated by using the Garfield++ toolkit. Results on avalanche production and spatial resolution are shown in Fig. 4 at the top and at the center respectively. Also the raw current signal induced by electrons on the anode strips was simulated thanks to the weighting fields produced by ANSYS software and the application of the Shockley–Ramo theorem. Fig. 4 (bottom) shows the raw signal induced by the avalanche electrons on a readout strip and the cross-talk signal produced on the adjacent strips.

2.2 Experimental set up

Laboratory tests were made to test the planar triple GEM detector, Fig. 5 shows the chamber during a measurement, while the charge signal produced by ionizing radiation is shown in Fig. 6. Here we report on the charge spectrum obtained with a ^{55}Fe (Fig. 7) source, while in Fig. 8 the gain obtained by irradiation with a ^{90}Sr source for different threshold values and different gas mixtures is shown.

Figure 5. Planar triple GEM detector during the tests.

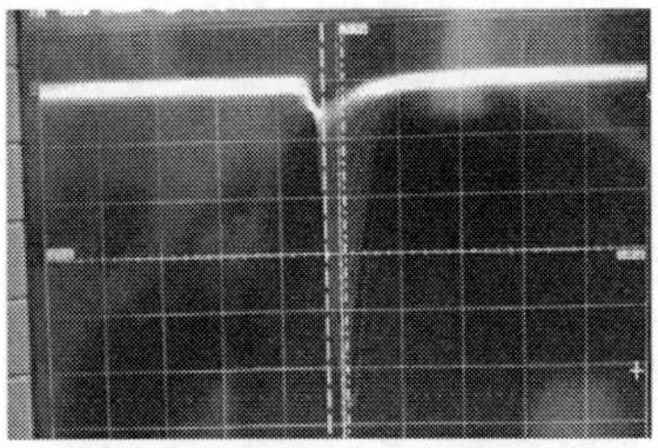

Figure 6. Triple GEM detector charge signal.

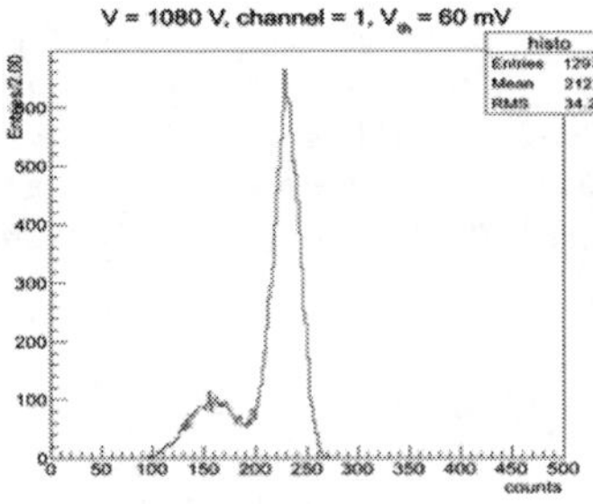

Figure 7. [55]Fe charge spectrum.

3 TPG detector prototype

A GEM based beam monitor has very good spatial and time resolution but a suitable beam monitor should also perturb as little as possible the beam characteristics. A TPC-GEM (TPG) detector, due to the low amount of material in the active volume, is "not invasive", therefore the beam characteristics are preserved, so minimizing the uncertainties on beam position, intensity and stability. For these reasons a TPG detector prototype has been designed. The TPG consists of the drift volumes of $40 \times 100 \times 100$ mm^3 with a uniform electric field produced by a field cage (100 μm diameter wires with a pitch of 2 mm).
The field cage has the shape of a cuboid, and is the first prototype which has the two transparent sides orthogonal with respect to the beam direction (to reduce the beam degradation as much as possible). The two others sides, parallel with respect to the beam direction, are printed circuits boards on which the wires are soldered. The detectors is made of three GEM foils, such that the two transfer

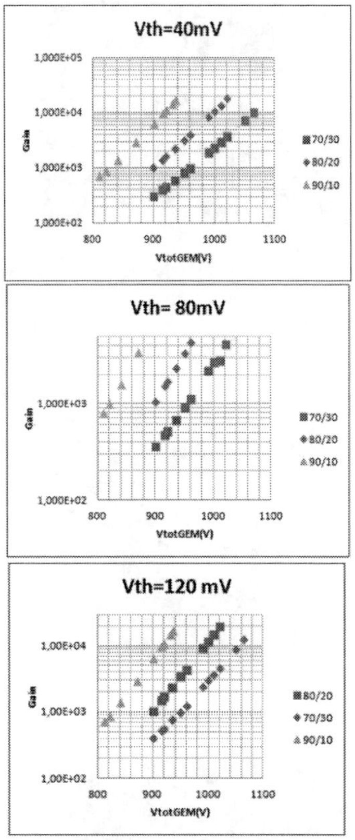

Figure 8. Gain as a function of the gas mixture (Ar and CO_2) and the threshold with a ^{90}Sr radioactive source.

gaps are 1.6 mm wide and the induction gap is 2.1 mm wide. The read-out electrode is divided into pads, there are two available geometries: 60×2 pads of 6×2 mm^2 and 30×4 pads of 3×3 mm^2. The detector is filled with a gas mixture of Ar and CO_2. In the future we will report also on tests with Ar and CF_4 gas mixtures.

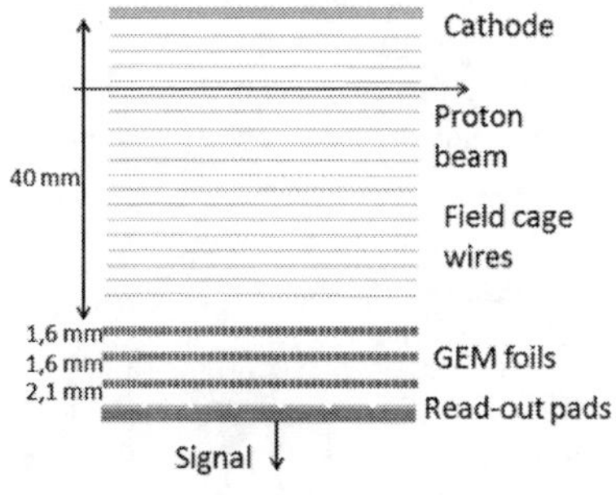

Figure 9. 13 TPG prototype schema. The beam cross the detector through two mylar windows.

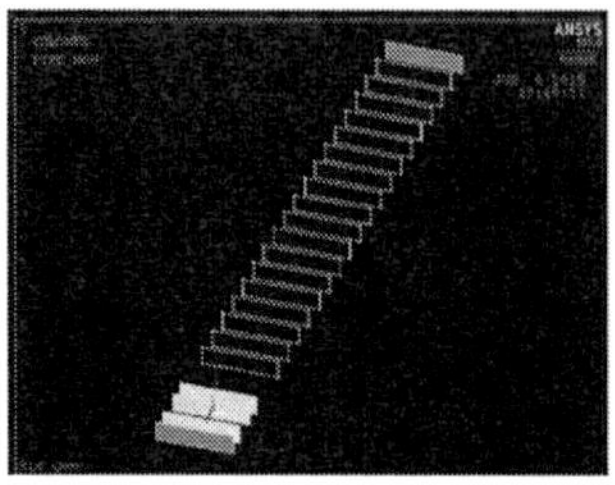

Figure 10. TPG ANSYS CAD model showing the cathode at the top, the field cage wires, the three GEM foils and the read-out plane at the bottom.

Figure 11. TPG prototype.

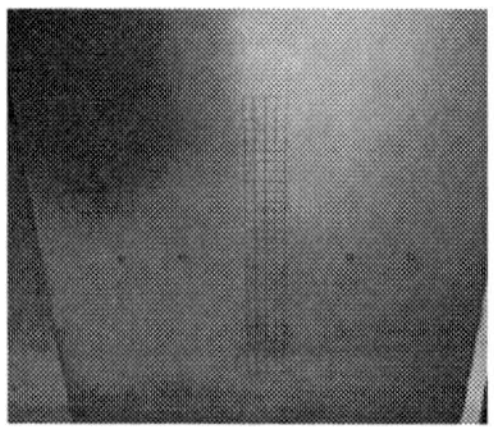

Figure 12. 3×3 mm^2 pad readout layout.

3.1 Monte Carlo simulation

The Monte Carlo simulation of the TPG prototype, as the simulation of the planar triple GEM detector, is based on the Garfield++ code and the engineering software ANSYS. The CAD model of the TPG implemented with ANSYS is shown in Fig. 10. The simulation of the electric field and detector performance are under development, and will be used in future studies to optimize the gas mixture and electric field.

3.2 Experimental set up

The TPG detector has been recently assembled and the experimental tests will start soon. Fig. 11 shows the TPG prototype, while in Fig. 12 the 3×3 mm^2 pad readout is shown. The very preliminary gain results with the use of an X-ray tube (30 kV, 130 μA, 2 mm collimator) are shown in Fig. 13. The distance between the TPG and the source is 30 cm, the drift region surface irradiated by the X-ray beam has a radius of 3 cm, the electric fields used are the following: 200 V/cm for the drift gap, 2 kV/cm for the two transfer gaps and 2.1 kV/cm for the induction gap.

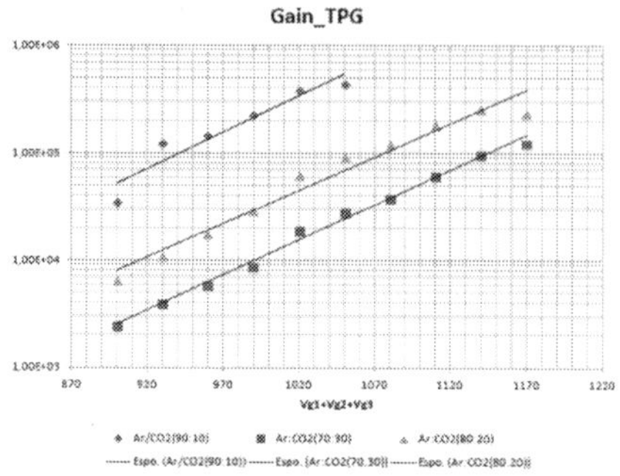

Figure 13. Preliminary TPG gain as a function of the gas mixture (Ar and CO_2) measured with an X-ray tube.

4 Conclusions

In the framework of AMIDERHA project the authors are studying and developing an innovative beam monitor based on MPGDs. Two main devices are under study: a Triple GEM based detector and a TPC-GEM prototype. Preliminary results from simulations and tests are encouraging and show that GEM based detectors are suitable for medical application. Next, we plan to test our prototypes, in particular the TPG detector, with a therapeutic proton beam.

Acknowledgements

The authors thank the Ministero dell'Istruzione, dell'Università e della Ricerca for the financial support. Particular thanks go to the Mechanical Service Design and Electronics Service of INFN-Bari for their precious support in the detector prototypes development.

References

[1] R. Orecchia et al., European Journal of Cancer **34**, 459-468 (1998).

[2] Th. Habereret al., Nuclear Instruments and Methods in Physics Research Section A, **330**, Issue 1-2, 296-305 (1993).

[3] H. Paganetti, *Proton Therapy Physics* (Taylor Francis & CRC Press, Massachusetts General Hospital and Harvard Medical School, Boston, USA), 1-704 (2011).

[4] F. Sauli, Nucl. Instr. and Meth. A, 386-531 (1997).

[5] Web site PON–REC AMIDHERA, http://www.ponrec.it/open-data/progetti/scheda-progetto?ProgettoID=5809

[6] H. Schindler and R. Veenhof, "Garfield++ - simulation of tracking detectors", http://cern.ch/garfieldpp

[7] ANSYS, http://www.ansys.com/.

Characterization of multilayer Thick-GEM geometries as ^{10}B converters aiming thermal neutron detection

H. Natal da Luz[1,a], F. A. Souza[2], M. Moralles[2], N. Carlin[1], R. A. N. Oliveira[1], M. Bregant[1], A. A. P. Suaide[1], J. F. D. Chubaci[1], M. Matsuoka[1], T. F. Silva[1], M. V. Moro[1], C. L. Rodrigues[1], and M. G. Munhoz[1]

[1] *Instituto de Física, Universidade de São Paulo, Brasil*
[2] *Instituto de Pesquisas Energéticas e Nucleares, IPEN-CNEN/SP, São Paulo, Brasil*

Abstract. Boron-based thermal neutron detectors have recently regained some attention from the instrumentation community as a strong alternative to helium-3 detectors. From the existing concepts exploiting boron layers in position sensitive detectors, the Cascade [1] is the one that takes full advantage of the 2D capabilities of gaseous detectors, with the position resolution not limited by the architecture of the detector. In this work, a proposal for the Cascade detector, based on Thick-GEMs is presented, together with some preliminary studies of the suitable pitch that optimizes the neutron conversion efficiency, while keeping the collection efficiency intact. The characterization of Thick-GEM prototypes produced in Brazil with hole pitch from 0.75 to 3 mm shows that these devices already present a stable performance at low gains, also resulting in fair energy resolution, when cascaded with a standard Kapton™ 50 µm GEM. Results of the first attempts of boron film depositions with Ion Beam Assisted Deposition and characterization by Ion Beam Analysis are also presented.

1 Introduction

Helium-3 has been for many years the most used thermal neutron converter in large detectors due to its high absorption cross section. Its recent unavailability has been well reported (e.g.: [2]) and triggered an intense research to find alternatives, turning the attention of detector developers back to boron-10.

Boron is in the solid state at NPT presenting a challenge in its deposition on surfaces, with a reasonable thickness. In addition, when increasing the thickness of the boron layer to improve the final detector efficiency, the self-absorption of the products of the nuclear reaction starts to play an important role. Therefore, there is an optimal thickness that maximizes the detection efficiency.

The use of many layers has been studied through simulations ([3, 4]) and confirmed by this work (figure 1), where it can be seen that the efficiency increases with the number of layers and the optimal thickness decreases. Furthermore, for more layers, the optimal efficiency becomes more dependent on the thickness of each layer. The deposition method must be able to deposit boron layers between 1 and 2.5 µm.

Several multi-layer geometries have already been developed, such as Multi-grid [5], Inclined detector [6], Jalousie [7] or Cascade [1] with efficiencies very close to those achieved with ^{3}He-based detectors. However, most of them achieve limited position resolutions due to the geometry of the neutron absorbers. Only the Cascade concept is able to fully exploit the position capabilitiy of a micropattern gaseous detector. However, for the stages dedicated to neutron absorption, working at unity charge gain, a simpler and more mechanically robust structure can be used. Even a detector with less stability against electrical discharges at high gains, can operate quietly when such a low gain is used.

In this work, a proposal based in the Cascade concept for the use of many boron layers is described, using Thick-GEMs (THGEM) as boron-coated neutron converters and electron transporters, together with a standard GEM-based charge amplification stage, as depicted in figure 2.

The THGEM was introduced in 2004 [8]. It is based on the Gas Electron Multiplier concept [9], with 0.5 mm holes drilled through a copper-coated FR4 board with a thickness of 0.5 mm. With the board immersed in a suitable gas mixture, by applying a high voltage between the two sides of the board, the electric field inside the holes is strong enough to multiply the primary electrons generated by the interaction of the ionizing radiation with the gas. THGEMs (like GEMs) can be used in a cascaded geometry to achieve higher gains at a more stable operation. For more details concerning GEM and THGEM operation see [10, 11]. By depositing ^{10}B films on both sides of a THGEM, the electrons generated near the surface by the absorption of the energy of the conversion reaction in the gas can be transported through several absorber layers and will be available for a final multiplication stage.

[a] e-mail: hugonluz@if.usp.br

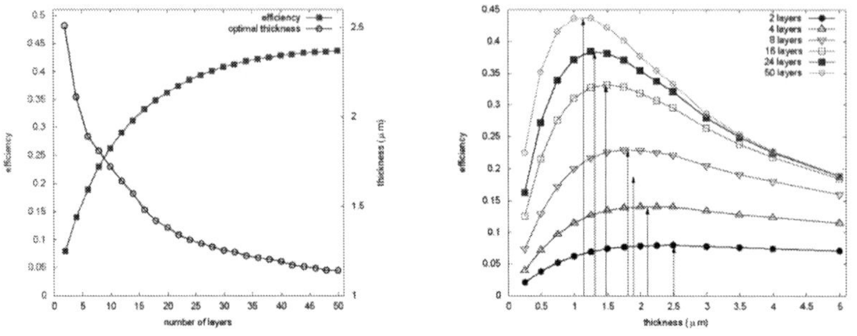

Figure 1. Simulations confirming the results of [3, 4]. Left: By increasing the number of layers, the final efficiency of the detector increases, whereas the optimal thickness decreases. Right: When the number of layers increases, the optimal thickness plateau becomes narrower. Simulation for 1.8 Å neutrons, without any energy threshold for particles leaving the ^{10}B layers.

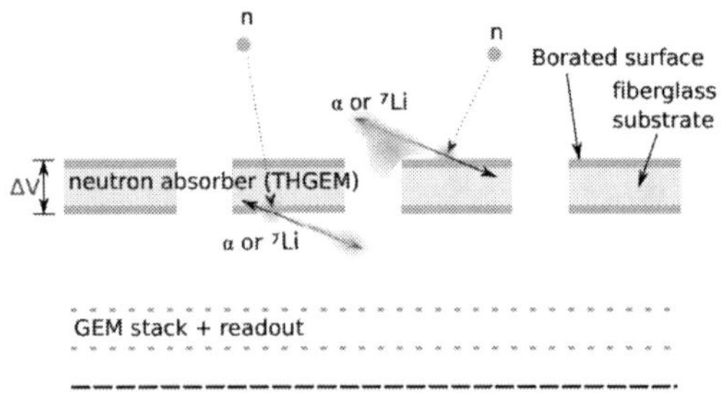

Figure 2. The THGEM based Cascade [1] concept: ^{10}B is deposited on both sides of the THGEMs that work as neutron converters and, while operating at unity gain, transport the electrons generated by the energy deposited by the conversion reaction to the multiplication stage. Only one THGEM layer is represented in the figure.

2 Experimental setup

The optical transparency of the THGEM also has an effect on the efficiency of the neutron converter. In fact, the surface available for ^{10}B deposition can increase from 60 % for a pitch of 0.75 mm up to 97 % for a pitch of 3 mm, considering a diameter of 0.5 mm for the holes. However this has a cost in the THGEM collection efficiency. It was understood soon after the introduction of the GEM that at high drift fields, the electric field inside the holes is not strong enough to push all the electrons in and some are lost on the THGEM surface, reducing the collection efficiency [12]. This effect increases with the drift field or when the distance between them becomes larger. Therefore, the hole pitch cannot increase infinitely, especially because the THGEMs serving as neutron converters will operate at very low gain, requiring low drift and transfer electric fields. Furthermore, the drift and transfer fields must be above the recombination threshold. All these features are competing with each other and a compromise between total collection efficiency, with maximum neutron conversion efficiency must be found by studying the THGEM behavior as a function of the hole pitch.

A THGEM prototype segmented in four sectors has been designed and produced using standard printed circuit board techniques. The four sectors had 0.75, 1, 2 and 3 mm hole pitch. The holes were drilled directly across the copper layers and the FR4, without any clearance between the copper and the hole. Different geometries were used in a chamber filled with an Ar/CO_2 (90/10) mixture at atmospheric pressure in an open flow of 6 L h^{-1}. The charge readout was made either from the bottom electrode of the THGEM, with a reversed induction field to avoid effects due to limitations in extraction efficiency, or with a $10 \times 10\,\mathrm{cm}^2$ pad readout consisting of a matrix of 10×12 rectangular pads. All the high voltage channels in the detector were biased independently. The drift gap was 6.2 mm thick.

3 Results

To assess the quality of the THGEMs produced, the gain of each sector was measured using an X-ray source (Amptek Mini-X, with a silver target) to generate a primary current. Figure 3 shows the gain as a function of the voltage across the holes for each of the four sectors for a single and double geometry. The gain curve for a standard single GEM is also plotted to give an idea of the scaling in voltage. The drift electric field was set to optimize the collection efficiency. The induction field was 600 V/cm and the transfer field for the double geometry was the same as the drift field. The transfer and induction gaps were 2 mm. As expected, no dependence of either the gain at each voltage or the maximum gain of each sector on the hole pitch was observed. The maximum gains achieved in single and double mode were around 150 and 1000, respectively. A clearance between the copper layer and the holes should increase the maximum gain, however, it is clear that the operation at gains around 1 will be stable.

To measure the collection efficiency, the same X-ray generator was used. A reversed induction field of 600 V/cm was applied, to allow collecting the charge at the bottom electrode of the THGEM. If the THGEM gain is kept constant, the current at the bottom electrode is proportional to the collection efficiency. Figure 4 shows the

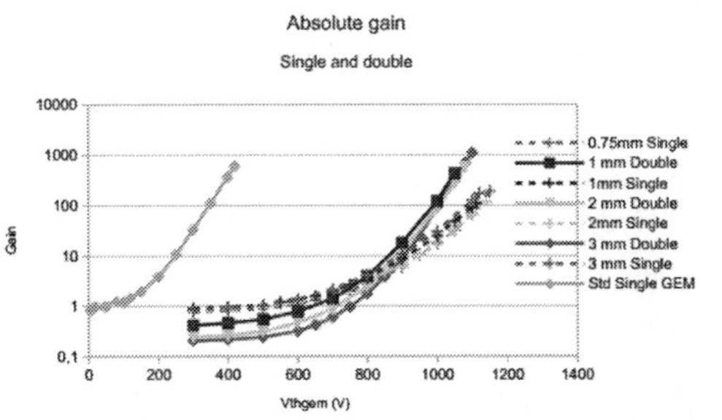

Figure 3. Gain curves for single and double geometries in each of the four sectors. The gain curve for a standard GEM is also plotted for comparison.

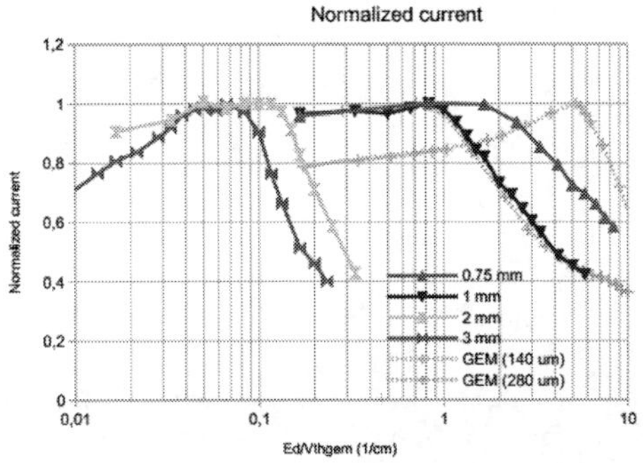

Figure 4. Current at the bottom electrode as a function of the drift electric field, normalized to the voltage across the holes of the THGEM. Each curve was normalized to its maximum value to have plateaus near 1, proportional to the collection efficiency.

current at the bottom electrode normalized to 1, as a function of the drift electric field (E_d) normalized to the voltage across the holes (V_{thgem}). For a comparison, the curves for two standard GEMs with pitch 140 µm and 280 µm are also plotted. The value of E_d/V_{thgem} at which the collection efficiency starts to drop depends on the hole pitch and is higher for smaller pitch. In the case of 2 and 3 mm pitch, this value is around 0.1 cm^{-1}, meaning that to operate at unity gain (between 200 and 400 V across the holes, according to figure 3), the drift field would have to be very low (20 to 40 V/cm) leading to higher difusion, and in the limit, to primary electron loss due to recombination.

A standard GEM with a thickness of 50 µm, holes with 70 µm diameter in the copper and a pitch of 140 µm was used as second multiplication stage to achieve a gain around 5000, high enough to provide energy spectra. The THGEM was operated at a gain around 100. The spectrum in figure 5 was taken from a ^{55}Fe radioactive source (5.9 keV) with a hole pitch of 2 mm. The energy resolution is 13.2 %. When a smaller pitch was used, the energy resolution improved, probably because the primary electron cloud spreads over more holes, reducing the influence of the non-uniformities in their size. Nevertheless, this en-

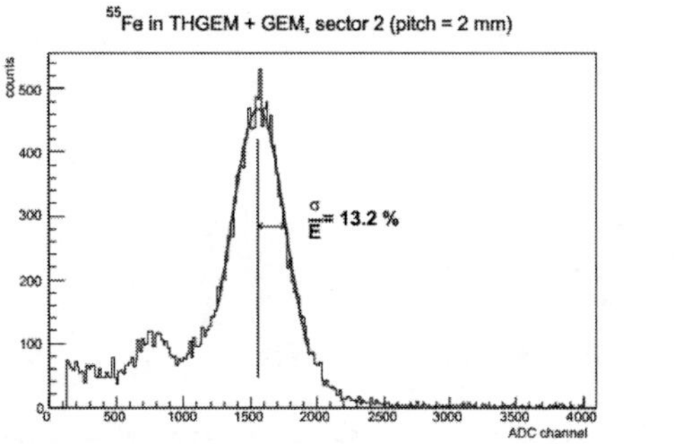

Figure 5. Pulse heigh distribution obtained with a ^{55}Fe radioactive source. On the left of the 5.9 keV peak, the argon escape peak is clearly visible.

ergy resolution is enough to allow a good discrimination of the signals when detecting thermal neutrons.

4 Boron deposition and ion beam analysis

The deposition of boron on copper surfaces poses a number of issues that have been studied and reported in other works (for example [13]). The ion beam assisted deposition (IBAD) technique [14] was used to prepare thin boron films on copper substrates. The IBAD apparatus used consists of a Kaufman-type ion gun and an electron-beam evaporator, adapted to a vacuum chamber. The base pressure in the vacuum chamber was about 7×10^{-5} Pa and 99.9 % pure boron evaporated using the electron-beam evaporator was deposited on a chemically cleaned copper substrate. The angle of incidence of the evaporated boron atoms on the substrate surface was 45° and the deposition rate of boron measured with a quartz crystal sensor was 0.054 nm/s. During the deposition, the substrate surface was normally exposed to an argon ion beam generated by the ion gun (the ion energy of 200 eV and the ion current density of 0.12 mA/cm2) to improve the adherence and densification of the film. The working pressure was around 2.4×10^{-2} Pa.

Three different Ion Beam Analysis (IBA) techniques for material characterization were performed on the same sample spot to check for the thin film composition. The selection of techniques was made regarding the complementarity of obtained information for the specific case of this study: the boron concentration was determined by Nuclear Reaction Analysis (NRA) using the $^{11}B(p,\alpha_0)^{8}Be$ nuclear reaction [15]; the oxygen content on the sample was determined by Elastic Backscattering Spectrometry (EBS); and Rutherford Backscattering Spectrometry (RBS) was used to provide additional information on the depth profile of elements. For the EBS modeling, the cross-section for the $^{16}O(p,p)^{16}O$ elastic scattering provided by SigmaCalc was used [16]. The set of spectra was processed self-consistently [17, 18], which means that all spectra were fitted simultaneously by the minimization of a χ^2 function that takes into account all spectra. The sample model that best describes all available data (minimum χ^2) is elected

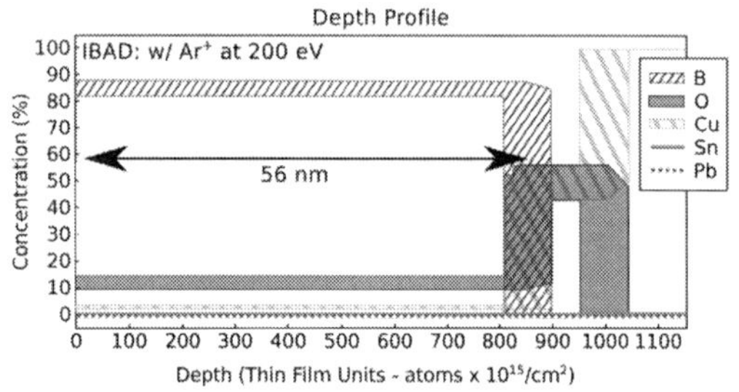

Figure 6. Element depth profile of the deposited boron film. Equivalent thickness for the boron film is 56 nm. Line thickness represents the calculated uncertainty.

to represent the sample. For a deeper description of the self-consistent approach, see Ref. [18].

Figure 6 presents the sample depth profile of elements, where the depth referred to the sample surface is expressed in 10^{15} at/cm^2 units. This model for the sample was obtained using the MultiSIMNRA software [18, 19], and the thickness of the lines represents the uncertainty obtained by a Monte-Carlo calculation to explore solutions with statistical significance on variations of the objective function. It is possible to observe a low-concentration of oxygen in the film, and a higher-concentration in the interface between the boron film and the copper substrate, suggesting the oxidation of the substrate surface prior the boron deposition. Trace-contaminants (tin and led) in low quantities (not visible in figure 6) were also found in the substrate, probably due the PCB manufacturing process.

5 Conclusions

The idea for a thermal neutron detector based in the Cascade concept, using cost effective THGEMs as boron-coated neutron converters and electron transporters was presented. The advantages of the THGEM with respect to the GEM are its mechanical robustness and cost effectiveness for large scale production.

Different distances between holes (pitch of 0.75, 1, 2 and 3 mm) were tested to determine a compromise between a larger area fraction covered with boron and a high collection efficiency. For the 3 mm pitch, the low drift electric field (around 20 V/cm) required to keep high collection efficiency is very close to the recombination threshold. The maximum gas gain (above 100) and energy resolutions ($\sigma \sim 13\,\%$) achieved are well within what is expected and needed for the operation of the detector. The position resolution of this system is dominated by the range of the conversion reaction products in the gas, the hole pitch of the THGEMs and the electronics used for charge collection in the anode. The range of the conversion products can be reduced by increasing the gas pressure in the detector. By reducing the hole pitch accordingly, the position resolution can be reduced down to less than 1 mm, with a cost in the detection efficiency.

Regarding the boron deposition, the IBA performed in the first samples made with IBAD reveals that the thickness of the boron film (56 nm) is still very small, compared to the $\sim$ 2 μm needed. Further depositions with a higher deposition rate and longer deposition times are foreseen to evaluate the limitations of this technique for larger thickness. Other deposition techniques are under study and will be tested in the near future.

References

[1] M. Klein, C.J. Schmidt, Nucl. Instr. Meth. A **628**, 9 (2011)
[2] D.A. Shea, D. Morgan, Tech. Rep. 7-5700, R41419, Congressional Research Service (2010)
[3] D. McGregor, M. Hammig, Y.H. Yang, H. Gersch, R. Klann, Nucl. Instr. Meth. A **500**, 272 (2003)
[4] J.C. Magdalena, Ph.D. thesis, Universidad de Zaragoza (2012)
[5] J.C. Buffet, J. Correa, P.V. Esch, B. Guerard, A. Khaplanov, F. Piscitelli, IEEE Nucl. Sci. Sym. 15 Conf. Rec. pp. 171–175 (2012)
[6] J. Buffet, J. Clergeau, R. Cooper, J. Darpentigny, A.D. Laulany, C. Fermon, S.Fetal, F. Fraga, B. Guérard, R. Kampmann et al., Nucl. Instr. Meth. A **554**, 392 (2005)
[7] C.J. Schmidt, M. Klein, ^{10}B – *Jalousie Detector*, http://www.n-cdt.com/
[8] R. Chechik, A. Breskin, C. Shalem, D. Mörmann, Nucl. Instr. Meth. A **535**, 303 (2004)
[9] F. Sauli, Nucl. Instr. Meth. A **386**, 531 (1997)
[10] F. Sauli, Nucl. Instr. Meth. A **805**, 2 (2016)
[11] A. Breskin, R. Alon, M. Cortesi, R. Chechik, J. Miyamoto, V. Dangendorf, J. Maia, J.D. Santos, Nucl. Instr. Meth. A **598**, 107 (2009)
[12] S. Bachmann, A. Bressan, L. Ropelewski, F. Sauli, A. Sharma, D. Mormann, Nucl. Instr. Meth. A **438**, 376 (1999)
[13] G. Nowak, M. Störmer, H.W. Becker, C. Horstmann, R. Kampmann, D. Höche, M. Haese-Seiller, J.F. Moulin, M. Pomm, C. Randau et al., Journal of Applied Physics **117**, 034901 (2015)
[14] M. Matsuoka, S. Isotani, N. Kuratani, K. Ogata, Surf. Coat. Technol. **202**, 3129 (2008)
[15] M.V. Moro, T.F. Silva, N. Added, M.A. Rizzutto, M.H. Tabacniks, J.B. Neira, J.B.F. Neto, AIP Conf. Proc. **1529**, 110 (2013)
[16] A.F. Gurbich, C. Jeynes, Nucl. Instr. Meth. B **119**, 270 (2014)
[17] T.F. Silva, M.V. Moro, G.F. Trindade, N. Added, M.H. Tabacniks, R.J. Santos, P.L. Santana, J.R.R. Bortoleto, Thin Solid Films **545**, 171 (2013)
[18] T.F. Silva, C.L. Rodrigues, M. Mayer, M.V. Moro, G.F. Trindade, F.R. Aguirre, N. Added, M.A. Rizzutto, M.H. Tabacniks, Nucl. Instr. Meth. B (2015), in press, http://dx.doi.org/10.1016/j.nimb.2015.10.038
[19] M. Mayer, Proc. of the 15th ICAARI, AIP Conference Proceedings **475**, 541 (1999)

Design, construction and quality control of resistive-Micromegas anode boards for the ATLAS experiment

F. Kuger[1,2,a], P. Iengo[1], and on behalf of the ATLAS Muon Group

[1] *CERN , Geneva (CH)*
[2] *Bayerische Julius Maximilians Universität Würzburg (DE)*

Abstract. For the upcoming upgrade of the forward muon stations of the ATLAS detector, 1280m^2 of Micromegas chambers have to be constructed. The industrialization of anode board production is an essential precondition. Design and construction methods of these boards have been optimized towards mass production. In parallel quality control procedures have been developed and established. The first set of large size Micromegas anode boards has finally been produced in industries and demonstrates the feasibility of the project on full-scale.

1 Introduction

The New Small Wheels (NSWs), an upgrade of the innermost stations of the ATLAS endcap muon detectors, will be installed during the LHC long shutdown in 2018-19 [1]. The 128 modules of Micromegas (MM) detectors [2], each one consisting of eight layers, to be installed on the two 10-meter-diameter wheels will comprise a total active detector area of 1280m^2.

The Micromegas anode boards, representing the key component of the detectors, are manufactured in industries, making the NSW Micromegas the first MPGD for a large experiment with its key component industrially produced. A total of 2048 anode boards, divided in 32 different types, each 0.5m wide and up to 2.2m long, are required for the construction of both NSWs.

This article summarizes the design of the NSW MM anode boards (Sec.2) and describes the steps of their production in industries with the corresponding quality requirements (Sec.3). The quality control (QC) scheme to be applied during this production is presented (Sec.4) and the results from a test production for NSW MM Module-0 are reported (Sec.5). Finally the status of large size Micromegas anode board production in industries is summarized (Sec.6).

2 Anode board design

The NSW Micromegas anode boards are a stacked structure of several layers, as shown in Fig. 1.

The copper readout structure comprises 1023 readout strips of 300μm width, covering the active area of the detector. The routing of the readout strips accounts for the positioning of assembly holes without strip cuts or interruptions (Fig.2d). Half of the strips are routed on the left or right to connector pads (Fig.2a) where the signal is transmitted to the front-end electronics via an elastomeric connector, avoiding soldering on the boards. Additionally the outer borders include, among other features, cross-mark targets for precision holes (Fig.2c), Rasnik Masks [4] for positioning and dimensions control (Fig.2b), alignment targets and a HV supply line including space for HV filters per side (Fig.2e).

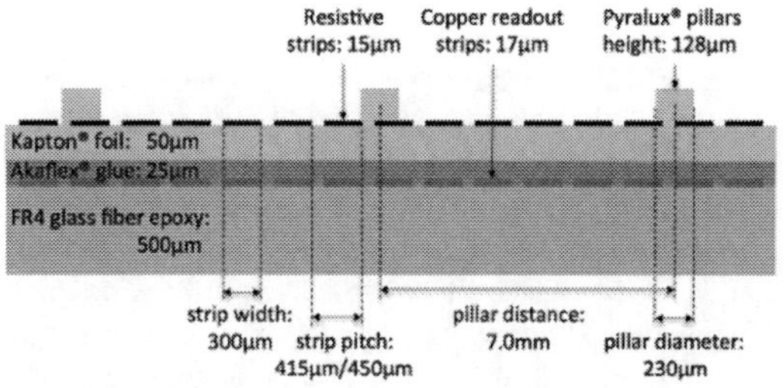

Figure 1: Schematic of the NSW Micromegas anode board layers and dimensions of its main structures (different strip pitch for small / large NSW sector PCBs). Not to scale.

A layer of strips printed with resistive ink on an insulating Kapton® foil follows the pattern of the readout strips. Interconnections between those lines are placed every 10 mm in an alternate configuration to achieve a homogeneous surface resistivity. The resistive layer provides protection for the readout electronics and spark suppression improving the high-rate capability of the detector [3]. High voltage is applied to the resistive layer from the copper pattern via a hole in the Kapton® foil. A line of silver polymer ink ensures a low impedance connection to the end of each strip (Fig.2e). The resistive strips are inter-

[a] e-mail: fabian.kuger@cern.ch

rupted in their middle (Fig.2d) to achieve a finer granularity in the high voltage distribution.

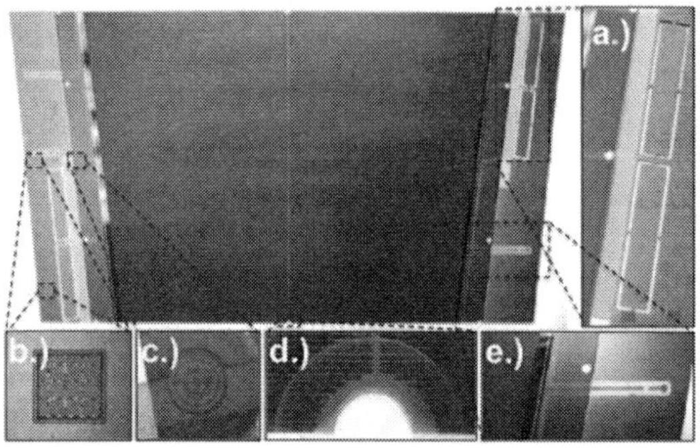

Figure 2: A NSW MM anode board with magnification of a.) the connector pad region including holes and the cut-out for electronic cooling; b.) a Rasnik Mask on various positions; c.) the precision target; d.) the routing of strips around a hole and resistive layer interruption in the board center; e.) HV supply point on the copper pattern and part of the silver conduct line.

Cylindrical pillars with a diameter of 230μm and a thickness of 128μm are arranged in a triangular array 7mm aside all over the detector area. They define, once the electrically grounded metallic mesh is applied on top, the amplification gap.

The boards are left-right symmetric to allow for installation of the same board types in a back-to-back doublet configuration. While on one doublet (eta) the strips will run perpendicular to the precision coordinate (and parallel to each other), the strips are inclined by $\pm1.5^o$ on the second doublet (stereo). This allows for the reconstruction of the precision coordinate with a spatial resolution O(100μm) in both configurations and of the second coordinate with a reduced resolution O(2-3mm) on a stereo doublet [5].

3 Production process and quality requirements

Although the design of the anode boards allows standardized industrial processes during production, the combination of these techniques, which are partially adopted from different branches of industry, and the large size of the boards result in a challenging production process. Additionally the physics performance and the mechanical requirements set tight quality constrains and acceptance criteria on the anode board.

The full production process is visualized in Fig.3 and each technique is shortly described in the following, pointing out the major quality requirements for the corresponding production step.

3.1 Copper pattern creation

The readout pattern is etched into the copper layer (17μm) on temperature stabilized glass fiber epoxy (FR4, 0.5mm

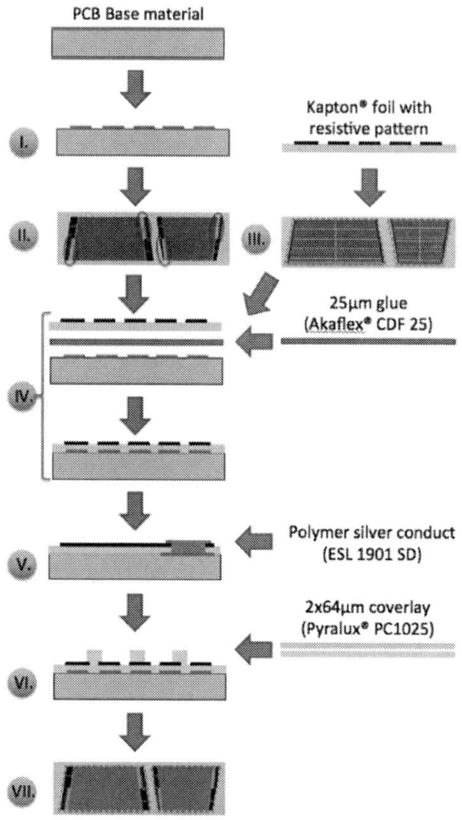

Figure 3: Process chart of the anode board construction, showing base materials and the different production steps: I. Copper pattern creation; II. Connector pad plating; III. Kapton® foil cutting; IV. Kapton®-PCB gluing; V. Silver conduct application; VI. Pillar pattern creation; VII. Cutting and drilling.

$\pm50\mu$m) via classical photo-lithographic processes [6] in a temperature and humidity controlled environment.

The required absolute accuracy is $\pm30\mu$m for the precision coordinate direction (short side) to guarantee accurate positioning of the readout strips in the detector, a precondition to achieve the required spacial resolution. On the long side $\pm100\mu$m/m accuracy are required. The local copper pattern inaccuracy, e.g. deviation in line thickness, is tolerated to $\pm20\mu$m. Shorts between two lines can be repaired and are only tolerated on <0.1% level, while up to 1% of non-repairable strip interruptions are accepted if not on neighboring strips.

3.2 Selective plating on the connector pads

To ensure a perfect conductivity between the connector pads and the elastomeric connectors, the copper has to be covered with a noble metal like Au, Ag or Pd using either electroless-/immersion-plating with a mediating Ni layer [7] or electroplating processes [6].

A complete and homogeneous coverage of the pads is required, where the layer thickness depends on the choice of the plating but ranges for all in O(1μm).

3.3 Kapton® foil preparation and cutting

The resistive pattern is applied on a Kapton® foil (50μm) via screen printing using a resistive ink (ESL D-RS 12115) [8], resulting in strips with a thickness of $\approx$15μm. This corresponds to an initial resistivity of 0.3-0.4MΩ/□, resulting in the target resistivity of $\approx$0.8MΩ/□, after an increase during the subsequent production steps.

The resistive layers are industrially produced in Japan and delivered to the PCB contractor. They have to be cut to size with moderate accuracy of $\pm$1mm and cleaned thoroughly.

3.4 Gluing of Kapton® foil on PCB

The Kapton® foil is glued on the readout side of the PCB with a 25μm thick Akaflex® glue layer under high pressure of 5-7kg/cm^2 at 170°C. An alignment accuracy of <0.5mm is required between the two layers.

Good cleanliness is essential to avoid enclosures between layers and dents on the surface. To reduce inhomogeneity in pressure, leading to air inclusion between the copper strips, additional layers of conformity material (Pacoflex® (0.27mm) or Pacotherm® (0.89mm)) are added to the stack before pressing and removed afterwards. A homogeneous, enclosure-free gluing with a flat surface is demanded, since small bumps on the surface of the anode could create discharges in the detector. The change in resistivity during the pressing has to be monitored.

3.5 Application of polymer silver conduct

The conductive line, connecting the HV input pad and the end of the resistive strips, is applied either manually or by screen printing using ESL 1901 SD silver conduct.

A coarse positioning of $\pm$1mm, flatness of the conduct line and a low impedance have to be guaranteed.

3.6 Coverlay and pillar pattern creation

After lamination with a double layer of 64μm Pyralux® PC1025, the pillar pattern is transferred via a photolithographic process [6] with a positioning accuracy of $\pm$1mm referring to the copper pattern.

The essential requirement is an (almost) complete pillar pattern, where missing pillars are only allowed if non consecutive and less than 0.1% in total. Neighboring missing pillars create the risk of discharges in the detector due to larger sagging of the metallic mesh between the supporting pillars. A maximum of ten extra structures and other defects on the coverlay are allowed if minor in size (<1x1mm^2). To obtain a homogeneous gain in the detector, a constant height of the pillars on <5μm level is required.

3.7 Board cutting and drilling of assembly holes

The finalized anode boards are milled parallel to the readout strips to size with a CNC machine on <100μm precision. The tight margin is necessary to avoid overlapping anode boards when joining them along the strips to a full detector plane.

Non-precise assembly holes are drilled with the same accuracy requirement along the outside borders of the detector plane.

4 Quality control and quality assurance

Given the complexity of the process a single quality control after finalization of the production proved to be insufficient to achieve a satisfying yield, limit material losses and trace systematic production-related errors. In close feedback with the involved industries we established a multi-step quality control plan, shown in Fig.4.

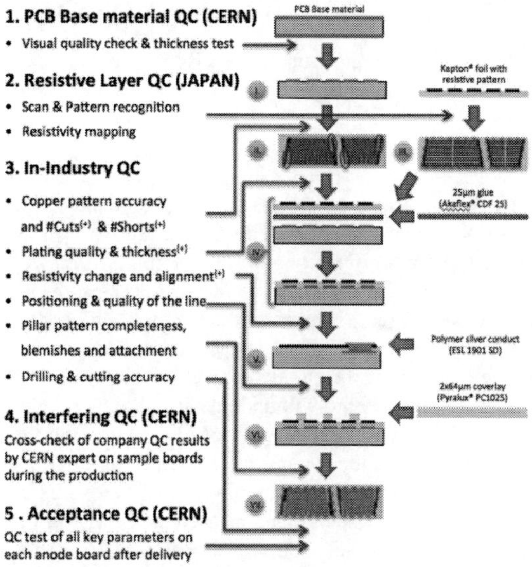

Figure 4: Visualization of the 5 stages of QC/QA during NSW MM anode board production, including the key parameters to be tested after each production step.

The relevant parameters of each production step are tested before further processing of the specific part. This minimizes material waste since unqualified boards are rejected immediately and only qualified components are further used.

Accordingly, the two supplies under responsibility of our collaboration, the PCB base material and the Kapton® foils with resistive pattern, undergo a QC procedure before being shipped to industries. This additionally enables us to track the change of material attributes during the production, like the resistivity increase, and distinguish between production faults and damages occurred during processing in industries.

Performing QC between each step of production requires involvement from industries and their operators. As shown in Fig.4 each key parameter of the different production steps will be tested immediately, where on some items (marked with $^{(+)}$) specified QC results shall be reported while on others compliance statements are considered sufficient. Facilitating these controls by e.g. integrated QC markers was one objective during the design phase. Two examples are rings on the copper pattern to verify drilling accuracy by simple visual control (Fig.5 top) and direct readable alignment targets (Fig.5 bottom).

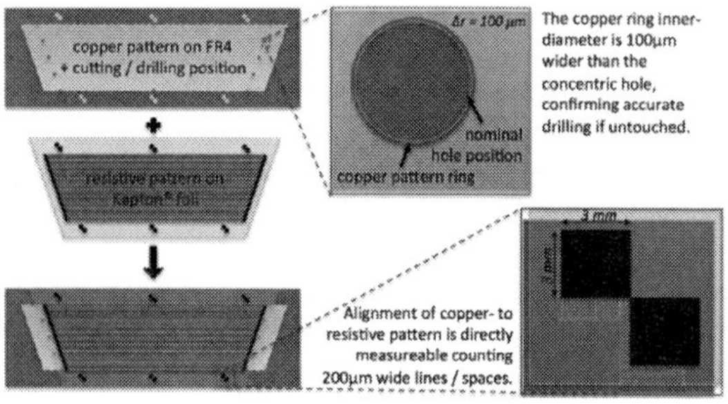

Figure 5: Left: Stacking scheme of FR4 with copper pattern, Kapton® with resistive layer and milling nominal positions. Right: Two kind of design-integrated QC markers.

Cross-checking these QC results obtained in industries by CERN personnel is foreseen during the full production. While a tight follow up during the production start-up will reduce the time for optimization of procedures and machine tuning, regular inspection during the running production are planned to maintain QC/QA standards and accordingly maximize the production yield.

Each finalized board will undergo an full quality control after delivery to CERN. Additionally to checking the key parameters, methods requiring dedicated tooling, not available in industries, can be applied at CERN. Examples are time efficient dimensional measurements (O(5μm)-accuracy) on a specially developed contact-CCD-table, O(1μm)-precise pillar height mappings and full surface resistivity measurements. Given the large quantity of boards to be tested time-efficiency, simplicity and repeatability of the methods where the main objectives during the development of the procedures and the tooling.

5 QC/QA results of first anode boards

The first anode boards to be used for the construction of NSW Micromegas full-scale prototype detectors (Module-0) have been produced in the summer of 2015 by two companies. The boards have undergone a control of quality at CERN to verify the compliance with the requirements. While some assumed critical parameters, like the homogeneity of the pillar height, were fulfilled on most of the boards, several common issues have been discovered in both productions:

· Deviations in the dimensions of the copper pattern

· Bubbles or dust between the Kapton® and the PCB

· Missing and weakly attached pillars

· Inaccurate and unclean edge cutting and hole drilling

These issues have been addressed with regular feedback and visits to the company's premises, leading to subsequent deliveries of higher quality boards. Although the overall yield of acceptable boards during Module-0 production is suffering from these start-up problems, the latest delivery included anode boards fully compliant with all requirement, demonstrating the feasibility of the production scheme and requirements in industrial processes.

6 Conclusion

The development work carried out at CERN to push the Micromegas technology to a new frontier is now coming to an end. The production of the first read-out boards for the upgrade of the ATLAS muon system demonstrates in full-scale the feasibility of the project.

A tender process to assign the production of the 2048 boards needed for the NSW project is ongoing and the mass production will start in the first half of 2016 and span over about 18 months.

PCB quality requirements are being finalized and quality control procedures in industries are established. The preparations for a final QC at CERN are in the last step of tooling preparation. Readiness of the PCB quality control procedures before the start of the series production is foreseeable.

References

[1] ATLAS Collaboration: ATLAS New Small Wheel Technical Design Report, CERN-LHCC-2013-006, ATLAS-TDR-020-2013, Geneva (2013).

[2] Y. Giomataris, et al: MICROMEGAS: a high-granularity position-sensitive gaseous detector for high particle-flux environments, Nucl. Instrum. Meth. A 376 (1996), 29-35.

[3] T. Alexopoulos, et al: A spark-resistant bulk-micromegas chamber for high-rate applications, Nucl. Instrum. Meth. A 640 (2011), 110-118.

[4] S. Aefsky, et al: The Optical Alignment System of the ATLAS Muon Spectrometer Endcaps, Journal of Instrumentation, Volume 3 (2008).

[5] M. Bianco, et al: Construction of two large-size four-plane micromegas detectors, arXiv:1511.03884 [physics.ins-det] (2015).

[6] S. Franssila: Introduction to Microfabrication, John Wiley & Sons, Ltd, 2nd Edition, Chichester (2010).

[7] J. Sudagar, J. Lian, W. Sha: Electroless nickel, alloy, composite and nano coatings - A critical review, Journal of Alloys and Compounds, Volume 571 (2013), 183-204.

[8] N. Serra: Epoxy-graphite Resistive Composites: Formulation, Characterization and Applications, EPFL Thesis No. 5346, Lausanne (2012). (doi:10.5075/epfl-thesis-5346)

Performance studies of resistive Micromegas chambers for the upgrade of the ATLAS Muon Spectrometer

Konstantinos Ntekas[1,2,a] on behalf of the ATLAS Muon Collaboration

[1] *National Technical University of Athens, Zografou Campus, 9, GR-159 73 Athens, Greece*
[2] *Brookhaven National Laboratory, Building 510, Upton NY 11973, US*

Abstract. The ATLAS collaboration at LHC has endorsed the resistive Micromegas technology (MM), along with the small-strip Thin Gap Chambers (sTGC), for the high luminosity upgrade of the first muon station in the high-rapidity region, the so called New Small Wheel (NSW) project. The NSW requires fully efficient MM chambers, up to a particle rate of $\sim 15\,\mathrm{kHz/cm^2}$, with spatial resolution better than $100\,\mu\mathrm{m}$ independent of the track incidence angle and the magnetic field ($B \leq 0.3\,\mathrm{T}$). Along with the precise tracking the MM should be able to provide a trigger signal, complementary to the sTGC, thus a decent timing resolution is required. Several tests have been performed on small ($10 \times 10\,\mathrm{cm^2}$) MM chambers using medium ($10\,\mathrm{GeV/c}$) and high ($150\,\mathrm{GeV/c}$) momentum hadron beams at CERN. Results on the efficiency and position resolution measured during these tests are presented demonstrating the excellent characteristics of the MM that fulfil the NSW requirements. Exploiting the ability of the MM to work as a Time Projection Chamber a novel method, called the μTPC, has been developed for the case of inclined tracks, allowing for a precise segment reconstruction using a single detection plane. A detailed description of the method along with thorough studies towards refining the method's performance are shown. Finally, during 2014 the first MM quadruplet (MMSW) following the NSW design scheme, comprising four detection planes in a stereo readout configuration, has been realised at CERN. Test-beam results of this prototype are discussed and compared to theoretical expectations.

1 The ATLAS New Small Wheel Upgrade Project (NSW)

The upgrade of the ATLAS [1] muon spectrometer is primarily motivated by the high background radiation expected during Run 3 (2020) and ultimately at $\mathcal{L} = 7 \times 10^{34}\,\mathrm{cm^{-2}s^{-1}}$ in HL-LHC (2025). Owing to this the detectors that occupy the innermost muon station called Small Wheel (SW), MDT, CSC & TGC, will go beyond their design luminosity limit of $\mathcal{L} = 1 \times 10^{34}\,\mathrm{cm^{-2}s^{-1}}$. In addition, the muon trigger rate will exceed the available bandwidth because of the fake endcap muon triggers (90% is coming from low energy particles, generated in the material located between the Small Wheel and the outer endcap muon station). The collaboration has decided to replace the SW with a NSW system [2], shown in Fig. 1, combining sTGC [3] & resistive MM detectors [4, 5]. Both detector technologies should provide tracking and triggering information for redundancy. The NSW will contribute to the suppression of fake triggers by reconstructing high quality ($\sigma_\theta \sim 1$ mrad) IP pointing segments online. Moreover, efficient & precise offline tracking ($\sigma_r \leq 100\,\mu\mathrm{m}$) even for the maximum expected rate of $15\,\mathrm{kHz/cm^2}$ is required.

[a] e-mail: konstantinos.ntekas@cern.ch

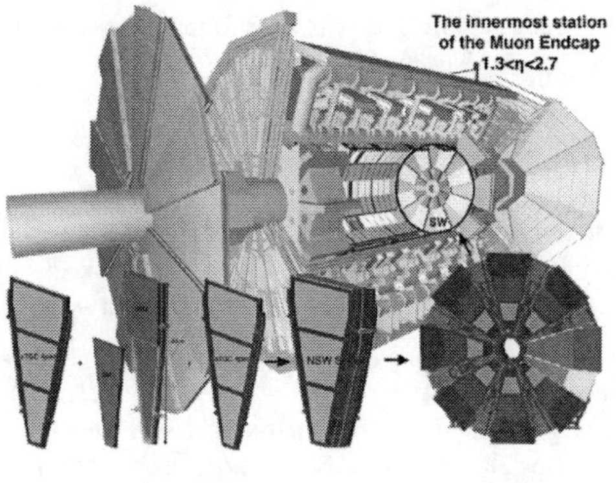

Figure 1. The muon system of the ATLAS experiment. The NSW will combine eight MM and eight sTGC detection layers per sector (eight small and eight large sectors per wheel).

2 Performance studies using test beams

In order to demonstrate the excellent performance of the MM technology and also optimise the design and operational parameters of the detector, several small and medium size chambers have been tested with medium and high momentum hadron beams at CERN. A graphical representation of the resistive MM internal structure and principle of operation is shown in Fig. 2. For the results that are presented, the front-end electronics were based on the APV25 ASIC (128 channels) [6], which will not be employed in the NSW, readout through the Scalable Readout System (SRS) developed by the RD51 collaboration [7]. The front-end ASIC providing the trigger and tracking primitives for both the MM and sTGC of the NSW will be the VMM [8] currently being produced in its second prototype version.

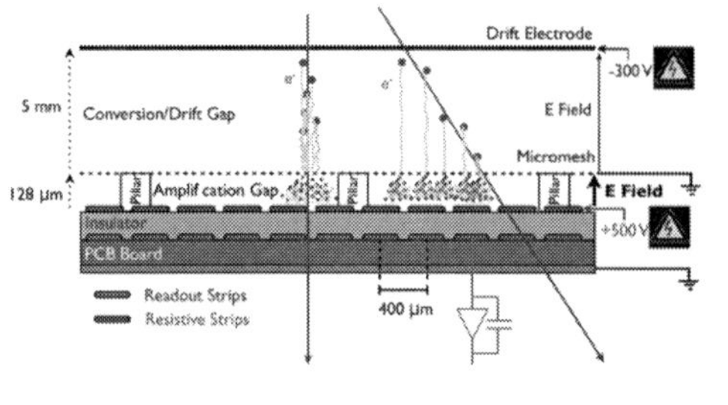

Figure 2. Cross-section of the resistive MM internal structure. A layer of resistive strips matching the read-out pattern makes the detector insensitive to discharges coming from massive ionisations.

3 Hit & Track reconstruction

The APV25 chip samples the charge every 25 ns over a predefined range ($\leq$ 675 ns). By analyzing the evolution of the integrated charge on each strip, as described in Fig. 3, the arrival time of each primary cluster along with the charge amplitude is calculated. Utilising the measured charge and time information, two different hit reconstruction techniques have been developed:

- **(Charge) Centroid :** Average of the strip position weighted by the strip charge.
- **µTPC** : 2D track reconstruction transforming the arrival time into distance. The hit is defined by the track interpolation in the middle of the drift gap. The track angle is measured using the slope of the reconstructed segment.

The charge centroid method reconstructs hits with high accuracy when the particle traverses the chamber plane perpendicularly. In this case the charge is shared among only a few strips allowing for a very precise charge interpolation. The spatial resolution determined with this method is well below 100 µm for perpendicular tracks. The µTPC method does not provide accurate results in this case as the pulse in each strip is the aggregation of pulses induced by more than one primary ionisation clusters generated at different heights within the drift gap. When a charged particle crosses the detector under an angle, the primary ionisation charge is distributed along several readout strips, with the signal induced in each strip is most probably coming from one primary cluster. In this occasion the µTPC method provides a very accurate measurement of the particle hit position on the detector. However, the strip signal amplitude becomes sensitive to the primary cluster charge fluctuations and thus the accuracy of the charge centroid method deteriorates with increasing track inclination angles.

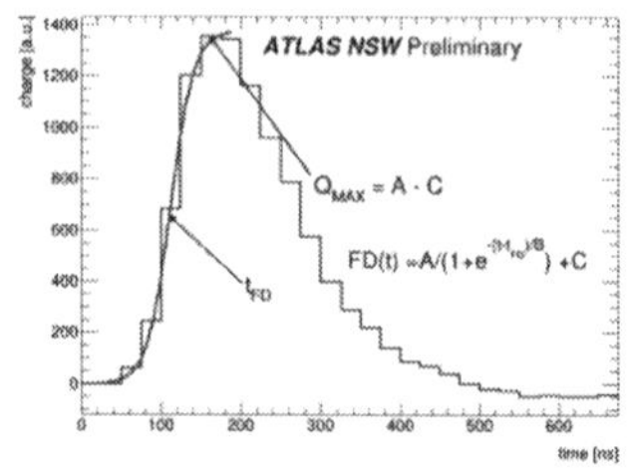

Figure 3. The integrated charge time evolution in a single strip over 25 samples of 25 ns each. A Fermi-Dirac function is used to describe the rising edge of the distribution and extract the time t_{FD} and the charge amplitude Q_{MAX}. (Taken from Ref. [9])

3.1 Refinement of the µTPC method

In the NSW the MM will be exposed to particle track angles between 8° and 30°. Thus, the performance of the µTPC track reconstruction method is of great importance for the ATLAS NSW MM project and a lot of effort has been dedicated in optimising the performance of this technique. An advanced clustering algorithm has been developed that properly treats events with more than one track, or events with noise and/or delta rays (~ 10% of the events). A pattern recognition, using the Hough transform, is used for the the track identification increasing the efficiency of the event selection and improving the accuracy of the reconstructed track by removing outlying strips [10].

Moreover, owing to the readout elements geometry, there is a non-negligible capacitance between adjacent strips causing the ionization charge to be shared among the neighbouring readout elements. The effect has been studied in simulation concluding that ~ 10% of the charge induced on a strip is capacitively coupled to its neighbours. This may bias the reconstructed track by artificially increasing the track footprint. A filtering algorithm has been developed identifying strips with charge coupled only from their neighbours discarding them from the track reconstruction procedure.

Another parameter that should be finely tuned is the assignment of the hit position along the strip pitch. In the offline reconstruction each strip position is usually assigned in the middle of the strip pitch. Actually, this is correct only for the middle strips of the cluster where the hit position is uniformly distributed along the strip pitch. For the edges of the cluster it is more probable for the hit position to be towards the strip pitch edge neighbouring to the cluster strips. The assignment of the hit in the middle of the strip pitch for the cluster edges results in biasing the reconstructed track towards larger inclination values, thus a method has been developed tuning the hit position on the cluster edges [11].

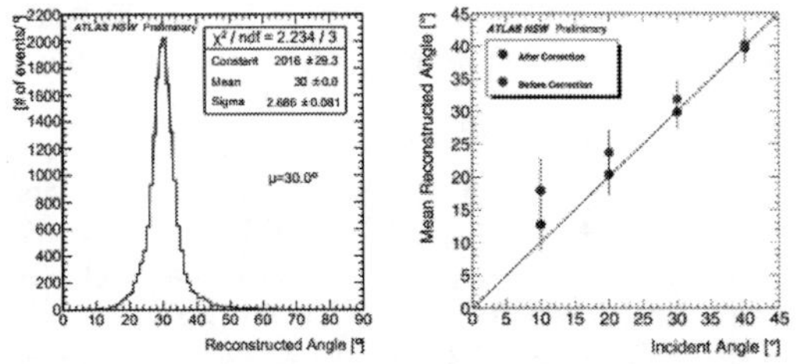

Figure 4. Left: Distribution of the track angle reconstructed with the μTPC method for a MM chamber inclined by 30° with respect to the beam. Right: The peak reconstructed angle as a function of the track inclination angle before and after the refinement of the μTPC method. (Taken from Ref. [9])

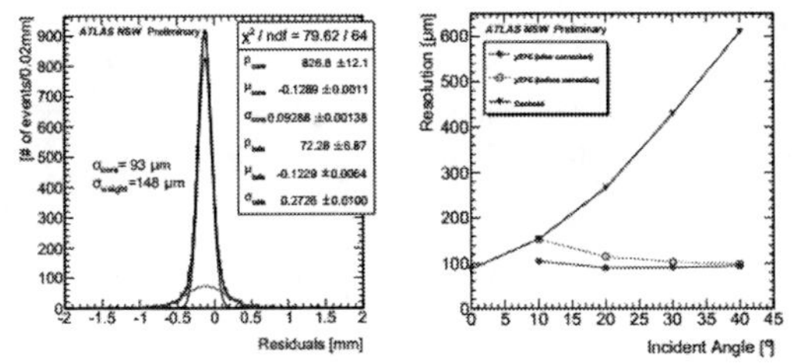

Figure 5. Left: Distribution of the position difference, divided by $\sqrt{2}$, between the reconstructed μTPC hits of two MM chambers inclined by 30° with respect to the beam. Right: The measured single spatial resolution for the centroid and the μTPC method, before and after the refinement, are shown as a function of the track angle. (Taken from Ref. [9])

The refined μTPC reconstruction algorithm was applied in test beam data with small size ($10 \times 10\,\mathrm{cm}^2$) prototypes exposed to a 120 GeV/c π^+ beam at CERN under different inclination angles. The reconstructed track angle and the measured spatial resolution with and without the refined algorithm are compared. The improvement in the accuracy of the μTPC method, using the techniques described above, is not only evident in the better agreement of the measured incident angle with the true one shown in Fig. 4, but also in the measurement of the spatial resolution which is displayed in Fig. 5. The impact is more significant for smaller incident angle values.

4 The effect of pillars in hit reconstruction

In addition to the precise hit position reconstruction the NSW MM detectors should be fully efficient even at the highest rates of $15\,\mathrm{kHz/cm}^2$ that are expected after the LHC upgrade. Along these lines, the efficiency of the small MM chambers has been studied in test beams in order to fully characterise and understand possible contributions to the detector inefficiency. Results for chambers with different characteristics and with different track inclination angles are presented.

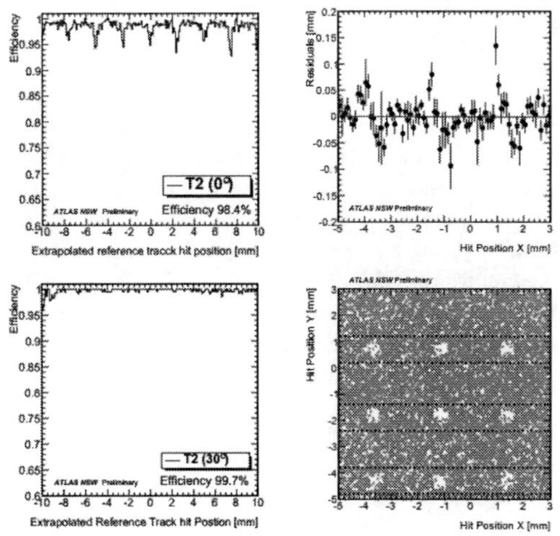

Figure 6. Left: The 300 μm in diameter cylindrical pillars placed every 2.5 mm supporting the mesh contribute 1.5% inefficiency in the case of perpendicular tracks. For inclined tracks the pillars are transparent to the efficiency owing to the larger number of strips fired per track. Right: Owing to the distortion of the electric field lines the reconstructed hit position is biassed in the the pillar region by as much as 150 μm. (Taken from Ref. [9])

Th efficiency is calculated by comparing the hits reconstructed in a single chamber with the reference chambers of the experimental set-up. The measured efficiency is plotted as a function of the reference track hit position for a MM chamber with a single readout layer in Fig. 6 (left column). In the top left plot, the case where the chamber plane is positioned vertically with respect to the beam axis is shown. Local efficiency drops, up to 5%, owing to the pillars appear every 2.5 mm. The total measured efficiency is rather high with only ~ 1.6% geometrical inefficiency attributed to the pillars. However, when the chamber is inclined by 30° with respect to the beam the pillars are no longer visible as shown in Fig. 6 (bottom left). The larger track footprint for inclined tracks makes the efficiency measurement insensitive to the pillars and an efficiency close to ~ 100% is reached.

Apart from affecting the MM efficiency, the presence of pillars locally distorts the electric field lines [12] biassing the reconstructed hit position in their vicinity. The effect has been measured for perpendicular tracks in a test beam, by calculating the residuals between a reference track and the reconstructed hit as a function of the reconstructed hit position. A MM chamber with 2D readout [13] was used and only hits reconstructed across "bands of pillars", as shown in Fig. 6 (bottom right), were included in the analysis. A bias in the reconstructed hit position around each pillar is evident in Fig. 6 (top right) with a maximum deviation of ~ 150 μm which is in qualitative agreement with what has been measured in simulation studies of a MM with different operational parameters [12].

5 Performance study of the first MM quadruplet

The performance of the recently developed MM quadruplet (MMSW) [14] was studied in a dedicated test beam at CERN. The MMSW was positioned in the beam line along with a hodoscope equipped with small MM chambers providing the 2D reference track. The operation of the MMSW, using the standard Ar+7% CO_2 gas mixture, was smooth during the whole period of irradiation with all the four layers being $\sim$ 99% efficient. For the results presented here the chamber was kept perpendicular to the beam axis.

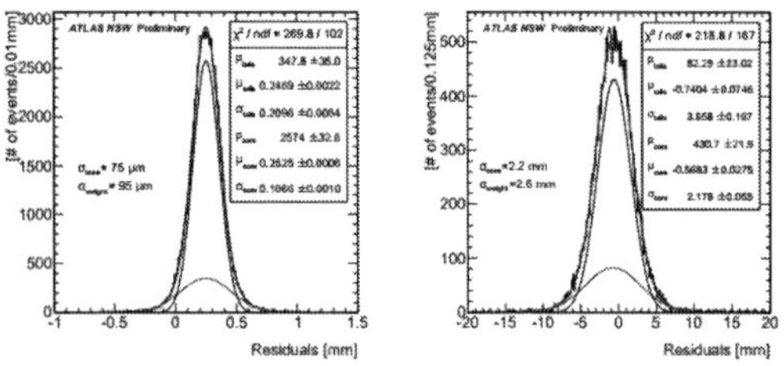

Figure 7. Left: Distribution of the hit position difference, divided by $\sqrt{2}$, between the precision coordinate layers of the MMSW. Right: Residual distribution of the MMSW second coordinate hit position with a second coordinate hit coming from one of the reference chambers. (Taken from Ref. [9])

By plotting the hit position difference between the first two layers of the MMSW, with readout strips parallel to each other, we were able to determine the spatial resolution of each of the two precision layers equal to 75 µm as shown in Fig. 7 (left). Exploiting the stereo readout strip geometry of the quadruplet, the second coordinate is also reconstructed. The chambers of the reference telescope are used to reconstruct the second coordinate of the extrapolated track on the MMSW frame. By comparing this with the second coordinate hit reconstructed combining the information from two stereo readouts the spatial resolution of the second coordinate is measured to be of the level of 2.2 mm, Fig. 7 (right) which is in accordance with theoretical expectations. Simulation studies of the stereo strips configuration [15] dictate, that by combining two stereo layers, with a 3° angle between them, the second coordinate can be reconstructed with an uncertainty of $\sim$ 2.2 mm, assuming single layer spatial resolution of $\sim$ 80 µm.

6 Conclusions

A series of performance studies in order to optimise the MM performance and operating conditions, in the context of the ATLAS NSW upgrade, are presented. The studies are mainly focussed on the development and optimisation of the µTPC reconstruction method that provides a precise segment reconstruction using a single detector gap. Moreover, studies on the contribution of the pillars in the measured efficiency and hit reconstruction accuracy are discussed. A non-negligible effect is measured for perpendicular tracks which is in qualitative agreement with simulation calculations. Finally, the MMSW quadruplet prototype, that follows the final NSW MM quadruplet layout, has been studied in test beam and the results presented follow the simulation and theoretical expectations.

Acknowledgements

The work was co-funded by the European Union (European Social Fund ESF) and Greek national funds through the Operational Program "Education and Life-long Learning" of the National Strategic Reference Framework (NSRF) 2007-2013 ARISTEIA-1893-ATLAS MICROMEGAS.

References

[1] ATLAS Collaboration, JINST 3 S08003 (2008)
[2] ATLAS Collaboration, CERN-LHCC-2013-006, ATLAS-TDR-020 (2013)
[3] V. Smakhtin *et al.*, Nucl. Instr. Meth. A 598 196-200 (2009)
[4] Y. Giomataris *et al.* , Nucl. Instr. Meth. A376, 29-35, (1996)
[5] T. Alexopoulos *et al.*, Nucl. Instr. Meth. A 640 110-118 (2011)
[6] M.J. French *et al.*, Nucl. Instr. Meth. A466, 359-365, (2001)
[7] S. Martoiu *et al.*, Nucl. Sc. Symp. and Med. Imag. Conf., IEEE, pp.2036,2038 (2011)
[8] G. De Geronimo *et al.*, IEEE Transactions on Nuclear Science, vol. 60, no. 3, (2013)
[9] ATLAS NSW Public Results : https://twiki.cern.ch/twiki/bin/view/AtlasPublic/NSWPublicResults
[10] T. Alexopoulos *et al.*, ATL-UPGRADE-PUB-2014-001 (2014)
[11] K Ntekas, ATL-MUON-PROC-2014-011 (2014)
[12] P. Bhattacharya *et al.*, Nucl. Instr. Meth. A 793, 41-48 (2015)
[13] M. Byszewski *et al.*, Journal of Instrumentation 7 02, C02060 (2015)
[14] M. Bianco *et al.*, arXiv:1511.03884 (2015)
[15] T. Alexopoulos *et al.*, ATL-MUON-PUB-2015-001 (2015)

Development of the GEM-TPC X-ray Polarimeter with the Scalable Readout System

Takao Kitaguchi[1,2,a], Asami Hayato[2], Wataru Iwakiri[2], Yoko Takeuchi[2,3], Megu Kubota[2,3], Kazuki Nishida[2,3], Teru-aki Enoto[4,2], and Toru Tamagawa[2]

[1] *Department of Physical Sciences, Hiroshima University, Higashi-Hiroshima, Hiroshima 739-8526, Japan*
[2] *RIKEN Nishina Center, 2-1 Hirosawa, Wako-shi, Saitama 351-0198, Japan*
[3] *Department of Physics, Tokyo University of Science, 1-3 Kagurazaka, Shinjuku-ku, Tokyo 162-8602, Japan*
[4] *Department of Astronomy, Kyoto University, Sakyo-ku, Kyoto 606-8502, Japan*

Abstract. We have developed a gaseous Time Projection Chamber (TPC) containing a single-layered foil of a gas electron multiplier (GEM) to open up a new window on cosmic X-ray polarimetry in the 2–10 keV band. The micro-pattern TPC polarimeter in combination with the Scalable Readout System produced by the RD51 collaboration has been built as an engineering model to optimize detector parameters and improve polarimeter sensitivity. The polarimeter was characterized with unpolarized X-rays from an X-ray generator in a laboratory and polarized X-rays on the BL32B2 beamline at the SPring-8 synchrotron radiation facility. Preliminary results show that the polarimeter has a comparable modulation factor to a prototype of the flight one.

1 Introduction

Recent progress in micro-pattern gas detectors enables us to sensitively and shapely track charged particles including photoelectrons with an energy down to 1 keV. The angular distribution of photoelectrons originated from the photoelectron effect is sensitive to the electric field vector (or polarization direction) of incident X-rays. Since the photoelectron effect is the dominant interaction of X-rays, the micro-pattern gas polarimeter is expected to open up a new window on cosmic X-ray polarimetry.

So far, the Crab Nebula is the only astronomical source that has been detected by X-ray polarization measurements since the 1970s [1]. In order to make a breakthrough in cosmic X-ray polarimetry, three satellite missions, each of which will carry micro-pattern gas polarimeters utilizing the photoelectric effect in orbit, were selected in 2015 as a promising candidate for launch in the 2020s. XIPE (X-ray Imaging Polarimetry Explorer) [2] is an ESA medium-class mission candidate, while IXPE (Imaging X-ray Polarimetry Explorer) and PRAXyS (Polarimeter for Relativistic Astrophysical X-ray Sources) [3] are NASA Small Explore finalists.

Micro-pattern gas polarimeters can detect the X-ray polarization by imaging a 2-D photoelectron track projected on to a plane perpendicular to the incident X-ray direction. They are mainly divide into two types according to the X-ray imaging capability. XIPE and IXPE will carry X-ray imaging polarimeters with a 2-D readout system [4, 5]. PRAXyS will have a nonimaging but higher sensitive polarimeter with 1-D strip electrodes using the time projection chamber (TPC) technique [6, 7]. In addition, the electric drift field direction is different: the imaging polarimeter has parallel electric field to the incident X-ray direction, whereas the nonimaging one has perpendicular field and therefore is unable to determine an interaction point along the electron drift direction. Instead, the latter polarimeter has more thicker absorption volume along the incident X-ray direction.

This paper deals with development of an engineering model of the micro-pattern TPC polarimeter combined with the Scalable Readout System (SRS) [8] produced by the RD51 Collaboration [9]. The polarimeter with SRS has been developed for ground tests to improve polarimetry sensitivity by quickly and flexibly changing detector parameters such as gas pressure and electric field strengths. Preliminary results of polarimeter performance evaluation is also described. However, more detailed results will be described elsewhere.

2 Overview of X-ray polarimeter

The detailed design and detection concept of the micro-pattern TPC polarimeter for cosmic X-rays onboard PRAXyS is described in [6]. A schematic view of the polarimeter with SRS is illustrated in Figure 1. The polarimeter chamber is sealed with dimethyl ether gas at low pressure of 190 Torr and consists of a thick (8 cm along the incident X-ray direction) drift/absorption region between the drift electrode and the gas electron multiplier (GEM) cathode and an induction region between the GEM anode and 128-ch readout strips with a pitch of 121 μm. These devices installed in the chamber for the engineering model

[a] e-mail: kitaguti@hiroshima-u.ac.jp

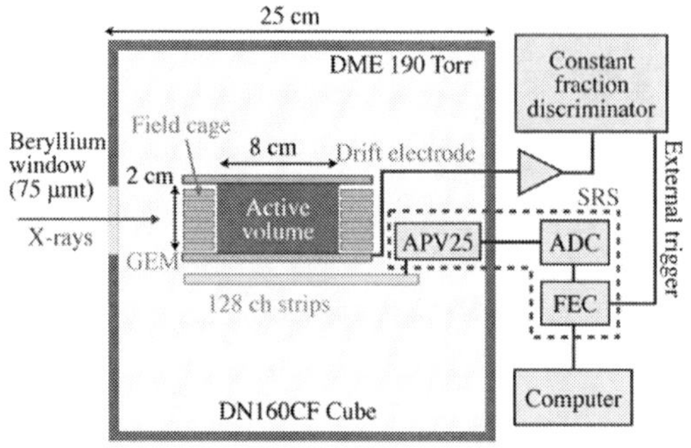

Figure 1. Schematic view of the polarimeter combined with SRS. The detector is self-triggered by using the GEM cathode signal in input to the trigger connector of SRS/FEC. The polarimeter chamber is placed on the X-Z stage to adjust the X-ray beam position.

Table 1. Difference of polarimeter parameters.

	Polarimeter	
	This work	Flight model
Sampling number	27	30
Sampling rate (MHz)	40	20
Drift field (V/cm)	374	196
Induction gap (mm)	1	0.25

are almost the same as those of the flight prototype. The main difference between them is a signal readout system: the flight prototype uses the APV25 ASIC [10] controlled by a NASA-programmed FPGA, whereas the polarimeter utilizes SRS with APV25. The other differences on detector settings are listed in Table 1. Because the ADC sampling rate of SRS is fixed at 40 MHz and twice as fast as that of the flight polarimeter, the electron drift velocity is set to be 4.84 μm ns^{-1} by adjusting the drift field strength so that the TPC image pixel size is 121 μm for both polarimeters.

An online data analysis software dedicated to X-ray polarimetry was developed based on the ROOT analysis framework [11]. The event-by-event analysis software consists of three threads: an event receiver, an analyzer/recorder, and a monitor. A quick-look monitor to display plots of a track image, spectrum, modulation curve, and so on as shown in Figure 2 allows us to check preliminary results of the polarization measurement as soon as data from SRS are taken. The data analyzer mainly performs track image processing which execute pedestal subtraction, common mode noise subtraction, electronics response deconvolution, and photoelectron angle reconstruction in sequence. Before X-ray beam irradiation, data of the pedestal and the impulsive electronics response are acquired with periodic internal triggers generated by FPGA on the SRS front-end card (FEC) and periodic test pulses from the on-chip calibration circuit on APV25, respectively.

3 Preliminary results of polarimeter performance

In order to characterize the polarimeter, it was irradiated with unpolarized X-rays in a laboratory and polarized X-rays on the BL32B2 beamline at the SPring-8 synchrotron facility. Unpolarized X-rays at energies of 4.5 and 6.4 keV separately generated by an X-ray generator were used to calibrate polarimeter parameters (gas pressure, high voltages, image processing parameters, and so on) so that the modulation curve, which is a distribution of photoelectron emission angles, becomes flat. The parameters derived from the unpolarized data were then used for polarized data. Polarized and collimated X-rays were generated at four energies (4.5, 5.5, 6.4, and 7.5 keV) at five detector positions or drift lengths from the GEM cathode (4.6, 7.6, 10.6, 13.6 and 16.6 mm). The polarized beam test at SPring-8 was performed in December 7–15, 2014.

Figure 3 shows 6.4 keV spectra at various detector positions. The energy resolution (FWHM) of all the spectra is 22%, which is larger than that of a prototype of the flight polarimeter, 16% [12]. The broader energy resolution is due to a non-uniform gain of the GEM foil [13] and could be improved by replacing it with a foil of improved quality. The spectral peak is shifted to the lower channel with the increase of the drift distance because of electron attachment of impurity gas caused by outgassing [14]. An electron attachment coefficient at the drift field of 374 V cm^{-1} is $\sim$ 0.05 cm^{-1}, derived by exponential fitting to the spectral peak as a function of the drift distance. The spectral peak shift due to electron attachment (not seen in the flight equivalent polarimeter [12]), can be reduced by long-term bake-out, which was performed, for the polarimeter chamber, with the temperature of 80°C for a week before the beam test.

Figure 4 shows photoelectron tracks with various energies at different X-ray irradiation positions. Some of them are curved due to large-angle Coulomb scattering with DME gas. The pixel region having a large amount of charge is the Bragg peak which represents the end point of the track. The track length increases with the increase of the incident X-ray energy. In addition, the track is more blurred with the increase of the electron drift length due to electron diffusion. The diffusion size for 1 cm drift length at the electric field of 374 V cm^{-1} is 140 μm or 1.2 pixel, which is numerically calculated with Magboltz [15].

In order to reconstruct the photoelectron emission angle, the major principal axis with the origin at the barycenter from the charge distribution of the track image is regarded as the reconstructed angle. The Bragg peak, which represents the track end point and is unnecessary to reconstruct the initial angle of photoelectron emission, is clearly seen in long and curved tracks with a high eccentricity (e.g. the top left panels of Figure 4). Therefore, for such high eccentricity tracks, the half of the charge distribution which is separated by the minor principal axis and contains the Bragg peak is cut off. Then, the major principal axis, which is regarded as the revised photoelectron emission angle, is calculated again with the remaining charge distribution. The Bragg peak cut improves the modulation

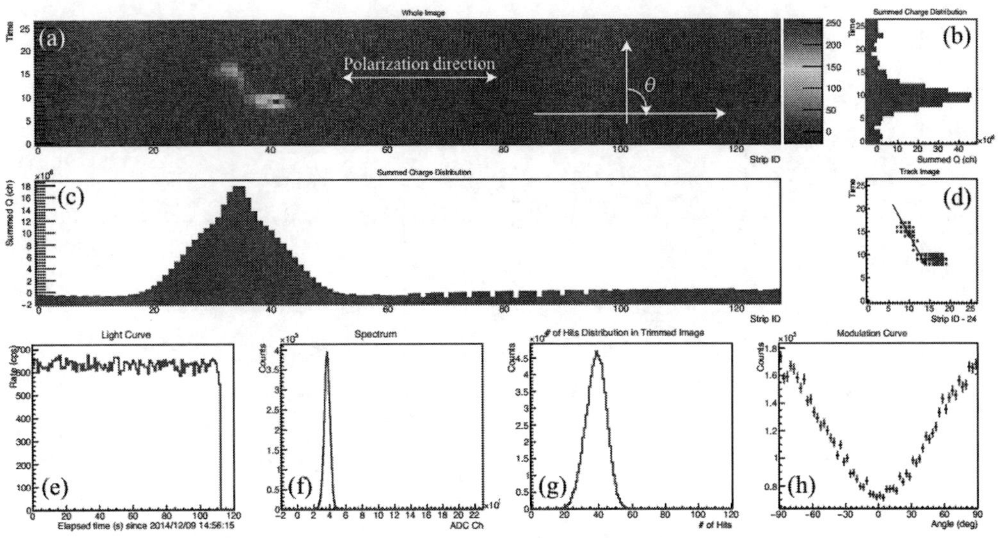

Figure 2. Online monitor to display polarization measurement data. Individual plots represent (a) full (27×128 pixels) track image with the θ angle definition, (b) and (c) stacked image projected to the Y-and X-axis of the track image, respectively, (d) trimmed (27 × 27 pixels) track image centering on the charge centroid with the blue arrow representing a estimated angle of photoelectron emission direction, (e) time variation of count rates, (f) ADC channel spectrum, (g) distribution of the number of hit pixels, and (h) modulation curve in the coordinate system that 0 deg points the top of the track image. The data were taken by irradiation of polarized X-rays with the energy of 6.4 keV at SPring-8/BL32B2.

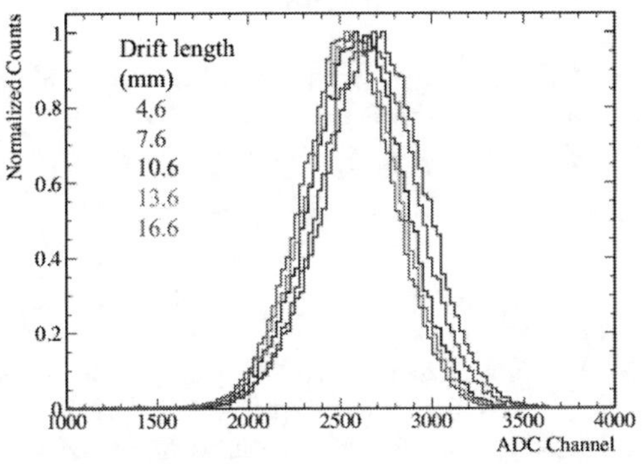

Figure 3. Measured spectra with 6.4 keV X-ray irradiation at different detector positions. The ADC channel of each event is total amount of charge in the trimmed track image with the size of 27 × 27 pixels.

factor of high-energy (> 5 keV) X-rays by approximately 10%. A preliminary result of the modulation factor for 6.4 keV X-rays at 7.6 mm drift length, which is near the optical axis of the polarimeter combined with an X-ray mirror, is approximately 50%, comparable to that of the flight equivalent polarimeter [12].

4 Summary

The engineering model of the micro-pattern TPC polarimeter for cosmic X-rays has been developed in combination with the SRS/AVP25 signal readout module for ground tests. The polarimeter was characterized with polarized X-rays at SPring-8/BL32B2. Preliminary results show that the developed polarimeter has a comparable modulation factor to the flight equivalent one.

Acknowledgements

The synchrotron radiation experiments were performed at the BL32B2 of SPring-8. We would like to thank Dr. Dai Takei and Dr. Yuki Takayama for their supports on X-ray beam adjustment. T.K. was supported by JSPS Grant-in-Aid for Young Scientists (B) (No. 24740185). This work was partially supported by MEXT KAKENHI Grant Number 24105007.

References

[1] M.C. Weisskopf, G.G. Cohen, H.L. Kestenbaum, K.S. Long, R. Novick, R.S. Wolff, Astrophys. J. Lett. **208**, L125 (1976)

[2] P. Soffitta, X. Barcons, R. Bellazzini, J. Braga, E. Costa, G.W. Fraser, S. Gburek, J. Huovelin, G. Matt, M. Pearce et al., Experimental Astronomy **36**, 523 (2013)

[3] K.M. Jahoda, J.K. Black, J.E. Hill, T.R. Kallman, P.E. Kaaret, C.B. Markwardt, T. Okajima, R. Petre, Y. Soong, T.E. Strohmayer et al., *X-ray polarization capabilities of a small explorer mission*, in *SPIE Conference Series* (2014), Vol. 9144, p. 0

[4] E. Costa, P. Soffitta, R. Bellazzini, A. Brez, N. Lumb, G. Spandre, Nature **411**, 662 (2001)

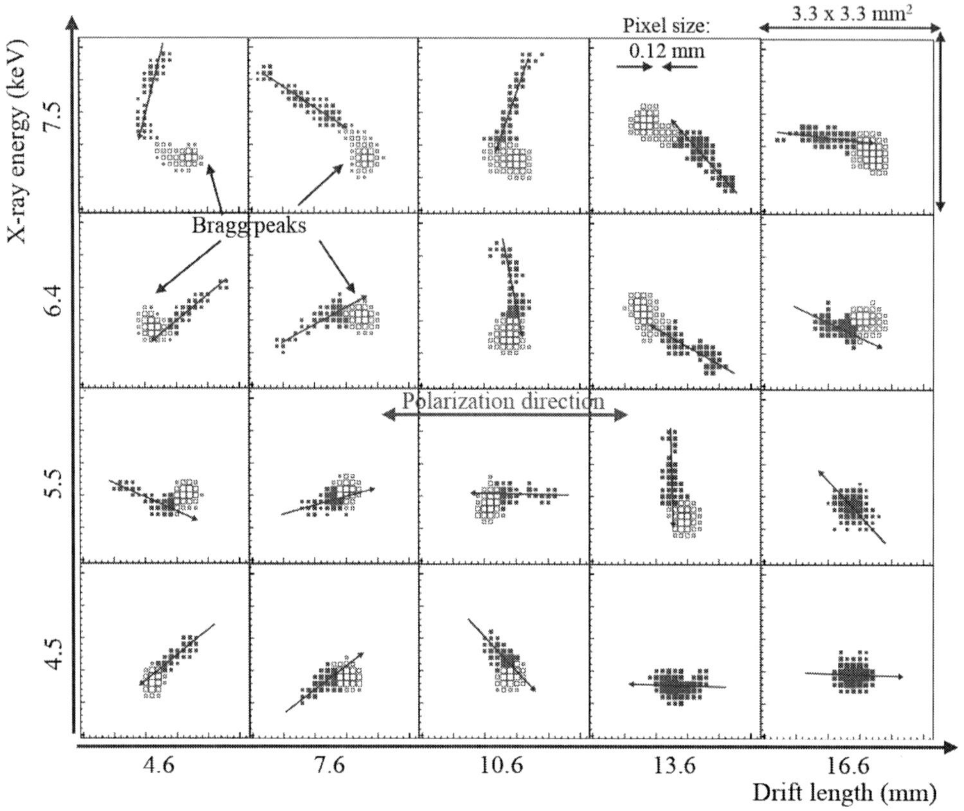

Figure 4. Zoo of photoelectron track images acquired by the polarimeter with SRS. Although the full image size is 128 (the number of strip electrodes) × 27 (sampling number), the above images are trimmed to a square image with a size of 27 × 27 centering on the charge centroid. The red box area is proportional to charge amount. The open red boxes around the Bragg peaks show unused pixels for angle reconstruction of photoelectron emission.

[5] R. Bellazzini, G. Spandre, M. Minuti, L. Baldini, A. Brez, L. Latronico, N. Omodei, M. Razzano, M.M. Massai, M. Pesce-Rollins et al., Nuclear Instruments and Methods in Physics Research A **579**, 853 (2007)

[6] J.K. Black, R.G. Baker, P. Deines-Jones, J.E. Hill, K. Jahoda, Nuclear Instruments and Methods in Physics Research A **581**, 755 (2007)

[7] J.E. Hill, J.K. Black, T.J. Emmett, T. Enoto, K.M. Jahoda, P. Kaaret, D.S. Nolan, T. Tamagawa, *Design improvements and x-ray performance of a time projection chamber polarimeter for persistent astronomical sources*, in *SPIE Conference Series* (2014), Vol. 9144, p. 1

[8] S. Martoiu, H. Muller, A. Tarazona, J. Toledo, Journal of Instrumentation **8**, C3015 (2013)

[9] S. Duarte Pinto (RD51), *Micropattern gas detector technologies and applications, the work of the RD51 collaboration*, in *Proceedings, 2010 IEEE Nuclear Science Symposium and Medical Imaging Conference (NSS/MIC 2010)* (2010), pp. 802–807

[10] M.J. French, L.L. Jones, Q. Morrissey, A. Neviani, R. Turchetta, J. Fulcher, G. Hall, E. Noah, M. Raymond, G. Cervelli et al., Nuclear Instruments and Methods in Physics Research A **466**, 359 (2001)

[11] R. Brun, F. Rademakers, Nuclear Instruments and Methods in Physics Research A **389**, 81 (1997)

[12] T. Enoto, J.K. Black, T. Kitaguchi, A. Hayato, J.E. Hill, K. Jahoda, T. Tamagawa, K. Kaneko, Y. Takeuchi, A. Yoshikawa et al., *Performance verification of the Gravity and Extreme Magnetism Small explorer (GEMS) x-ray polarimeter*, in *SPIE Conference Series* (2014), Vol. 9144, p. 4

[13] Y. Takeuchi, T. Kitaguchi, A. Hayato, T. Tamagawa, W. Iwakiri, F. Asami, A. Yoshikawa, K. Kaneko, T. Enoto, K. Black et al., *Properties of the flight model gas electron multiplier for the GEMS mission*, in *SPIE Conference Series* (2014), Vol. 9144, p. 4

[14] J.E. Hill, J.K. Black, L. Brieda, P.L. Dickens, K. Montt de Garcia, D.L. Hawk, A. Hayato, K. Jahoda, J. Mohammed, *Lifetime estimation of a time projection chamber x-ray polarimeter*, in *SPIE Conference Series* (2013), Vol. 8859, p. 0

[15] S.F. Biagi, Nuclear Instruments and Methods in Physics Research A **421**, 234 (1999)

EPJ Web of Conferences 174, 01016 (2018) https://doi.org/10.1051/epjconf/201817401016
MPGD 2015

Characterization of a hybrid GEM-Micromegas detector with respect to its application in a continuously read out TPC

Viktor Ratza[1], Markus Ball[1], M. Liebtrau[1], and Bernhard Ketzer[1] for the ALICE TPC Collaboration

[1] Helmholtz-Institut für Strahlen- und Kernphysik, Nussallee 14-16, 53115 Bonn, Germany

Abstract. In the context of the upgrade of the LHC during the second long shutdown the interaction rate of the ALICE experiment will be increased up to 50 kHz for Pb-Pb collisions. As a consequence, a continuous read-out of the Time Projection Chamber (TPC) will be required. To keep the space-charge distortions at a manageable size, the ion backflow of the charge amplification system has to be significantly reduced. At the same time an excellent detector performance and stability of the system has to be maintained. A solution with four Gaseous Electron Multipliers (GEMs) has been adopted as baseline solution for the upgraded chambers. As an alternative approach a hybrid GEM-Micromegas detector consisting of one Micromegas (MM) and two GEMs has been investigated. The recent results of the study of the hybrid GEM-Micromegas detector will be presented and compared to measurements with four GEM foils.

1 Introduction

A Time Projection Chamber (TPC) [1] is used in many ongoing experiments like STAR [2] or ALICE [3] as central detector. With its low material budget and excellent pattern recognition capabilities it is an ideal device for three-dimensional tracking and identification of charged particles. The working principle of a TPC is based on ionization and gas amplification: Once a charged particle traverses the gas-filled detector the gas molecules will be ionized along the particle's trajectory. Owing to an electric field within the detector volume, the produced electrons will drift to the endplates where they get amplified and finally detected at a space-resolved readout. With a gas volume of $90\,\mathrm{m}^3$ the ALICE TPC is the largest TPC on the world. Up to now the readout is done by 72 Multi-Wire Proportional Chambers (MWPC) [4] with a total of 550000 readout cathode pads. In order to prevent ions to drift back to the active drift volume, a bipolar Gating Grid (GG) is used which limits the theoretical readout rate to the order of 3 kHz. Together with the present readout system, the rate capability for central Pb-Pb collisions is limited to 300 Hz. As the interaction rate of the Large Hadron Collider (LHC) will be increased up to 50 kHz for RUN3, the expected amount of events can not be handled by the current readout technology which demands a complete re-design of the TPC. In Ne-CO_2-N_2 (90-10-5) and at a total gain of 2000 the energy resolution must not exceed $\sigma/E = 12\,\%$ for ^{55}Fe which translates to a $\mathrm{d}E/\mathrm{d}x$ resolution of about 5.5 % and 7 % in pp and central Pb–Pb collisions. The ion backflow should be lower than 1 % in order to be able to completely correct all effects from space-charge distortions [3]. Therefore the gated and MWPC-based readout technology will be replaced by a continuous read-out technology [5] based on Micro-Pattern Gaseous Detectors (MPGD). In previous investigations it has been shown that at least four GEM foils [6] with different pitches (see table 1) are required in order to cope with the aspired goal [7]. An alternative approach is given by the hybrid GEM-Micromegas detector which consists of only two standard GEM foils and a bulk Micromegas [8].

Geometry	p / µm	d_i / µm	d_o / µm
small pitch (SP)	90	40	60
standard pitch (S)	140	50	70
medium pitch (MP)	200	50	70
large pitch (LP)	280	50	70

Table 1. Different GEM geometries (p pitch, d_i inner diameter, d_o outer diameter).

2 Hybrid GEM-Micromegas detector

Figure 1 shows the setup of the investigated hybrid GEM-Micromegas detector: Due to the geometric and electrostatic structure of the Micromegas its ion backflow suppression capability is superior to a single GEM. The two GEM stages in front of the Micromegas are installed to keep the discharge probability of the device at a reasonable level while the energy resolution can be maintained. Furthermore the ion backflow can be reduced. Each standard GEM foil has a hexagonal hole structure according to table 1 and consists of a Polyimide substrate (50 µm) which is coated with a thin layer of copper on both sides (5 µm). To prevent unwanted overlapping effects of the holes, both foils are rotated by 90 degrees against each

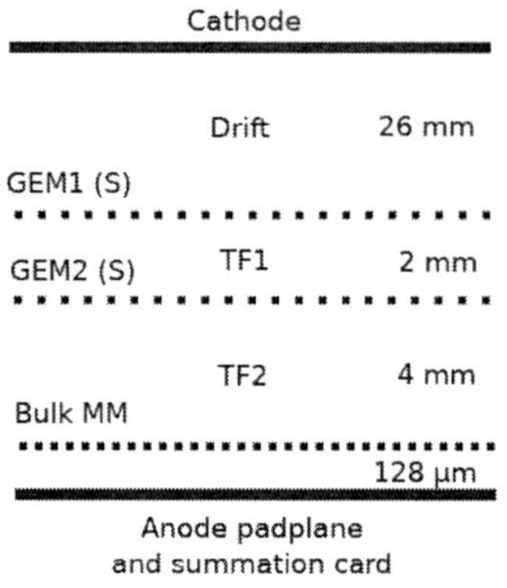

Figure 1. Schematic setup of the investigated hybrid GEM-Micromegas detector.

other. The investigated bulk Micromegas was manufactured at CERN and has a gap of 128 μm which separates a small mesh (160 lines / cm) from the readout padplane (10 cm x 10 cm). Finally a summation card has been designed to sum up all 128 pads as four concentric rings. All rings can be read out individually or in an arbitrary combination. The setup has been assembled within a gas tight vessel with a volume of 6 liters. A Kapton window on the lid allows an external irradiation of the drift region with an X-ray or an ^{55}Fe source. The currents at the cathode, the different GEM stages or the anode can be measured by picoammeters which were developed at the TU Munich [9] and which are sensitive for currents down to 20 pA before systematic errors become dominant. In order to reduce electromagnetic influences and to shield the system, the whole detector and the picoammeters are placed in a Faraday cage made of copper and lead.

3 Results

3.1 Gain contributions

Depending on the field configuration above and below a GEM, electrons can get lost by drifting to the copper or the Kapton layer, i.e. the measured effective gain (equation 1) is a function of the field-dependent extraction– and collection efficiency ϵ_{coll} and ϵ_{extr} [10] of the GEM.

$$G_{\text{eff}} = \epsilon_{\text{coll}} \cdot G_{\text{abs}} \cdot \epsilon_{\text{extr}} \quad (1)$$

Finally the total gain of the detector can be expressed as the product of the effective gain of its individual amplification stages (equation 2).

$$G_{\text{total}} = G_{\text{eff,GEM1}} \cdot G_{\text{eff,GEM2}} \cdot G_{\text{eff,MM}} \quad (2)$$

In order to understand the contribution of the different amplification stages, the gain of each component has been measured. Figure 2 shows the absolute gain of the Micromegas as a function of the applied voltage: Since the ^{55}Fe source has a certain solid angle, the full padplane (all pads) needs to be read out in order to cover the avalanches

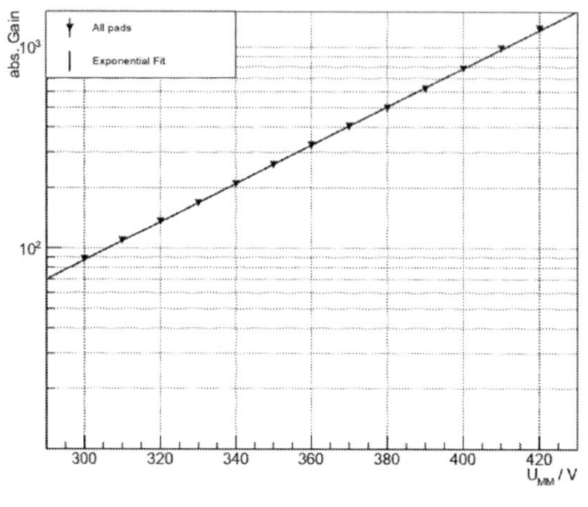

Figure 2. Absolute gain (=effective gain) of the Micromegas for different voltages (^{55}Fe source in Ne-CO_2 (90-10), Drift 400 V/cm, TF1 3 kV/cm, TF2 400 V/cm, GEM1 250 V, GEM2 250 V, Error x 5).

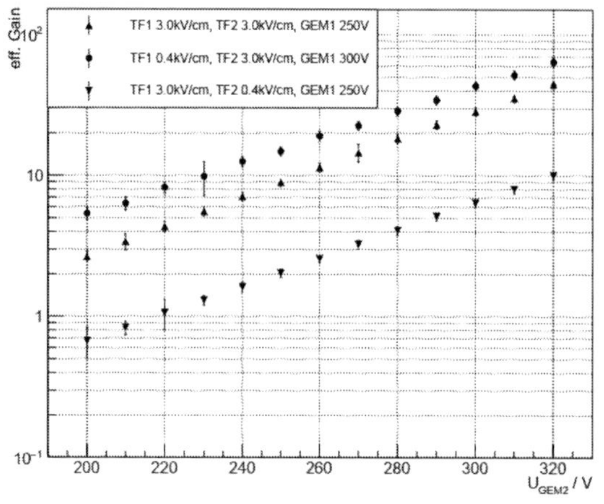

Figure 3. Effective gain of GEM2 for different field configurations (^{55}Fe source in Ne-CO_2 (90-10), Drift 400 V/cm, MM 0 V).

from all converted photons. The gain rises exponentially and is in the order of several 10^2 – 10^3 for common Micromegas voltages. Since the primary charge above the first amplification stage is below the sensitive range of the picoammeters, the gain contribution of the first GEM can only be estimated by the second GEM (see figure 3): A field configuration of TF1 0.4 kV/cm and TF2 3.0 kV is commonly used for the first amplification stage and thus represents the effective gain of the first GEM. Accordingly field configurations like TF1 3.0 kV/cm and TF2 0.4 kV/cm are common settings for the second GEM. Most of the total gain of the hybrid GEM-Micromegas detector is contributed by the Micromegas (order 10^2 – 10^3), followed by the first GEM (order 10^1 – 10^2). Commonly the second GEM is operated at an effective gain around 1

https://doi.org/10.1051/epjconf/201817401016

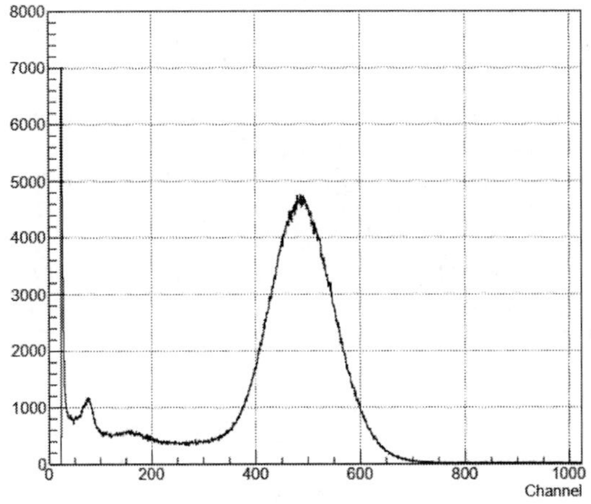

Figure 4. Example MCA spectrum for ^{55}Fe source in Ne-CO_2 (90-10) showing the central $K_{\alpha/\beta}$ at 5.9 keV.

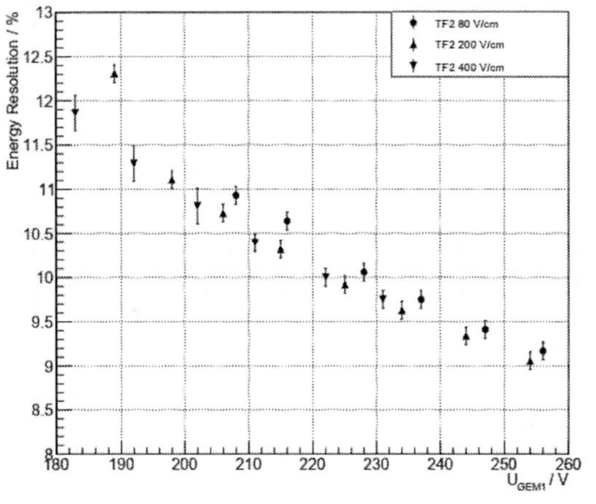

Figure 5. Energy resolution as a function of the GEM1 voltage (GEM1-MM scan in Ne-CO_2 (90-10), Drift 400 V/cm, TF1 3 kV/cm, GEM2 250 V, gain 2000, error energy resolution x 10)

to maintain the energy resolution and to act as a further stage to block the ions from the Micromegas.

3.2 Energy resolution and ion backflow

For the planned upgrade of the ALICE TPC the ion backflow has to be less than 1 % while the energy resolution (σ/E) should not exceed 12 % for ^{55}Fe (at a total gain of 2000 in Ne-CO_2-N_2 (90-10-5)) [3]. The energy resolution and the ion backflow of the hybrid GEM-Micromegas detector have been determined within a GEM1-Micromegas scan, i.e. the applied voltage of the first GEM was successively increased or decreased while the total gain of the detector was kept at a constant level of 2000 by adjusting only the voltage of the Micromegas. While the electrons from avalanches drift to the readout padplane, some of the produced ions may drift back to the cathode. Thus the ion backflow can be defined as the fraction of the currents at the readout padplane (anode) and the cathode according to equation 3, where ϵ is the number of back drifting ions per incoming electron coming from the amplification region. For the ion backflow measurements an Amptek MiniX has been used to irradiate the detector (30 kV, 5 μA) and the whole padplane was read out.

$$\mathrm{IB} = \frac{1+\epsilon}{G_{\mathrm{eff}}} = \frac{I_{\mathrm{cathode}}}{I_{\mathrm{anode}}} \tag{3}$$

All energy resolutions were determined with an ^{55}Fe source and the innermost pads (2x2) were read out. An example MCA spectrum is shown in figure 4: The prominent peak at the center belongs to $K_{\alpha/\beta}$ photons which converted in the drift region while the lower energetic peaks appear due to conversion processes between the amplification stages (GEM1-GEM2 and GEM2-Micromegas). To determine the energy resolution the $K_{\alpha/\beta}$ peak was fitted by a simple Gaussian. Due to the non-negligible background under the peak a more refined fit model will be used in future investigations. Within the GEM1-Micromegas scan the energy resolution of the detector can be tuned by changing the amplification at the first GEM stage as the number of avalanche electrons will change and influence the energy resolution statistics. Figure 5 shows that the energy resolution is mostly dominated by the first amplification stage (GEM1) and only weakly depends on TF2 or the Micromegas. Only for low TF2 (80 V/cm) the energy resolution gets worse: Due to a small extraction efficiency of the second GEM electrons get lost within the transfer process to the Micromegas and hence its gain fluctuations become more dominant. Furthermore a change of the amplification at the first GEM strongly influences the ion backflow since the number of ions (which are immediately injected into the drift region) will be altered (figure 6). In contrast to the energy resolution the ion backflow is also influenced by TF2 and the Micromegas as most of the ions are produced within the last amplification stage (highest gain contribution). Thus the energy resolution and the ion backflow are not independent of each other: An optimization of the energy resolution leads to a degradation of the ion backflow and vice versa. Figure 7 illustrates this behavior for different TF2 and additionally indicates that specific field configurations exist which fulfill the requirements of the upgrade of the ALICE TPC. A comparison to the 4-GEM baseline configuration (S-LP-LP-S) reveals that the hybrid GEM-Micromegas detector can be considered as an alternative approach.

4 Summary and Outlook

A hybrid GEM-Micromegas detector consisting of two GEM foils (standard pitch) and a bulk Micromegas has been assembled at Bonn University. First measurements reveal that an energy resolution less than 12 % (σ/E with ^{55}Fe) and an ion backflow better than 1 % is possible in Ne-CO_2 (90-10) at a total gain of 2000. Within the detector most of the gain is contributed by the Micromegas

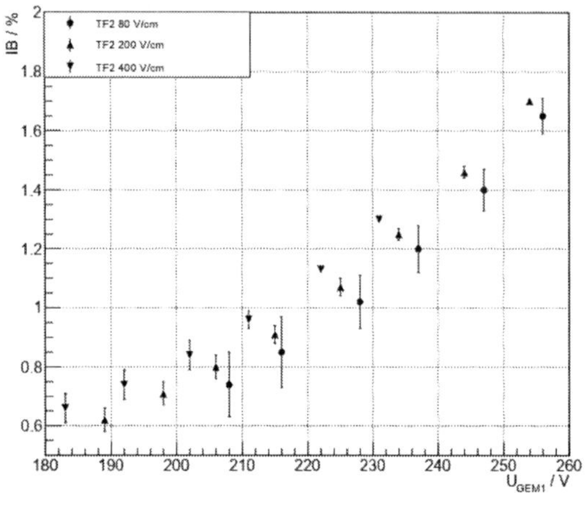

Figure 6. Ion backflow as a function of the GEM1 voltage (GEM1-MM scan in Ne-CO_2 (90-10), Drift 400 V/cm, TF1 3 kV/cm, GEM2 250 V, gain 2000)

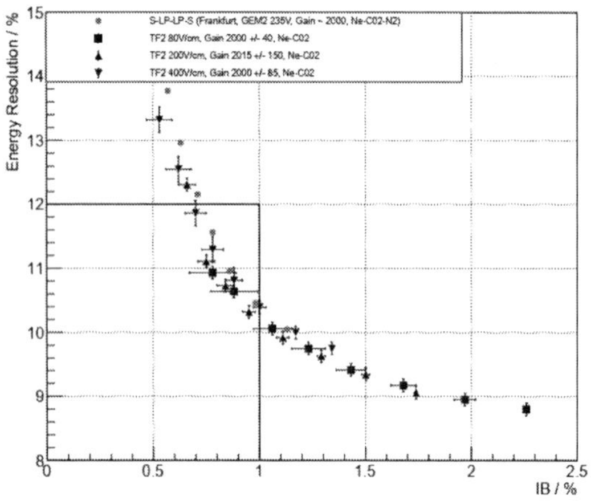

Figure 7. Energy resolution against ion backflow for the hybrid GEM-Micromegas detector (GEM1-MM scan in Ne-CO_2 (90-10), Drift 400 V/cm, TF1 3 kV/cm, GEM2 250 V, gain 2000, error energy resolution x 10) and the 4-GEM baseline configuration for the ALICE TPC (Frankfurt measurements, GEM1-GEM2 scan in Ne-CO_2-N_2 (90-10-5)).

(order 10^2 – 10^3), followed by the first GEM stage (order 10^1 – 10^2). The second GEM stage is used in order to prevent ions from drifting back to the drift region while the energy resolution can be maintained. With respect to the ion backflow and the energy resolution the capability of the hybrid GEM-Micromegas detector can be compared to the 4-GEM baseline configuration and shows a similar performance. As Ne-CO_2-N_2 (90-10-5) will be used for the ALICE TPC in RUN3 all previous measurements will be repeated in this gas mixture. Yet the full possible parameter space of the hybrid GEM-Micromegas detector has not been scanned in order to find an optimized configuration with respect to the ion backflow and the energy resolution. To optimize the energy resolution measurements, a veto setup for the innermost readout pads has been established and will be used in further investigations. Several measurements have shown that the quadruple S-LP-LP-S GEM configuration fulfills the specified requirements and thus has been chosen for the upgrade of the ALICE TPC. At the time of the decision the hybrid GEM-Micromegas detector was not studied as extensive as the multiple GEM setup and further stability measurements are still necessary to get a complete understanding of the hybrid GEM-Micromegas detector.

References

[1] D. R. Nygren, J. N. Marx, Phys. Today **31N10**, 45-53 (1978)

[2] K. H. Ackermann et al., Nucl. Phys. A **661**, 681-684 (1999)

[3] ALICE Collaboration, *TDR for the Upgrade of the ALICE TPC* (CERN/LHCC, Geneva, 2014)

[4] G. Charpak et al., Nucl. Instr. Meth. **62**, 262 (1968)

[5] B. Ketzer, Nucl. Instr. Meth. A **732**, 237-240 (2013)

[6] F. Sauli, Nucl. Instr. Meth. A **386**, 531-534 (1997)

[7] M. Ball et al., JINST **9**, C04025 (2014)

[8] Y. Giomataris, Ph. Rebourgeard, J. P. Robert and G. Charpak, Nucl. Instr. Meth. **A376**, 29-35 (1996)

[9] C. Höppner, Ph.D. thesis (Tech. U., Munich, 2012)

[10] M. Killenberg et al., Nucl. Instr. Meth. A **498**, 369-383 (2003)

Development of Resistive Micromegas for Sampling Calorimetry

T. Geralis[1,a], G. Fanourakis[1], A. Kalamaris[1], D. Nikas[1], A. Psallidas[1], M. Chefdeville[2], I. Karyotakis[2], I. Koletsou[2], and M. Titov[3]

[1] INPP, NCSR Demokritos, Athens, Greece
[2] LAPP, Université de Savoie, CNRS/IN2P3, Annecy-Le-Vieux, France
[3] IRFU, Saclay CEA, Gif-sur-Yvette, France

Abstract. Resistive micromegas is proposed as an active element for sampling calorimetry. Future linear collider experiments or the HL-LHC experiments can profit from those developments for Particle Flow Calorimetry. Micromegas possesses remarkable properties concerning gain stability, reduced ion feedback, response linearity, adaptable sensitive element granularity, fast response and high rate capability. Recent developments on Micromegas with a protective resistive layer present excellent results, resolving the problem of discharges caused by local high charge deposition, thanks to its RC-slowed charge evacuation. Higher resistivity though, may cause loss of the response linearity at high rates. We have scanned a wide range of resistivities and performed laboratory tests with X-rays that demonstrate excellent response linearity up to rates of (a few) times $10MHz/cm^2$, with simultaneous mitigation of discharges. Beam test studies at SPS/CERN with hadrons have also shown a remarkable stability of the resistive Micromegas and low currents for rates up to $15MHz/cm^2$. We present results from the aforementioned studies confronted with MC simulation

1 Introduction

Resistive Micromegas was proposed [1] for the ATLAS muon detector upgrade for the HL-LHC, in order to cope with the high particle fluxes of (a few) times $10kHz/cm^2$ of minimum ionising particles (MIPs) and mitigate the problem of discharges at high flux radiation environments. The goal of our research program is to develop micromegas detectors with resistive pads suitable for use in Particle Flow (PF) calorimetry at high rates. The expected rates are of the order of $1MHz/cm^2$ - (a few) times $10MHz/cm^2$ MIPs at high pseudorapidity regions. The discharge rate diminishes as the resistivity increases at the cost of loss of linearity at high rates. We have scanned five order of magnitude in resistivities, in order to optimise for these counter acting effects.

2 Resistive Micromegas design

The anode pads in resistive micromegas are covered with dielectric material on top of which a resistive layer is overlaid. The resistive coating prevents the fast evacuation of the charge that is produced by the primary electrons cascade. The semi-static charge, accumulated on the resistive layer, deforms the electric field, lowering the gain and finally quenching the sparks. The parameters controlling this effect are the capacitance between the resistive layer and the anode pad and the resistance to ground. The charge

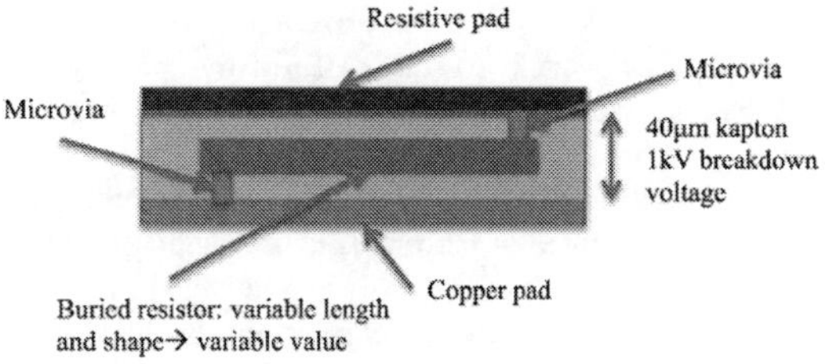

Figure 1. Resistive micromegas pad schematic. A dielectric is overlaid on the anode pad, on top of which the resistive layer is laid. Inside the dielectric a resistor is laid with vias connecting it to the anode and the resistive pad respectively.

can be evacuated sideways through a net of resistive elements or by directly connecting every pad via a resistor to ground. A new technique with buried resistors in line with the second option, proposed and implemented by Rui De Oliveira, is shown schematically in Figure 1. We have adopted the buried resistor technique for large area detectors to avoid the accumulation of charges on the full extend of the detector surface. The buried resistor can be altered by varying the shape and the length between the connecting vias. We have also used different resistive pastes ranging from $1 - 100k\Omega/\square$.

Using different buried resistor shapes and different resistive pastes [2] we have scanned five orders of magnitude in resistivitiy, from $4k\Omega$ to $40M\Omega$. The prototypes consisted of a $10 \times 10cm^2$ active area, divided in pads of $1cm^2$ with

[a]Corresponding author, e-mail: Theodoros.Geralis@cern.ch

a bulk mesh kept at a distance of 128 μm. The drift gap was 3mm and all the measurements were performed with Ar/CO_2 93 - 7.

3 Linearity study using X-rays

Rate scans were performed with X-rays at 3 keV (rhodium K-Line) and at 8 keV (copper K-Line) provided by X-ray tube beams collimated to an area of $8mm^2$. The event rate is calibrated operating in pulse-counting mode. The measurements were performed at gains of up to about 5×10^3.

Figure 2 shows the mesh current response of one of the resistive prototypes (with $R_{Buried} = 1.5M\Omega$) up to rates of $10MHz/mm^2$, while Figure 3 shows a zoom at the lowest part of the scale of Figure 2 up to rates of $10MHz/cm^2$. The response in Figure 2 deviates from linearity because of the charging up of the resistive pad that is more pronounced at the highest rates. For small voltage drops, up to 10 V (i.e. gain drop less than 25%), the following parametrization is valid to a few percent accuracy, for the mesh current i:

$$i(f) = \frac{q_0 f}{1 + BRq_0 f} \qquad (1)$$

where q_0 is the anode charge per event in absence of charging-up, B is the slope of the gain curve, f is the X-ray source flux and R the pad to ground resistance. The slope is precisely known so R is easily fitted to the data points and yields $R = 1.5M\Omega$, as expected from the geometry and the embedded resistor. All resistive prototypes apart the ones with lowest resistivities ($4k\Omega$) operated up to rates close to $10MHz/mm^2$ with saturation depending on the resistivity as expected. The non-resistive prototypes could not be operated at high rates above gains of about 500. In order to ensure the linearity at relatively low rates, which are more realistic for an experiment, we have performed the same tests for rates up to $10MHz/cm^2$ (or up to $0.1MHz/mm^2$ zooming at the very beginning of the x-axis in Figure 2). The linearity in that region is excellent.

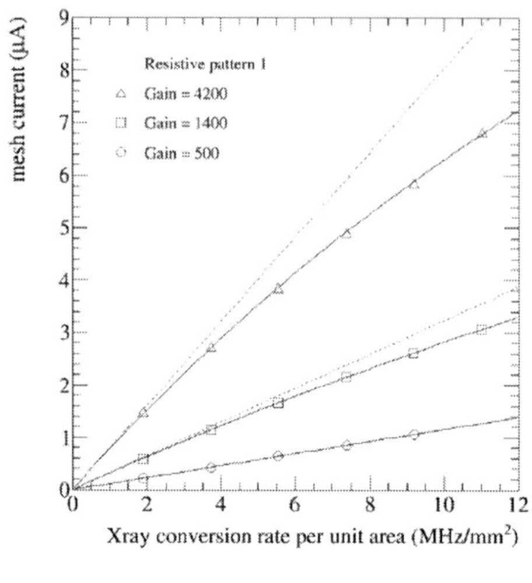

Figure 2. Micromegas linearity at high rates for $R = 1.5M\Omega$. Continues lines are a fit of Eq.1 to data points, while the dashed lines present the theoretical response.

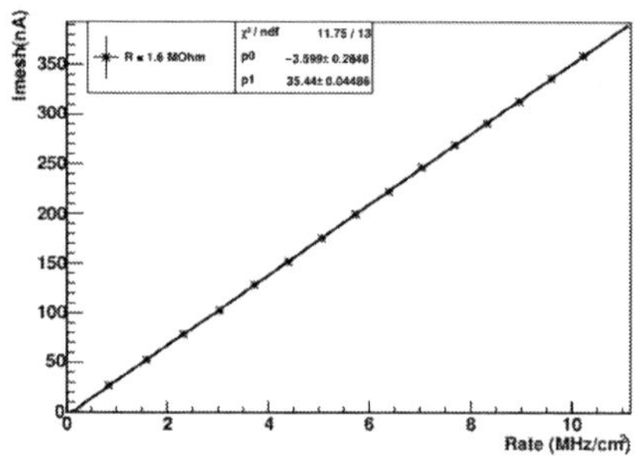

Figure 3. Micromegas linearity for rates up to $10MHz/cm^2$

4 Discharge rate measurements with X-rays

In order to measure discharges that could cause a global voltage drop of the mesh HV we have connected a differentiator circuit with a high time constant $RC = 0.1s$ to the mesh electrode. Simulation on spice circuit simulator [3] shows that for signal frequencies higher than 30 Hz there is no attenuation of the signal amplitude, thus any voltage drop caused by a spark can be counted and its pulse height recorded. We have acquired the spectra of discharges for long periods of irradiation with X-rays at high rates on the same detector as the one used for the linearity plots (Figures 2 and 3) with $R_{buried} = 1.5M\Omega$. Table 1 shows the discharge rates for gain spanning from $(2-6) \times 10^3$ at a constant rate of $11MHz/cm^2$ for a period of 24 hours. The column labeled "Discharges (I)" refers to rates with voltage drop higher than 30 mV while the column "Discharges (II)" refers to discharge rates with voltage drop (approximately greater than 0.5V) causing gain drop greater than 1%.

Table 1. Discharge rates at $11MHz/cm^2$

Gain	Maximum HV drop (V)	Discharges (I) $sprks/cm^2/s$	Discharges (II) $sprks/cm^2/s$
2000	0.25	$5.5x10^{-4}$	$< 1.3x10^{-4}$
3000	0.80	$1.4x10^{-3}$	$2.8x10^{-4}$
4000	0.80	$2.7x10^{-3}$	$6.8x10^{-4}$
6000	4.90	$1.8x10^{-1}$	$1.4x10^{-1}$

5 Test beam at SPS

Seven resistive micromegas together with two non-resistive for reference were tested in the SPS/H4 test beam in order to study their stability and discharge rate at high particle fluxes and also their efficiency, energy response and linearity. Two campaigns were performed, in Decem-

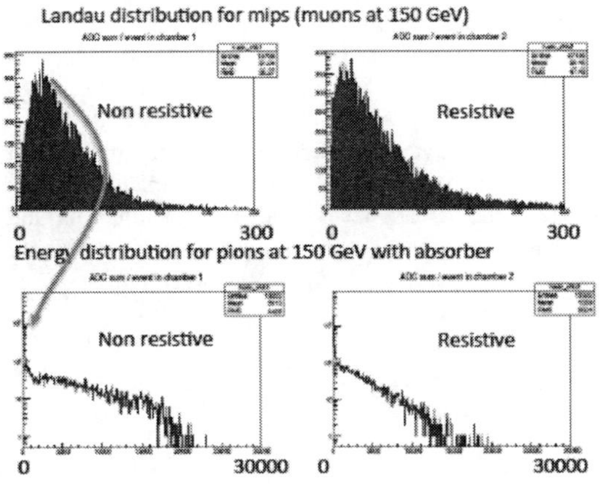

Figure 4. Landau distributions (top) and pion shower distributions (bottom) for resistive and non-resistive micromegas.

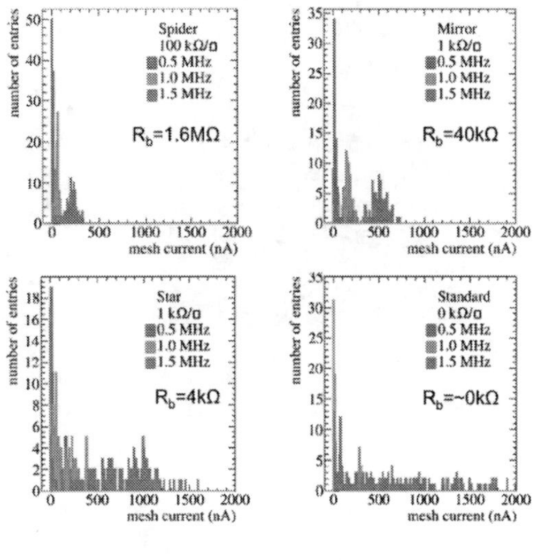

Figure 5. Micromegas mesh current with pion showers.

ber 2014 and in July 2015 using muon, hadron (pions) and electron beams provided by the SPS at energies ranging from 30 GeV to 200 GeV. The DAQ was based on VME, running in C++ with the Gassiplex FE cards that provide energy information deposited on every pad. The maximum acquisition rate was about 1.4 kHz thus we have recorded about 15 million useful events in both test beam periods.

The detectors efficiency was measured with muon beams at 200 GeV. Three detectors were used as telescope to identify a track and measure the efficiency of the resistive detector under study. Efficiencies are measured at the level of 96% for most detectors at moderate gains of about 3000. We have obtained Landau distributions that are shown in Figure 4 for a non-resistive and a resistive micromegas. The same Figure shows the energy distribution of 150 GeV pions converted on an iron block of 1.5λ, where λ is the interaction length, placed in front of the detector. In those plots the Landau peak is visible, showing the large dynamic range of the detector and also the possibility for calibration using the Landau distribution.

The iron absorber of 1.5λ was used to monitor the mesh current as a function of the pion beam rate for every resistive detector. Maximum rates of more than 1 MHz/spill (5s) with pions corresponding to a particle flux of $15MHz/cm^2$ after the cascade produced currents of about 500 nA on the mesh electrode and no voltage drop, for resistivities down to $40k\Omega$ (see Figure 5). It is important to notice that the resistive micromegas with buried resistor of $1k\Omega$ behaves almost as the non resistive that draw currents up to $2\mu A$ and frequently suffer mesh voltage trips.

The first series of tests with electrons were performed by placing every resistive detector at the shower maximum in order to expose the detectors on high charge deposition in every single event. The layout consisted of an iron block of $2.2X_0$ followed by a non resistive detector and a second iron block of $3.5X_0$ followed by the resistive detector. We have placed all resistive prototypes one by one after the second block and performed individual tests, using

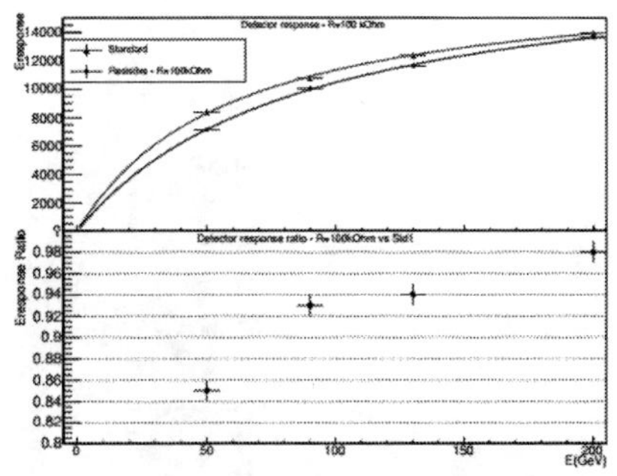

Figure 6. Comparison of the Detector response of Non Resistive (red curve, triangle marker) vs Resistive (blue curve, square marker) as a function of the energy. Their ratio is presented in the bottom plot.

the non-resistive as reference and as preshower and veto in order to clean up the beam from pion contamination. The electron beam energies used are 50 GeV, 90 GeV, 130 GeV and 200 GeV at low rates. Detailed simulation of the layout using Geant ([4]), shows that the peak energy deposited on the resistive detector is 150 keV, 240 keV, 330 keV and 450 keV at the above energies correspondingly.

Figure 6 shows that the detector response rate (Resistive vs Standard) doesn't vary significantly as a function of the energy. A slight drop at 50 GeV remains to be understood.

The second series of tests refer to a full containment small prototype calorimeter consisting of 6 resistive micromegas layers and a total of about $20X_0$. Figure 7 shows a schematic of the prototype calorimeter (top), a picture of the experimental apparatus (middle) and a MC Geant simulated event in the same apparatus.

Figure 8 shows the spectra of the energies (on identical arbitrary scale) deposited on every calorimeter layer and the corresponding hit map for 200 GeV electrons. We

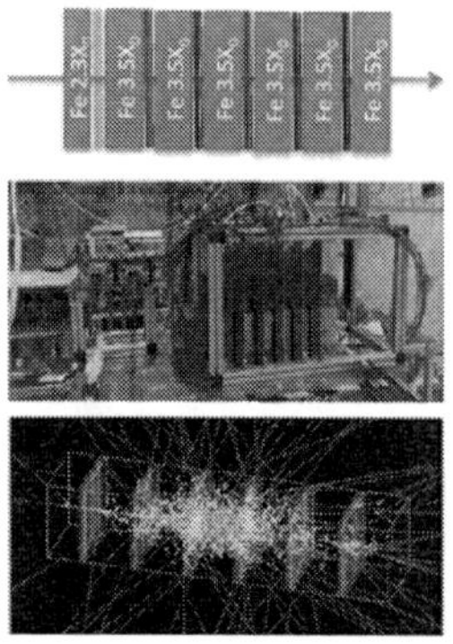

Figure 7. Prototype Calorimeter: setup schematic (top), picture of the apparatus (middle) and Geant simulated event (bottom).

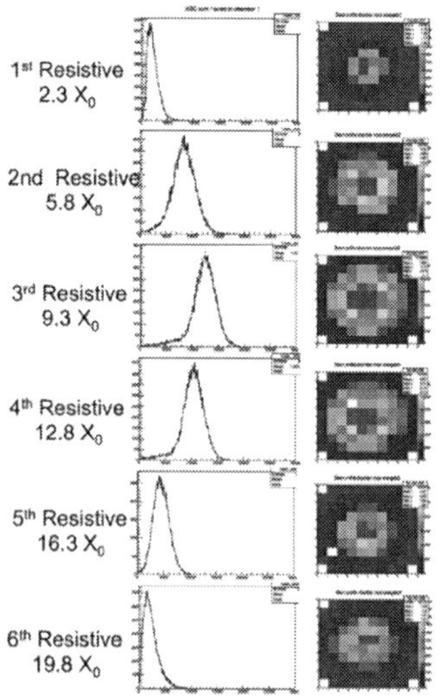

Figure 8. Spectra of energy deposited on every calorimeter layer (left) and the corresponding hit map (right) for 200 GeV e^-.

can see the development of the cluster from the conversion layer to its maximum and the total absorption up to the last layer. The analog energy response (Figure 9) saturates due to poor sampling (only 6 layers) and the low energy sampling fraction (10^{-4}). In the same Figure the sum of the ADC sum resolution is shown for the data (top - red curve) and for the MC simulation (bottom - blue curve). The stochastic term is in excellent agreement for data and MC, while the constant term is not included in the MC.

6 Conclusions

We have performed an *R&D* on resistive micromegas with pads for sampling calorimetry and scanned resistivities spanning 5 orders of magnitude. We have reached the extreme limits on which the detector cannot operate (low R) due to discharges and at the opposite side (high R) were saturation of the order of 30% takes place at extremely high rates. Discharges are suppressed while linearity is preserved to a high degree for buried resistors of the order of $1M\Omega$ at rates of about $10MHz/cm^2$. For this resistivity and flux the sparking rate and the mesh voltage drop are totally suppressed while linearity is excellent at the few per mille level. High charge deposition does not seem to affect the detector response. We work on a model to understand the charge evacuation phenomenon and also to measure parameters like the effective capacitance and resistivity at the cascade level. The current studies and the results are very encouraging and promising that particle flow calorimetry can profit from these micromegas detector developments.

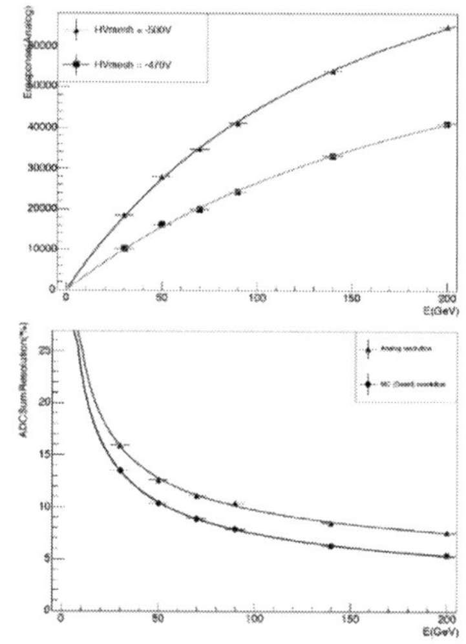

Figure 9. The detector Energy response (top) and the Energy resolution (bottom), when we consider as Energy the sum of all the detectors' ADC values.

7 Acknowledgements

This research has been co-financed partially by the EU - ESF and Greek national funds through the Operational Program "Education and Lifelong Learning" of the NSRF - Research Funding Program MIS-448332-ORASY (NSRF 2007 - 13 ACTION, KRIPIS) at INPP/NCSR Demokritos and by the THALIS-DIBOSON program.

References

[1] T. Alexopoulos et al., A spark-resistant bulk-micromegas chamber for high-rate applications, Nuclear Instruments and Methods A 640 (2011) 110-118

[2] M. Chefdeville et al., Resistive Micromegas for sampling calorimetry, a study of charge-up effects, Elba Conf., May 2015, Nuclear Instruments and Methods A.

[3] The Spice Page: http://bwrcs.eecs.berkeley.edu /Classes/IcBook/SPICE/

[4] S. Agostinelli et al., Geant4: a simulation toolkit, Nuclear Instruments and Methods A, 506, Nr. 3, (2003) 250 - 303

https://doi.org/10.1051/epjconf/201817402001

Large Area Coverage of a TPC Endcap with GridPix Detectors

Jochen Kaminski[1,a] for the LCTPC Collaboration

[1] University of Bonn

Abstract. The Large Prototype TPC at DESY, Hamburg, was built by the LCTPC collaboration as a testbed for new readout technologies of Time Projection Chambers. Up to seven modules of about 400 cm^2 each can be placed in the endcap. Three of these modules were equipped with a total of 160 GridPix detectors. This is a combination of a highly pixelated readout ASIC and a Micromegas built on top. GridPix detectors have a very high efficiency of detecting primary electrons, which leads to excellent spatial and energy resolutions. For the first time a large number of GridPix detectors has been operated and long segments of tracks have been recorded with excellent precision.

1 Introduction

The International Linear Collider (ILC) is a future accelerator colliding electrons and positrons at center-of-mass energies of 250-500 GeV. There are two detector concepts foreseen to take data alternatingly at the collision point. One of the detector concepts, the International Large Detector (ILD), foresees a large volume TPC as a main tracking device. The most stringent requirement is derived from the Higgs-recoil measurement and is given as $\Delta\frac{1}{p_t} \sim 9 \cdot 10^{-5}/\text{GeV/c}$ for the TPC. This results in spatial resolution requirement of better than 100 μm over the complete drift distance, which can not be achieved by a conventional wire-based readout. The LCTPC collaboration studies, therefore, several different MPGD technologies which could fulfill the requirements of the ILD physics program. To test and compare the readout technologies, a test setup with a fully operational TPC prototype (called Large Prototype, LP) in a $B = 1$ T magnetic field and an $E = 6$ GeV electron test beam was set up at DESY, Hamburg, by the collaboration. The LP can be equipped with up to seven readout modules of about 400 cm^2 each. Micromegas and Gas Electron Multiplier with a pad based charge collection plane are studied and have demonstrated that the requirements can be fulfilled.

2 GridPix Detectors

As an alternative technology option we are studying also GridPix detectors, which consist of a Micromegas gas amplification stages on top of highly pixelized Timepix readout ASICs [1]. The pillars and grid of the Micromegas are fabricated in photolithographic processes [2] ensuring a very good alignment of each grid hole with a pixel of the ASIC (see Fig. 1). When a single primary electron enters a grid hole, it is amplified in the gap and the complete charge avalanche is collected on one pixel. Since typical gas amplifications of 2,000-3,000 surpass significantly the threshold level of 700-1,000 electrons, the detection efficiency of primary electrons is very high. Provided the diffusion in the drift region is sufficient to spread the primary electrons over different grid holes, this has significant advantages both for tracking applications as well as for energy determination. The track reconstruction profits from the low occupancy because of the small pixel sizes as well as from the large amount of information available for each track. For example, the effect of delta electrons, multiple scattering kinks or radiation of photons can be corrected for. Also, the angular pad effect does not exist anymore. The measurement of the energy deposited by a particle along its track can be done by counting the number of pixels associated with the track. Thus, the fluctuations intrinsic to the gas amplification can be avoided narrowing the dE/dx-distribution to a theoretical limit and improving the particle identification.

Figure 1. SEM picture of an InGrid, where the grid is partially removed for better visibility.

[a]e-mail: kaminski@physik.uni-bonn.de

 https://doi.org/10.1051/epjconf/201817402001

3 Challenges of Large Area Coverage

The current design of the ILD-TPC foresees an outer radius R_{max} = 1.739 m of the active area resulting in a total area of about 9 m^2 per endcap. Each endcap is made up of 240 single modules with a keystone shape with maximum dimensions of 17 cm in radial direction and 22 cm in ϕ-direction (see Fig. 2). The GridPix detector features a small active area of about 2 cm^2 only, which results in a demand of 45,000 detectors per endcap. To demonstrate the feasibility of such a proposal, we have prepared a setup of three LP modules with a total of 160 GridPix detectors summing up to 10.5 million readout channels. 96 GridPixes were placed on the central module, which corresponds to a 50 % coverage. Much tighter packaging can not be reached, because of the wire bondings and HV-distribution. The 32 GridPixes on each of the outer modules were arranged to maximize the recorded track length of each particle. Fig. 3 shows the position of the GridPixes.

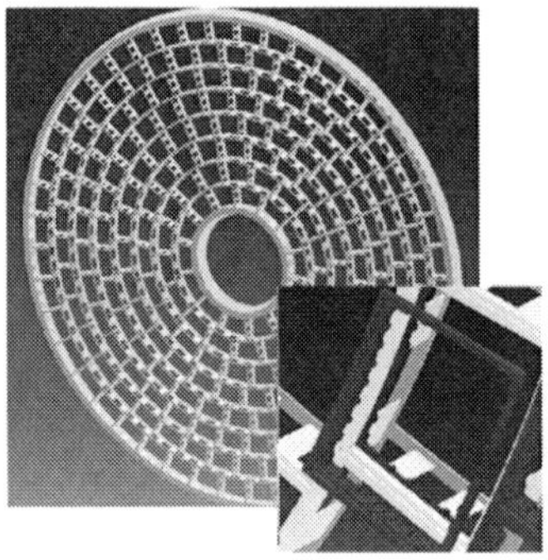

Figure 2. Drawing of the proposed end-plate for the ILD-TPC. In the insert a backframe which is supporting the actual readout module, is shown. [3]

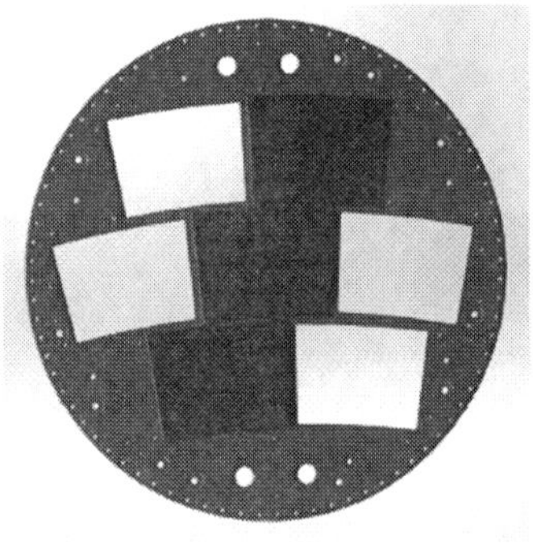

Figure 3. Endcap of the Large Prototype with three out of seven openings filled with GridPix-modules.

During this project a number of challenges had to be overcome:

- Production of a sufficient number of GridPixes
- Simultaneous readout of 160 GridPixes
- LV supply for 160 GridPixes
- Cooling of modules

These challenges are discussed in the following subsections.

3.1 GridPix Production

The production of GridPixes was pioneered by the University of Twente [2]. But because of limited machine sizes only 1 to 9 GridPixes could be produced at a time. Taking into account the long production times, this could not satisfy the demand. We have, therefore, established a new production process at the Fraunhofer Institute IZM at Berlin. This process allows a wafer-based production of 107 GridPix detectors at a time [4]. The crucial step turned out to be the deposition of a Si_xN_y-layer, which is applied on the ASIC to protect it from possible discharges. To keep the wire bond pads free of this resistive layer, a photoresist is applied on them beforehand. After the Si_xN_y deposition is completed, the photoresist is removed along with the Si_xN_y deposited on the pads. Because of rather high layer thicknesses the standard processes routinely applied in MEMS technology have to be modified and both machines and materials have to be used in a border area of their specifications. Six productions runs with one to three wafers were performed with various successes. However, the last two runs yielded a sufficiently high number of good GridPix detectors to start the construction of the modules.

Eight GridPixes were mounted on each chip carrier called octoboard. These were tested both electronically and for HV stability. Finally, the octoboards were mounted on the module.

3.2 Readout System

At the beginning of this project none of the available readout systems could handle more than eight ASICs. We developed therefore a new readout system based on the Scalable Readout System (SRS) [5] of the RD51 collaboration [6]. It is very flexible and designed for R&D purposes, where a few ASICs up to several hundred ASICs can be easily accommodated. The central part of the SRS is the Front-End Concentrator (FEC), which hosts a Virtex FPGA and both a fiber connector for communication with a PC and PCI connector for communication with adapter boards. Because of the high flexibility of the system, different readout ASICs can be easily implemented by exchanging the adapter board and the FPGA code. Both hardware adapter as well as the FPGA firmware were developed to handle up to 32 Timepix ASICs per FEC. Thus, only 5 FECs and adapter were necessary. Five HDMI cables were used to connect each adapter board to the module, where four octoboards were grouped in a subdivision. The HDMI cables were used for slow control commands and data exchange including chip readout. Fig. 4 shows a FEC with adapter card and a prototype module for one octoboard. Since the eight chips on the octoboards had to be readout serially, the readout frequency was limited by

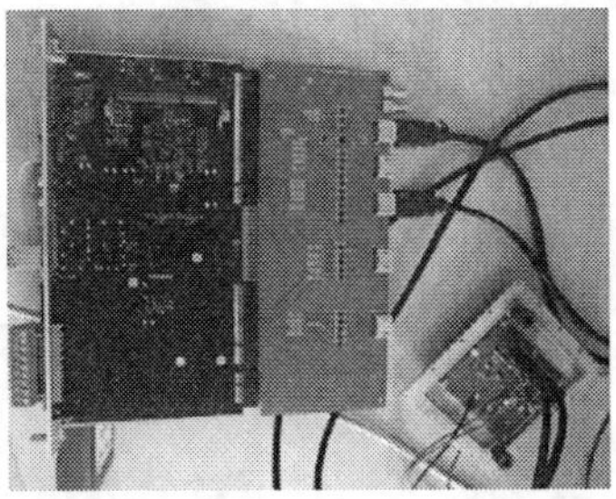

Figure 4. FEC with adapter card connected via two HDMI cables to a prototype module.

this process and a maximum rate of about 5 Hz could be reached, when the trigger was fast enough.

3.3 LV Power Supply

Detailed studies on the power consumption of an octoboard were performed. The two power supply lines (for the analog and the digital part of the ASIC) behave differently. While the digital part draws only very little current when no signal is recorded, it can sink up to 4.5 A when a large fraction of pixels is activated. The analog part in contrast has a higher power consumption also in the stand by mode, but rises to 2.3 A only, when most pixels are activated. Even though a large number of pixels was not expected to be hit under standard operation, the LV system was laid out for the most extreme case of up to 136 A taking into account both supply lines.

To avoid a large voltage drop over the supply cables to affect the readout electronics, low drop out regulators were foreseen on the modules with large buffer capacitors of $C = 0.47$ F. This design was implemented on piggyback PCBs, which were mounted perpendicular on the modules (see light green boards in figure 5). The 3.3 V and 5 V lines

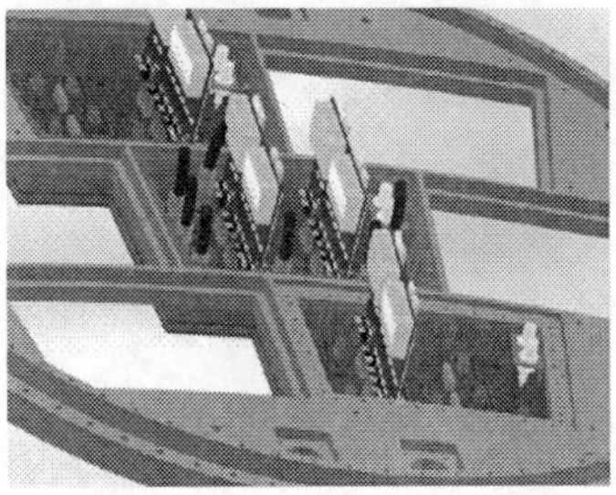

Figure 5. CAD-drawing of endplate with 3 GridPix modules and LV boards mounted.

of standard ATX power supplies were used to supply the LV-board with currents. Since the setup was operated in a magnetic field of up to $B = 1$ T the supplies were installed at a safe distance and 15 m long cables with cross sections of 6-16 mm^2 were used.

3.4 Cooling

To remove the dissipated power water cooling of the modules was foreseen. For this, a 3 mm thick cooling plate made of aluminum was introduced between the module and the octoboards. The plate consisted of two parts, a thicker one, where cooling channels were milled into and a thin cover plate, which was diffusion welded on the thicker plate at the research center Jülich. The welding was helium and vacuum tight. During the setup at DESY, the plates were connected to tap water.

4 Test Beam

A first test beam was performed in 2013 to test several components of the setup [7], but only one octoboard on a single module was used. In March and April 2015 the final setup was tested including one week of beam time. The setup can be seen fully cabled in figure 6. During the test

Figure 6. Test beam setup at DESY, Hamburg, with three GridPix modules mounted and cabled.

beam 1.5 million events were recorded in various detector conditions. The voltage on the grid, drift distance, electron momentum, magnetic field and the track inclination in the readout plane were varied. One track segment recorded by a single octoboard can be seen in figure 7.

Figure 7. Detail (1 octoboard) of the online event display with a single track and a curling delta electron.

No problems occurred during the test beam campaign. However, several GridPix detectors either stopped transferring data or the noise level became so high, that no data could be taken. From the 160 GridPixes installed 15 were not useful initially, while at the end of the test beam, this number had increased to 33. The largest fraction of the failing chips were on the same wafer, but also the last production cycle of the GriPix detectors showed significantly larger failure rates both right after mounting on the octoboards as well as during the test beam. SEM examinations revealed frequent damages, of the Si_xN_y-layer suggesting an increased vulnerability to discharges. However, this does not explain the high failure rate before HV was applied. We, therefore, continue to study potential explanations.

5 Data Analysis

The data is analyzed with MarlinTPC, a C++ based simulation reconstruction and analysis tool developed by the LCTPC collaboration to unify the software efforts by different groups. In a first step the noise counts were reduced by masking ASICs known to malfunction and removing single noisy pixels. The three dimensional position of each hit was calculated. Then tracks were reconstructed by gathering the hits, which belong together, with the help of a Hough transformation. These tracks were subsequently analyzed. For example the transverse spatial resolution is determined by histograming the distance between the hits and the reconstructed tracks. In Fig. 8 the width of this distribution is shown in dependence on the drift distance z. One can see, that the width increases as $\sigma = \sqrt{\sigma_{xy,0}^2 + D_T^2 z}$ and is close to the diffusion limit of single electrons. The constant offset σ_0 combines detector related effects, which have not been corrected for. In particular the $E \times B$ effects at the GridPix borders have not been addressed yet and lead to a noticeable degradation of the performance.

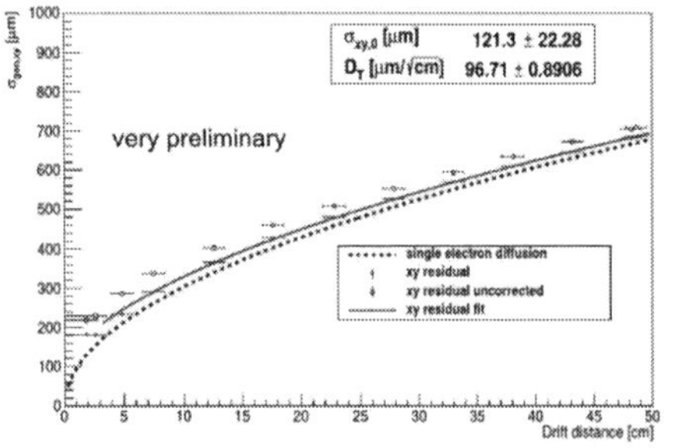

Figure 8. Transverse spatial resolution in dependence on the drift distance in a magnetic field $B = 1$ T

Other performance checks have been done and are within expectations considering the early state of the analysis. The longitudinal spatial resolution is rather poor, because of the time walk effect, which can not be corrected for with the Timepix ASIC, since it can not measure simultaneously the arrival time and the collected charge for each pixel. A track typically has about 3000 hits, which is well in line with expectations from primary ionizations because of dE/dx and indicates an efficiency of detecting primary electrons with more than 90 %. Very preliminary energy resolution studies give $\sigma_E/E \approx 10$ % for 50 cm of tracks (including insensitive areas).

6 Summary

We have designed and built a TPC readout based on 160 GridPix detectors. The new detectors were tested in the Large Prototype TPC of the LCTPC collaboration at DESY. A large amount of data was taken during a test beam campaign and analysis results show an excellent performance.

7 Acknowledgement

The measurements leading to these results have been performed at the Test Beam Facility at DESY Hamburg (Germany), a member of the Helmholtz Association.
This work was supported by the Federal Ministry of Education and Research of Germany (BMBF) under the contract numbers 05H09PD7 and 05H12PD6.
This work is also supported by the Commission of the European Communities under the 6th Framework Programme "Structuring the European Research Area", contract number RII3-026126, and under the FP7 Research Infrastructures project AIDA, grant agreement no. 262025. The information herein only reflects the views of its authors and not those of the European Commission and no warranty expressed or implied is made with regard to such information or its use.
This material is based in part upon work supported by the National Science Foundation. Any opinions, findings, and conclusions or recommendations expressed in this material are those of the author(s) and do not necessarily reflect the views of the National Science Foundation.
We would also like to thank the SRS group at CERN in particular Hans Muller for their support.

References

[1] X. Llopart, et al., Nucl. Instr. and Meth. A **581**, 485-494 (2007); Erratum ibid **585**, 106-108 (2008)
[2] M. Chefdeville, et al., Nucl. Instr. and Meth. A **556**, 490-494 (2006)
[3] Ties Behnke, et al., The International Linear Colloder, Technical Design Report Volume 4, ISBN 978-3-935702-78-2
[4] W. Koppert, et al., Nucl. Instr. and Meth. A **732**, 245-249 (2013)
[5] S. Martoiu, et al., Nuclear Science Symposium and Medical Imag- ing Conference (NSS/MIC), 2036 (2011)
[6] http://rd51-public.web.cern.ch/RD51-Public/
[7] M. Lupberger, et al., JINST **9** C01033 (2014)

Fast Timing for High-Rate Environments with Micromegas

Thomas Papaevangelou[1,a], Daniel Desforge[1], Esther Ferrer-Ribas[1], Ioannis Giomataris[1], Cyprien Godinot[1], Diego Gonzalez Diaz[2], Thomas Gustavsson[3], Mariam Kebbiri[1], Eraldo Oliveri[2], Filippo Resnati[2], Leszek Ropelewski[2], Georgios Tsiledakis[1], Rob Veenhof[2], and Sebastian White[2,4]

[1] *CEA, IRFU, Centre d'Etudes Nucléaires de Saclay, Gif-sur-Yvette 91191, France*
[2] *European Organization for Nuclear Research (CERN), Genève 1211, Switzerland*
[3] *Laboratoire Interactions, Dynamiques et Lasers (LIDyL) CEA, CNRS, Université Paris-Saclay, Gif sur Yvette 91191, France*
[4] *Princeton University, Princeton, New Jersey 08544, USA*

Abstract. The current state of the art in fast timing resolution for existing experiments is of the order of 100 ps on the time of arrival of both charged particles and electromagnetic showers. Current R&D on charged particle timing is approaching the level of 10 ps but is not primarily directed at sustained performance at high rates and under high radiation (as would be needed for HL-LHC pileup mitigation). We demonstrate a Micromegas based solution to reach this level of performance. The Micromegas acts as a photomultiplier coupled to a Cerenkov-radiator front window, which produces sufficient UV photons to convert the ~100 ps single-photoelectron jitter into a timing response of the order of 10-20 ps per incident charged particle. A prototype has been built in order to demonstrate this performance. The first laboratory tests with a pico-second laser have shown a time resolution of the order of 27 ps for ~50 primary photoelectrons, using a bulk Micromegas readout.

1 Introduction

From 2025 onward, it is planned to operate the LHC with typically 140 collisions per proton bunch crossing (HL-LHC phase), which will greatly complicate the analysis, particularly in the forward regions, where it will be very difficult to link the tracks with the primary vertex (associated with the only interesting collision) and thus to prevent the formation of fake jets created by the random stacking or pile-up of tracks from several secondary vertices. A solution proposed to fight against this pile-up effect is to measure very accurately the time of arrival of the particles [1]. A time resolution of a few tens of picoseconds with a spatial granularity better than 10 mm (depending on pseudo-rapidity) will be needed to obtain a fake jet rejection rate that is acceptable for physics analyses. Such performances should be obtained with a detector capable of withstanding the very important radiation levels expected in the HL-LHC era.

The ongoing R&D on timing for charged particles is approaching a level of 10 ps [2] but is not primarily directed at sustained performance at high rates and under high radiation (as would be needed for HL-LHC pileup mitigation). Several technologies are presently under evaluation within the LHC collaborations: Silicon detectors with some built-in amplification, Silicon detectors without amplification and gaseous detectors. Gaseous detectors have the potential of very good timing performance, provided the amplification gap is small enough. They have also the advantages of being low-cost, simple to operate, and are potentially insensitive to irradiation effects.

Modern photo-lithographic technology has enabled the development of a series of novel Micro Pattern Gaseous Detectors (MPGD): Micro-Strip Gas Chamber (MSGC), Gas Electron Multiplier (GEM), Micro-mesh Gaseous Structure (Micromegas), and many others revolutionizing cell size limits for many gas detector applications. The Micromegas detector fulfils the needs of high-luminosity particle physics experiments, and compared to classical gas counters, it offers intrinsic high rate capability, excellent spatial resolution and excellent time resolution [3, 4].

2 Micromegas for fast timing

2.1 Detection principle

The concept of a Micromegas detector coupled to a front window that acts as Cerenkov-radiator, equipped with a photocathode, is examined in order to address its timing performance. The radiator should produce sufficient UV photons and should be equipped with an appropriate photocathode in order to create the necessary number of electrons to convert the expected ~100 ps single-photoelectron time jitter into a timing response of 10-20 ps per incident particle. The photocathode has to be spark-resistant and must have good aging properties, appropriate for the high particle flux environment of HL-LHC.

In gaseous detectors the time resolution is typically of the order of several ns because of the time jitter of primary ionization (for high energy MIPs) in the drift gap.

[a] e-mail: thomas.papaevangelou@cea.fr

Our strategy to improve the time response is to minimize this jitter by creating simultaneously all primary electrons at about the same distance from the amplification region and by minimizing the drift space. The primary electrons are created by the Cerenkov light which is emitted by the charged particle when it crosses a 2-3 mm MgF_2 crystal. A thin CsI photocathode is deposited on the micromesh (reflective mode) or on the bottom of the crystal (semi-transparent mode) as shown in figure 1, and it is used to convert UV photons to electrons. The time spread of the Cerenkov light (~10 ps) is too small to limit the projected time accuracy.

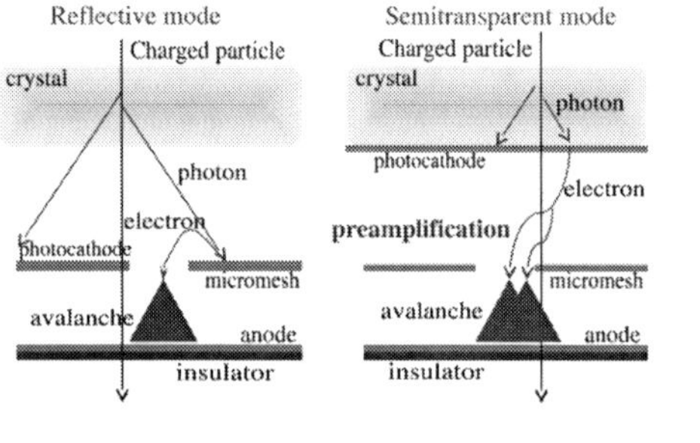

Figure 1. Schematic view of the charged particle detection principle, in the two modes; Reflective mode: a thin CsI layer is deposited on top of the micromesh; Semitransparent mode: CsI is deposited on the bottom surface of the UV crystal, defining a drift gap with the micromesh.

In the semitransparent mode the photocathode is deposited at the bottom face of the MgF_2 crystal at a distance of 100-300 μm from the micromesh to limit longitudinal diffusion. A strong field can be applied in the drift gap allowing some pre-amplification. The advantages of this configuration are the high single-electron sensitivity due to higher total gain, the possibility to use woven mesh Micromegas (bulk technology [5]) and the fabrication simplicity. However, the partial exposure of the photocathode to the ion backflow may cause aging and robustness issues. In the reflective mode the photocathode is deposited on the Micromegas mesh and no strong drift field is needed. The main advantages of this method are the very effective protection of the photocathode from the ion backflow (better aging properties, robustness) and the complete suppression of gas ionization in the drift region. However, this mode is limited to flat mesh detectors (Microbulk technology is the more appropriate [6]), while the internal reflection of the Cerenkov light on the crystal-gas interface might limit the efficiency.

2.2 Photoelectron production

A MIP passing through a crystal will emit:

$$\frac{d^2N}{dEdx} = \frac{\alpha^2 z^2}{r_e m_e c^2}\sin^2\theta_c \approx \frac{370}{eVcm}\left(1-\frac{1}{n^2(E)}\right)$$

photons. If L is the thickness in cm, $T(E)$ is the transmission of the crystal and $QE(E)$ the quantum efficiency of the photocathode, the number of produced photoelectrons will be:

$$N_{p.e.} \approx 370L\int T(E)QE(E)\left(1-\frac{1}{n^2(E)}\right)dE$$

For a MgF_2 crystal and a CsI photocathode with $QE(E)$ taken from [7], $T(E)$ from [8] and $n(E)$ from [9], we expect $370 \times 0.89 = 330$ p.e./cm (figure 2). However, this QE concerns operation in reflective mode. In the semitransparent mode it is expected to be significantly lower (factor 2-3), so, for a 3 mm MgF_2 crystal we would expect 30-50 photoelectrons.

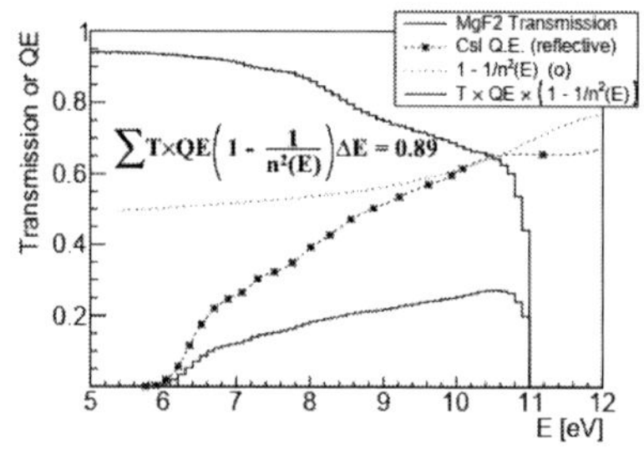

Figure 2. Transmission, refractive index of MgF_2 and quantum efficiency of CsI in reflective mode which were used for the calculation of the number of photoelectrons.

2.3 Diffusion in the drift region

Since the time jitter in photoelectron creation from the Cerenkov light is small (<10 ps), the dominant effect limiting the time resolution will be the diffusion of the photoelectrons in the drift region.

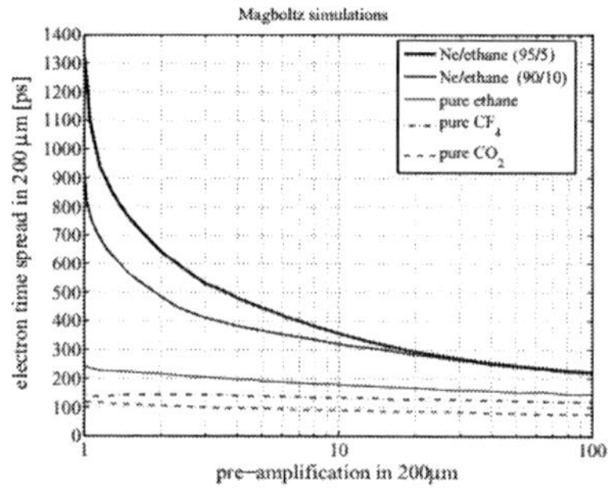

Figure 3. Electron time spread due to diffusion for several gases as a function of the achieved pre-amplification in a 200 μm gap.

In the case of the reflective photocathode, the jitter in the drift path will reflect the difference in the path depending on the creation point on the mesh and its distance to the amplification gap, so fine pitch meshes are preferable. Simulations were carried out to investigate the semitransparent mode. Magboltz simulation results are shown in figure 3 where the time spread due to electron diffusion in a 200 μm drift gap is plotted as function of the preamplification for several gas mixtures. A first assessment based on Magboltz thus indicates that a single photoelectron time resolution of 100-200 ps could be reachable for relatively small (or even in the absence of) pre-amplification.

Compared to noble-gas admixtures, pure quenchers show potentially a smaller time spread, but demand stronger electric fields in order to achieve the same gain. The optimum selection will be a tradeoff between the minimum time jitter and the maximum signal to noise ratio.

2.4 The Micromegas prototype

A prototype detector has been constructed (figure 4) in order to study the timing properties of the proposed scheme. It has been designed to be compatible with operation in both semitransparent and reflective mode.

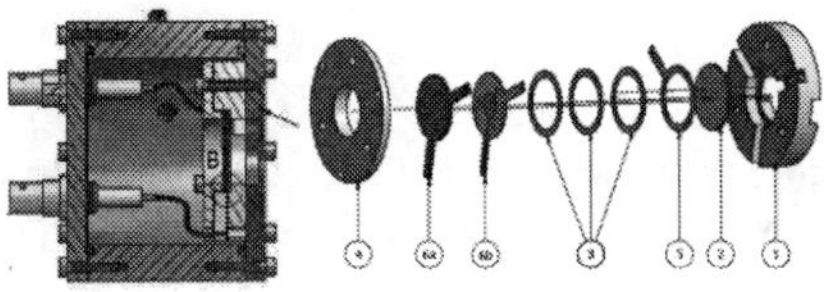

Figure 4. Schematic view of the detector chamber and expanded view of the Micromegas - spacers - crystal assembly.

The Micromegas was constructed using both Bulk and Microbulk technologies [5, 6]. It was a single-anode detector of 1 cm diameter (bigger than the maximum pixel size of a possible tracker). The amplification gap was 50 μm for the Microbulk and 128 μm for the bulk. Kapton rings, 50 μm thick, were used in order to support the MgF_2 or Quartz crystal on which the photocathode is deposited and define the drift gap of 200 μm. The top ring was metalized on the top surface, in order to apply the bias voltage to the photocathode. A thin (10 nm) Al layer was deposited on a quartz crystal and served as photocathode. This simplified configuration is sufficient to study the timing properties of the detector, using a very fast UV laser. For charged particle detection, the quartz has to be replaced with MgF_2, while the photocathode will be a CsI layer (100-300 nm) deposited on a metallic substrate (5-50 nm). The exact configuration will be optimized in terms of aging properties and of the maximization of the number of produced photoelectrons.

The Micromegas-crystal assembly is encased in a stainless-steel chamber. A quartz window, transparent to the UV laser, is mounted on the top surface of the chamber. The chamber and the components were properly chosen and special precaution was taken while mounting the detector in order to be appropriate for operating without gas circulation, to facilitate thus the measurements at laser and particle beam facilities.

2.5 Performance of the prototype

The gas mixture was chosen to be 90% Ne plus 10% Ethane, a mixture that gives high gain with good energy resolution and has shown good behavior for sealed mode operation [10], though it is not optimal for timing. We used a pulsed deuterium lamp in order to monitor the gain stability without gas circulation. Data were taken over one week showing only random fluctuations (arising from the lamp intensity) and no significant drop of the gain, demonstrating the possibility to perform measurements without renewing the gas.

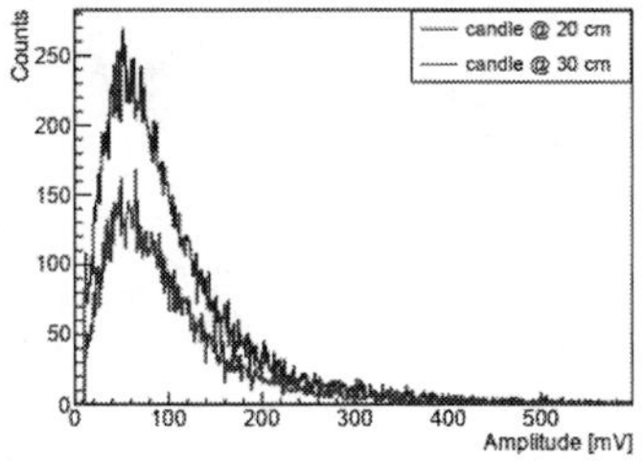

Figure 5. Pulse amplitude spectra for a candle placed at ~20 cm and ~30 cm distance from the detector with a charge preamplifier. The shape of the spectrum does not change, a clear indication for single electron pulses.

The sensitivity of the detector to single photons was verified using a candle flame, which is emitting continuously light from the IR down to the VUV, the spectrum stopping at ~200 nm due to the absorption of the air [10]. The quantum efficiency of the aluminium is sufficient to have several photoelectrons per second when the distance to the flame is not very long. Two spectra taken at a distance of 20 and 30 cm for the same time duration are shown in figure 5. The amplitude distribution does not change with the distance, while the rate does, proving that the pulses correspond to single photons. The resulting very good energy resolution is due to the single stage amplification in the Micromegas gap.

3 Time resolution measurements

The time response of the bulk Micromegas detector in semitransparent mode was studied at LIDyL laboratory

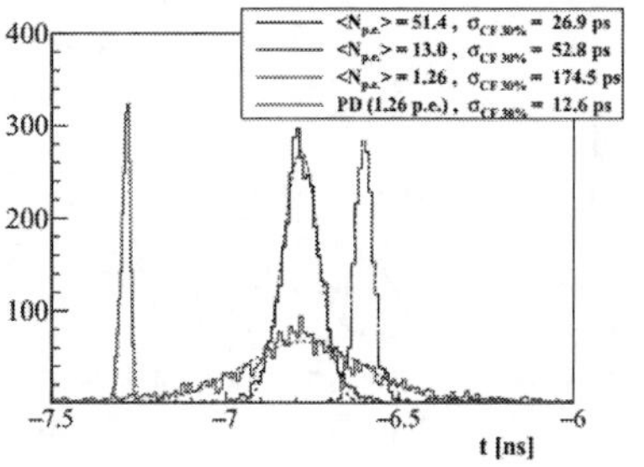

Figure 6. Distribution of the time at the 30% fraction of the amplitude for the Micromegas pulses, for different mean number of photoelectrons. The time spread of the photodiode is also shown.

(CEA/Saclay). An Optical Parametric Oscillator (Coherent MIRA-OPO), pumped by a Ti:sapphire laser (Coherent MIRA 900) provided 120 fs pulses at 550 nm. The repetition rate was varied between 9 kHz and 4.75 MHz by using a Pulse-Picker (Coherent 9200). The second harmonic at 275 nm was generated in a 5 mm BBO crystal. The typical output energy was 0.15-1 pJ for a beam diameter < 1 mm.

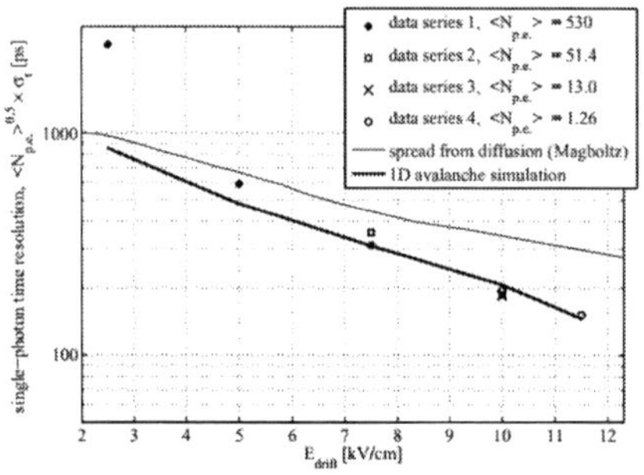

Figure 7. Dependence of the measured time resolution with the drift field, scaled to the single photoelectron case. The difference between the thin and thick lines indicates the improvement due to pre-amplification, according to a stochastic 1D avalanche model.

Though the extraction efficiency for the Al photocathode at those wavelengths is small, the intensity of the laser (up to 10^6 photons/pulse) allows for the production of a sufficient number of photoelectrons (up to 1000 p.e./pulse for 10 nm Al, depending also on the beam tuning and the applied Micromegas drift field). The number of photoelectrons per laser pulse was controlled by using a series of light attenuators: fine electroformed nickel meshes (100-2000 LPI) with optical transmission varying between 10% - 25%, giving attenuation factors of 4, 5, 10 and their combinations. The pitch of the meshes was varied in order to avoid systematic effects when combining several of them.

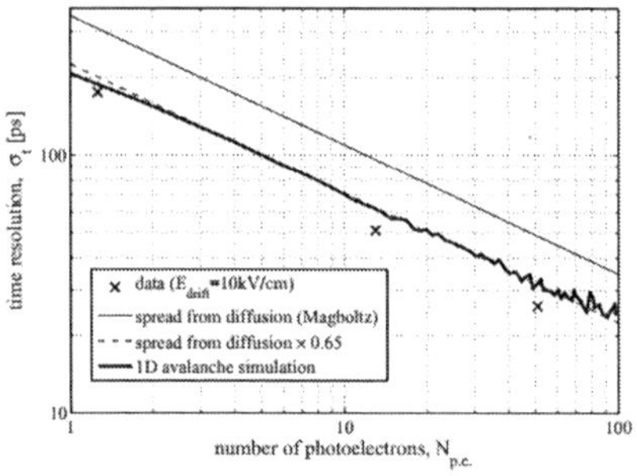

Figure 8. Dependence of the measured time resolution with the mean number of photoelectrons, for fixed amplification and drift fields. A resolution of 200 ps per single photoelectron and 27 ps for 50 photoelectrons has been achieved.

The laser beam was split in two, part of it illuminating our prototype and the other part an ultra fast Si detector (Thorlabs DET10A/M). Both signals were recorded by a 2 GHz oscilloscope operating at 10 GS/s. The time jitter of the photodiode was of the order of 13 ps, being appropriate as trigger and as reference for the measurement. The signal from the Micromegas was filtered for high frequency noise using a moving average. The photocathode and the filtered Micromegas signal times were calculated by taking a constant fraction of each signal. Best results were obtained for a 30% fraction of the amplitude (figure 6).

Data were taken for different values of the drift field in combination with the attenuators, in order to maintain the signal within a reasonable dynamic range. The time resolution improves with the drift field intensity due to the decrease of the longitudinal diffusion but also due to the preamplification at high field values. The results are summarized in figures 7 and 8, in comparison with a stochastic 1D avalanche simulation (that will be detailed in a forthcoming publication).

4 Conclusions

A Micromegas photodetector, equipped with a UV-transparent crystal and a metallic photocathode deposited on the bottom of the crystal has been constructed in order to investigate the potential of MPGDs for fast timing tracking of minimum ionizing particles. The aim of the first tests was the investigation of the intrinsic time characteristics of the detector. A very fast UV laser (time spread 120 fs) was used to produce the photoelectrons in a controlled way. With a gas mixture consisting of 90% Ne and 10% Ethane we measured a time resolution of the order of 200 ps for single photoelectrons and 27 ps for 50 photoelectrons. Simulations show that this already satisfactory result can be further improved by optimizing the gas mixture. A more exhaustive description of these results will be published elsewhere. Tests on a particle beam, with a MgF_2 crystal and CsI photocathode will be made in 2016.

The current project is carried out within the framework of the RD51 common Project "Fast Timing for High Rate Environments: A Micromegas Solution".

References

[1] S. White, proc Picosecond Workshop, Clermond-Ferrand 2014, arXiv:1409.1165 [physics.ins-det].

[2] C. Williams et al., Nucl.Instrum.Meth. **A594** (2008) 39-43.

[3] Y. Giomataris, P. Rebourgeard, J.P. Robert, G. Charpak, Nucl.Instrum.Meth. **A376** (1996) 29.

[4] I. Giomataris, Nucl.Instrum.Meth. **A419** (1998) 239.

[5] I. Giomataris et al., Nucl.Instrum.Meth. **A560** (2006) 405-408

[6] S. Andriamonje et al., JINST **5** (2010) P02001

[7] J. Séguinot et al., Nucl.Instrum.Meth. **A297** Issues 1-2 (1990) 133-147.

[8] http://www.korth.de/index.php/162/items/21.html

[9] P. Laporte et al., Optical Society of America Journal (ISSN 0030-3941), **73**, (1983) 1062-1069.

[10] A. Peyaud et al., Nucl.Instrum.Meth. **A787** (2015) 102-104.

First result of large size Scintillating Glass GEM imager

Takeshi Fujiwara[1,a], Yuki Mitsuya[2], Hiroyuki Takahashi[2], and Hiroyuki Toyokawa[1]

[1] *National Institute of Advanced Industrial Science and Technology (AIST), Research Institute for Measurement and Analytical Instrumentation, 1-1-1 Umezono, Tsukuba, Ibaraki, 305-8568 Japan*
[2] *The University of Tokyo, Japan, Department of Nuclear Engineering and Management, 7-3-1 Hongo, Bunkyo, Tokyo, 113-8656 Japan*

Abstract. A large size x-ray imaging gaseous detector, which has 280 × 280 mm^2 effective area has been successfully developed and x-ray imaging has been demonstrated. The imaging system consists of a chamber filled with Ne/CF_4 scintillating gas mixture, inside of which Glass GEM (G-GEM) is mounted for gas multiplication. In this system electrons are generated by the reaction between x-rays and the gas, and visible photons by excited Ne/CF_4 gas molecules during the gas electron multiplication process in the G-GEM holes. These photons are simply detected with CCD-camera and a radiograph is formed. Here, we report on the properties of large size scintillation G-GEM and the results of using it as a digital x-ray imager with a large sensitive area.

1 Introduction

Gaseous detectors have been playing important role for various types of radiation measurements. Gas electron multiplier (GEM) is one of the most popular and successful types of micro-patterned gaseous detector (MPGD). [1–4] However, building a large size detector with MPGDs needs a lot of effort to install and maintain. In addition, signal readout is another big issue for operating MPGDs. For imaging with GEMs, readout pads needs to be carefully designed and so as the many channels of readout circuits which goes up to hundreds to thousands of channels. On the other hand, not all radiation measurements are required to be precisely measured. For instance, x-ray radiograph can be formed with simple integrating circuits such as CCDs and CMOS. [5, 6] From that point of view, gaseous scintillation with GEMs has been studied in past works, in order to simplify the signal readout for imaging. [7–10] Scintillation light caused during electron avalanche process with scintillation gas (such as CF_4) can be easily detected with optical camera. However, amount of photons produced is limited with conventional GEM's avalanche process and it is difficult to be detected with conventional optical camera. This is mostly due to the gas gain is not enough with single GEM ($\approx 10^3$). On the other hand, we have been developing Glass GEMs (G-GEM), which is promising electron multiplier with high gas gain and low charge ups [11, 12]. Gas gain of G-GEM reached up to 10^5 with a single G-GEM structure may contribute to high scintillation yield. In addition, we have fabricated large size G-GEM with 300 × 300 mm^2 glass substrate, which has 280 × 280 mm^2 effective area [13]. In this paper, we report on the development of large area x-ray imager using G-GEM and gaseous scintillation. Development of our first large-area Scintillating G-GEM detector is shown. We indicate the measured results of gas gain and optical signal to show that it worked properly. We then show the first result of gain x-ray imaging result with large-area Scintillating G-GEM.

[a] e-mail: fujiwara-t@aist.go.jp

2 Properties of large size G-GEM and initial experiment

2.1 Principle of the setup

We have fabricated large size G-GEM using 300 × 300 mm^2 photo-etchable glass substrate, using fabrication process of HOYA Corporation. This glass is called PEG3, and the G-GEM has 280 × 280 mm^2 effective area. Properties of the large are G-GEM especially focusing on comparison with G-GEM fabricated in our past works [10, 12] which has 100 × 100 mm^2 effective area are shown in table 1. The outlook of the large area G-GEM is shown in Figure 1. As seen in the figure, the G-GEM are stiff enough and has a self-supporting structure, which results in easy to handle and needs no support for mounting.

On the other hand, CF_4 gas is known as a scintillation gas, and to achieve high enough yield of scintillation photons for imaging, gas gain needs to be high enough also. The mechanism of scintillation light yield during G-GEM's avalanche process is shown in Figure 2. Ne/CF_4 gas mixture is attractive gas for high gas amplification with low applying voltage. In this study, we first studied gain properties and scintillation light from Ne/CF_4 gas for x-ray imaging.

Table 1. Comparison with past reported Glass GEM and large size Glass GEM fabricated in this study.

	Past reported G-GEM[10, 12]	Large size G-GEM
Effective area	100 mm square	280 mm square
Substrate size	145 mm square	300 mm square
Hole size	180 μm	170 μm
Hole pitch	280 μm	280 μm
Substrate thickness	680 μm	700 μm
Number of holes	141,417	1,154,423

Figure 1. Outlook of the large size G-GEM (280 × 280 mm^2 effective area), fabricated with 300 × 300 mm^2 photo-etchable glass substrate. As the G-GEM is 700 μm thick, and has a perfectly self-supporting structure.

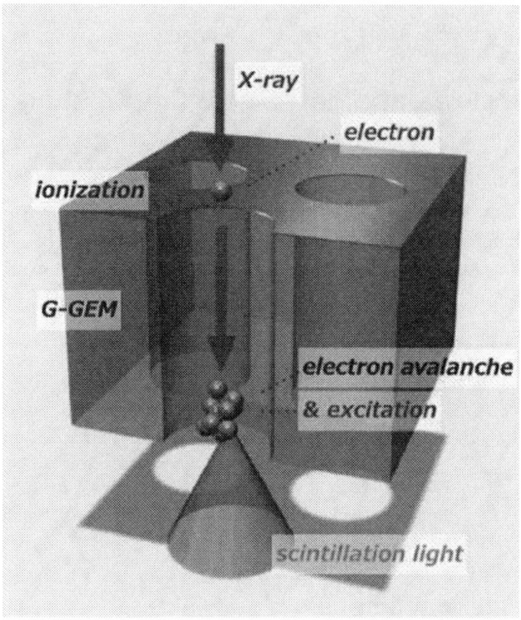

Figure 2. Diagrammatic sketch of scintillation light yield from G-GEM hole during electron avalanche process in scintillation gas.

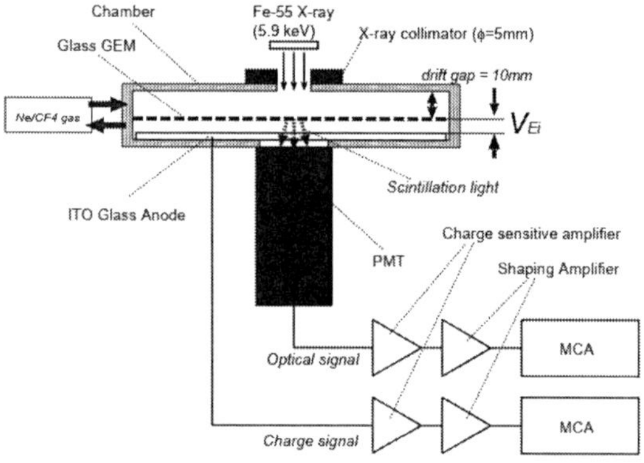

Figure 3. Experiment setup for measuring the basic performance of the scintillating G-GEM detector coupled with the photomultiplier tube (Hamamatsu R329) are described. ^{55}Fe x-ray (5.9 keV) source was used in this study, and the x-ray was collimated to ϕ =5mm at the entrance window of the chamber. Both the charge and the optical signal were recorded with MCAs.

2.2 Gain properties with Ne/CF$_4$ gas and Scintillation light measurement

In order to investigate scintillation light from G-GEM with Ne/CF$_4$ gas, initial experiments were carried out using ^{55}Fe source and photo multiplier tube (PMT). Figure 3 is the experimental setup. We installed a G-GEM in a gas filled chamber with drift cathode 10 mm , and a readout 2 mm gap from the optical window which plays a role as a readout anode. The chamber was flushed with an Ne/CF$_4$ (90:10) gas mixture at 50 mL/min of flow rate. High voltages to the G-GEM's electrodes were applied with individual voltage supply. The drift field was 1 kV/cm, and induction field was 2 kV/cm. In this setup, charge and scintillation light produced with G-GEM were read at the same time using two MCAs. First, charge signal was read out from a charge sensitive preamplifier connected to the anode electrode placed in the bottom of G-GEM. The output from charge sensitive amplifier output was connected to a shaping amplifier and a MCA for taking pulse height spectra (figure 4). At the same time, scintillation light from G-GEM were detected with PMT placed right under the chamber. PMT was also connected to charge sensitive amplifier, shaping amplifier and MCA, and pulse height spectra of ^{55}Fe were recorded at the same time. The spectra obtained in this measurement in Figure 5. Due to the efficiency of light collecting was not good enough in this setup, we could not observe clear peak of 5.9 keV x-ray, beside the energy spectra obtained with charge signal. However, the result indicates high yield of scintillation photons were produced with G-GEM and Ne/CF$_4$ gas. Gas gain curve of this G-GEM is shown in Figure 6. Gas gain reached up to 7 × 10^3 stably with single G-GEM and Ne/CF$_4$ gas.

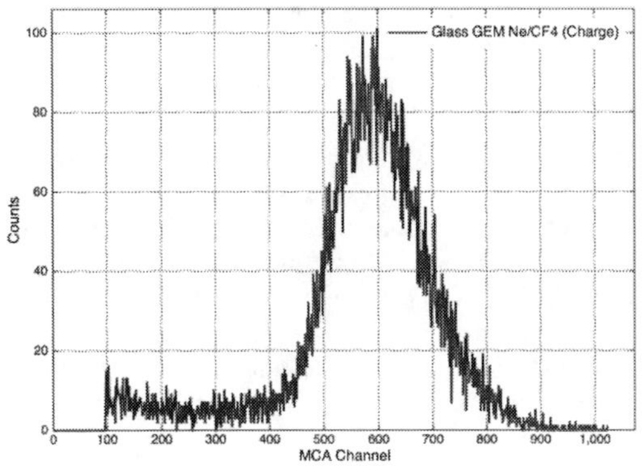

Figure 4. Measured pulse height spectrum with ^{55}Fe X-ray source (5.9 keV). Energy spectrum of charge readout is shown. Energy resolution is 29% in FWHM.

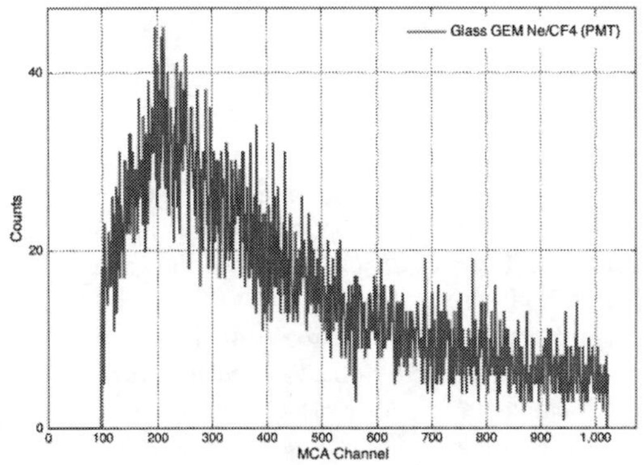

Figure 5. Energy spectrum taken at the same time using a photomultiplier tube directly coupled under optical window placed in the bottom of the gas chamber.

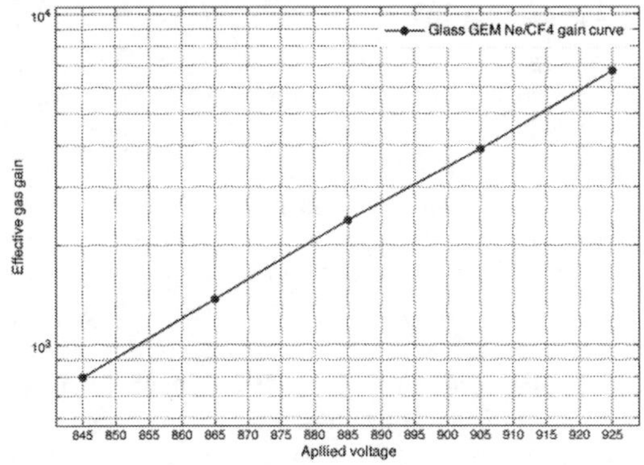

Figure 6. Gas gain curve achieved with G-GEM and Ne/CF4 (90:10) gas. Maximum gain reached up tp 7×10^3, applying 925 V to a single G-GEM.

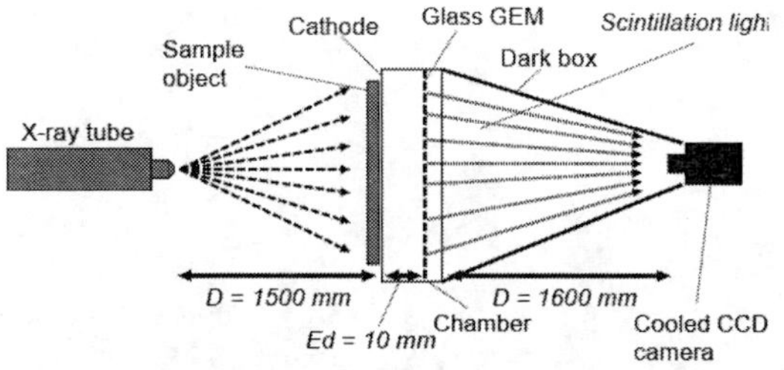

Figure 7. Schematic and detection principle of scintillating G-GEM detector. Electrons created in reaction with x-rays are multiplied by the G-GEM in the chamber. Scintillation light emitted during the electron avalanche process is detected with a CCD camera placed behind the chamber, and form an x-ray transmission image.

3 Imaging test of large size Scintillating Glass GEM

Figure 7 illustrates the schematic and detection principle detection principle of x-ray imaging with a scintillating G-GEM detector. For the x-ray window of the chamber, a 20 μ m thick Kapton film coated with 20 nm Aluminum layer was used. In this experiment, gas gain of G-GEM were set to approximately 3×10^3. Using this setup, we investigated the first x-ray imaging test with the large size scintillating G-GEM detector. Detailed experimental conditions and equipment used in this study is shown in table 2. The sample object for demonstating x-ray imaging is shown inf figure 8. A tool box made out of plastics are filled with screws and connectors. X-ray tube were placed in front of gas chamber, and in the same line, optical cooled CCD camera (Bitran BH-50L) were placed behind the chamber. Camera is looking towards the optical window of the chamber surrounded with dark box. First demonstrating x-ray transmission image taken by large size Scintillating G-GEM is shown in figure 9. X-ray image of 280 × 280 mm^2 are successfully taken in 5 seconds of exposure time with simple optical camera.

Table 2. Detailed setup and equipment used in imaging test with large size scintillating G-GEM Detector

Electron multiplier	280 mm square G-GEM
Scintillation gas	Ne/CF_4 (90:10)
Gas gain	3×10^3
Optical camera	BITRAN BH50L(300,000pixels)
Lens	Fujinon 25 mm F0.85
x-ray source	Amptek Mini-x (25kV, 50μA)

4 Summary

A large size scintillating G-GEM detector that 280 × 280 mm^2 in size was developed and operated successfully. The gain properties and gas scintillation yield with Ne/CF_4 gas were quantitatively studied, and succeeded in achieving fine X-ray transmission images with the scintillating

Figure 8. A photograph of tool box filled in with screws an connectors, used as a sample object for x-ray imaging demonstration of scintillating G-GEM detector.

Figure 9. X-ray transmission image of a tool box obtained by large size scintillating G-GEM detector with an x-ray tube voltage of 25 kV, current 50 μA. Effective area of the x-ray image is 280×280 mm^2. The image was taken with an exposure time of 5 seconds (gas gain: 3,000).

G-GEM detector. Using the high yield photon emission from the Ne/CF_4 gas mixture, the scintillation light from the high gain G-GEM can be easily detected with a commercially available CCD camera coupled to a conventional lens system. Although the data acquisition rate is restricted due to the speed of the CCD readout, the optical readout is very simple and attractive. Moreover, the CCD camera can provide fine imaging due to the very large number of channels (up to 3×10^6). Therefore, we conclude that our detector has the potential to be an alternative to flat-panel detectors, imaging plates, x-ray films, and CMOS sensors for digital x-ray imaging.

Acknowledgement

The authors wish to acknowledge Takashi Fushie (HOYA Corporation), an expert on fabricating Glass GEMs with photo-etchable glass, for his great help in interpreting the significance of the results of this work.

References

[1] Sauli F 1997 GEM: A new concept for electron amplification in gas detectors Nuclear Instruments and Methods in Physics Research Section A: Accelerators, Spectrometers, Detectors and Associated Equipment 386 531–4

[2] Sauli F 2016 The gas electron multiplier (GEM): Operating principles and applications Nuclear Instruments and Methods in Physics Research Section A: Accelerators, Spectrometers, Detectors and Associated Equipment 805 2–24

[3] Breskin A, Alon R, Cortesi M, Chechik R, Miyamoto J, Dangendorf V, Maia J M and Santos Dos J M F 2009 A concise review on THGEM detectors Nuclear Instruments and Methods in Physics Research Section A: Accelerators, Spectrometers, Detectors and Associated Equipment 598 107–11

[4] Coimbra A E C, Breskin A and Santos Dos J M F 2012 THGEM operation in high pressure Ne/CF 4 J. Inst. 7 C02062–2

[5] Lumb, D.H.; Nousek, J.A., "Energy and time response of CCD X-ray detectors," in Nuclear Science, IEEE Transactions on , vol.39, no.5, pp.1379-1383, Oct 1992

[6] Cho, Min Kook; Kim, Ho Kyung; Graeve, T.; Seung Man Yun; Lim, Chang Hwy; Dong Hyun Cho; Kim, J.-M., "Measurements of X-ray Imaging Performance of Granular Phosphors With Direct-Coupled CMOS Sensors," in Nuclear Science, IEEE Transactions on , vol.55, no.3, pp.1338-1343, June 2008

[7] Fraga F, Margato L, Fetals S, Fraga M, Marques R and Policarpo A 2003 Luminescence and imaging with gas electron multipliers Nuclear Instruments and Methods in Physics Research Section A: Accelerators, Spectrometers, Detectors and Associated Equipment 513 379–87

[8] Fraga F A F, Margato L M S, Fetal S T G, Fraga M M F R, Ferreira Marques R and Policarpo A J P

L 2001 Optical readout of GEMs Nuclear Instruments and Methods in Physics Research Section A: Accelerators, Spectrometers, Detectors and Associated Equipment 471 125–30

[9] Fraga F A F, Fetal S T G, Margato L M S, Ivaniouchenkov I, Marques R F and Policarpo A J P L 1999 Imaging detectors based on the gas electron multiplier scintillation light 1999 IEEE Nuclear Science Symposium. Conference Record. 1999 IEEE Nuclear Science Symposium and Medical Imaging Conference vol 2 (IEEE) pp 829–33

[10] Fujiwara T, Tanaka S, Mitsuya Y, Takahashi H, Tagi K, Kusano J, Tanabe E, Yamamoto M, Nakamura N, Dobashi K, TOMITA H and Uesaka M 2013 Development of a scintillating G-GEM detector for a 6-MeV X-band Linac for medical applications J. Inst. 8 C12020–0

[11] Takahashi H, Mitsuya Y, Fujiwara T and Fushie T 2013 Development of a glass GEM Nuclear Instruments and Methods in Physics Research Section A: Accelerators, Spectrometers, Detectors and Associated Equipment 724 1–4

[12] Fujiwara T, Mitsuya Y, Takahashi H, Fushie T, Kishimito S, Guerard B and Uesaka M 2014 The performance of Glass GEM J. Inst. 9 P11007–7

[13] Mitsuya Y, Fujiwara T, Fushie T, Maekawa T and Takahashi H 2015 Development of large-area glass GEM Nuclear Instruments and Methods in Physics Research Section A: Accelerators, Spectrometers, Detectors and Associated Equipment 795 156–9

Development of a transparent single-grid-type MSGC based on LCD technology

Hiroyuki Takahashi[1], Hiroaki Miyoshi[2], Lian Xuan[1], Yuki Mitsuya[1], Kenji Shimazoe[1], Masashi Ohno[1] and Takeshi Iimoto[3]

[1]*Departmenf of Nuclear Engineering and Management, Graduate School of Engineering, The University of Tokyo, 7-3-1 Hongo, Bunkyo-ku, Tokyo 113-8656, JAPAN*
[2]*SHARP Co. Ltd, 22-22 Nagaike-cho, Abeno-ku, Osaka 545-8522, JAPAN*
[3]*Division for Environment, Health, and Safety, The University of Tokyo, 7-3-1 Hongo, Bunkyo-ku, Tokyo 113-8654, JAPAN*

Abstract. We are developing a multi-grid-type MSGC based on the liquid crystal display (LCD) technology, which enables a large area and fine structure. Single-grid-type MSGC using transparent electrodes has been fabricated and successfully operated in several different gas mixtures. The use of LCD technology allows us to integrate some simple electronics using thin film transistors. Such an integrated device is our next target. The successful operation of Single-grid-type MSGC is the very important first step for us.

1 Introduction

We are working on the use of microfabrication techniques to the radiation detection for almost 20 years. In this period, we studied on MSGC (MicroStrip Gas Chamber), Glass GEM (Gas electron multiplier), and dedicated ASIC (Application Specific Integrated Circuit). Based on our experience in these individual components of present-day's gaseous radiation detectors, we are becoming aware of the importance of more integrated devices. Many principles of the signal readout method are invented but the difficulty of the application of each principle lies in the interface between the sensor part and the readout electronics. New important topics may even exist in the interfacing itself. Now we re-think about the MSGC. People are almost forgetting about the MSGC [1], however, during these decades, big progress has been made in the microfabrication techniques on the glass, in particular, in the field of liquid crystal display (LCD).
Because an essence of the LCD technology is the formation of very precise electrode patterns on a large area glass plate, the LCD technology can readily be applied to the MSGC fabrication. One problem of the conventional MSGC was related to the substrate. The glass substrate isolates the anode and the cathode, however the high resistivity surface of the glass substrate causes a charge-up problem. Although the small-gap MSGC can solve this problem, it cannot operate at high gas gain since the anode and the cathode are too close in this geometry and the surface discharge is unavoidable. We have already shown a new configuration of MSGC using the multi-grid-type electrode structure [2], where the electrodes are not just the anode and the cathode but intermediate electrodes (grids) are inserted in between to stabilize the detector and increase gas gain.

[a] Corresponding author: leo@n.t.u-tokyo.ac.jp

Also, we demonstrated that the MSGC with transparent electrodes can be operated in an Ar/CF_4 gas mixture. Scintillation light is emitted in this gas mixture [3] in the course of avalanche process, which can be seen through the transparent electrode and the glass substrate. The use of optical signal for position sensing was demonstrated [4]. The LCD technology relies on the transparent electrode such as indium tin oxide (ITO) and indium zinc oxide (IZO), which is favourable for the optical readout. Also, we could integrate some simple electronics with thin film transistors (TFTs) on the glass as already done in most of LCD panels. Thus, in principle, we can integrate all the necessary functions for a gaseous radiation detector with readout electronics on just one glass plate as shown in Figure 1.

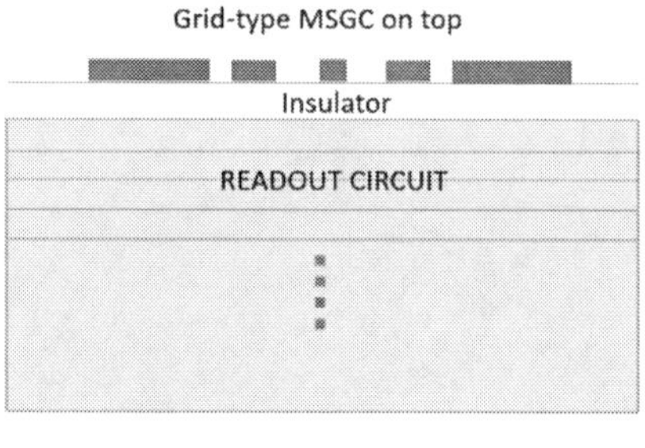

Figure 1. Concept of new generation of MSGC based on LCD technology.

This is definitely compact, efficient, and cost-effective. Besides the optical readout, we can also use induced charge [5], or direct coupling to the MSGC cathode. Based on such an idea, we are intensively working on the development of new generation MSGC. Radiation signals

https://doi.org/10.1051/epjconf/201817402004

can be transferred to the integrated readout circuit by visible photons, induced charges, or direct current (see Figure 2). The simplest readout circuit is just a TFT array with integrated capacitance, which is applied to the flat panel detector for medical applications, however, many other possibilities are being considered in our collaboration team. In particular, the signal readout electronics for the gas detector operated at high voltage often suffers from damages caused by spark events. If we use optical signals as the interface from micropattern gas detector to the readout circuit, we are completely free from the damage problem. Besides this, we could also utilize the charge signal.

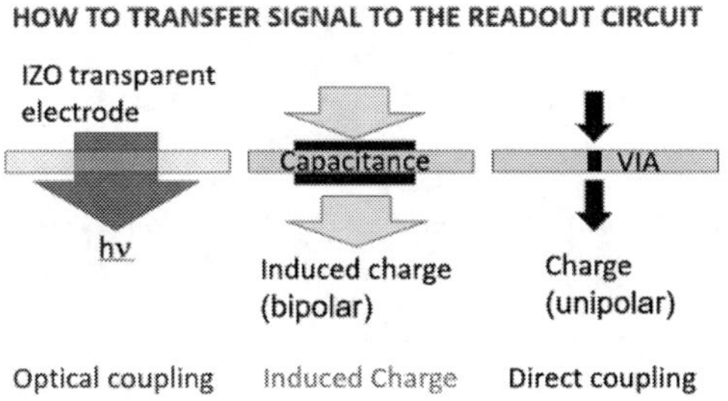

Figure 2. Three different signal readout schemes through the insulation layer. 'VIA' means electrical connection through the insulation layer.

Based on such an idea, we are fabricating test MSGCs as the first step to the integrated device. We report some of the initial results in this paper.

2 Design and fabrication of transparent MSGC with a single grid

We have designed a new single-grid type MSGC (S-MSGC) using IZO electrode. Aiming at low energy X-ray imaging applications, we selected a fine pitch of 150 μm. We adopted a single grid inserted between the anode and the cathode as illustrated in Figure 3. The anode and cathode widths were set to 8 μm and 62 μm, respectively. It is equipped with 20 μm wide grid between the anode and the cathode. The grid is biased at an intermediate potential between the anode potential and the cathode potential. The gap widths between neighbouring electrodes are set at 10 μm, maximizing the surface covered with conductive electrodes. All grid strips are connected together to apply a single bias voltage. Figure 4 shows the fabricated test device. The effective area of the test device was 15 mm x 23 mm.

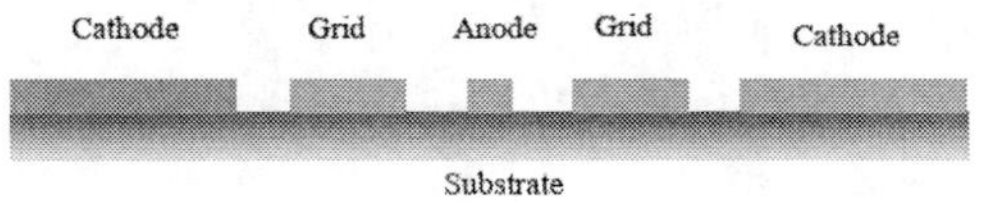

Figure 3. Cross sectional image of the electrode pattern of the designed S-MSGC.

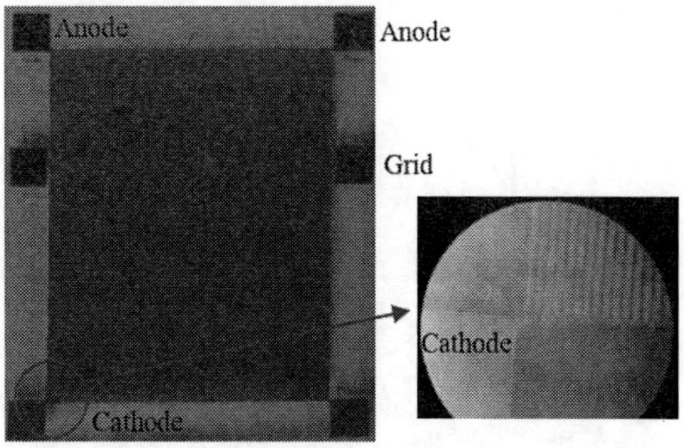

Figure 4. Left: Overall layout image of MSGC electrodes (For visualization purpose, opaque electrodes were used). Right: Microscopy reflection images of IZO electrodes (Near the cathode pad shown in the red circle in the left image). Contrast is adjusted and white part shows the electrode regions.

3 Test of the transparent S-MSGC

3.1 Charge signal measurements

The S-MSGC was tested under different gas mixtures. We have successfully operated our transparent S-MSGC in Ar (90 %) / CH_4 (10 %), Kr (90 %) / CO_2 (10 %), and Ar (90 %) / CF_4(10 %). The applied voltage was around 400-550 V to the anode and 100-250 V to the grid. The anode strips were connected together and read out by a charge sensitive amplifier. The maximum gas gain was ~500. A clear photo peak was observed for an ^{55}Fe X-ray source. Figure 5 shows a result from Kr/CO_2 gas, where the anode was biased at 530 V and the grid voltage was 240 V. Figure 6 shows a result from Ar/CF_4 gas where the anode voltage was 500 V and the grid voltage was 250 V. This is a reasonable performance for the single-grid design.

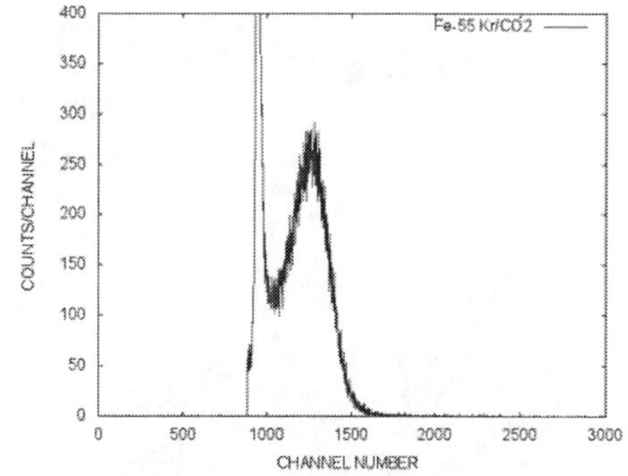

Figure 5. Observed pulse height spectrum for an ^{55}Fe source. The fabricated IZO S-MSGC was operated in a Kr/CO_2 gas mixture. The electric charge was read out from the common anode line.

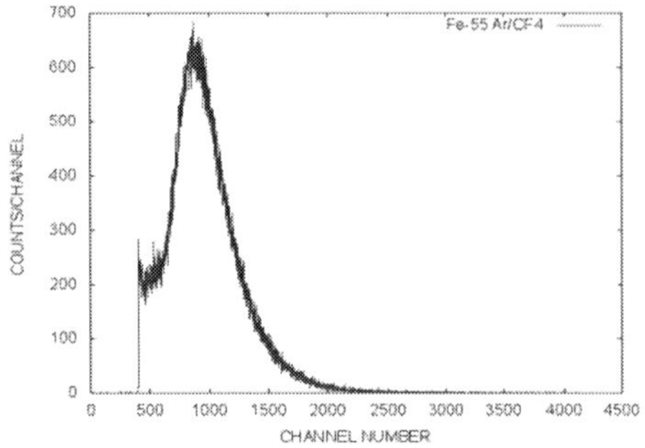

Figure 6. Observed pulse height spectrum for an ^{55}Fe source. The IZO S-MSGC was operated in an Ar/CF_4 gas mixture.

Kr gas shows a higher mass attenuation coefficient between 17 keV and 30 keV than Xe gas. Thus, X-ray diagnostics applications are considered with the Kr gas mixture.

3.2. Optical signal readout

Visible light photons are emitted during the avalanche process in Ar/CF_4 gas. The red light at wavelength around 600 nm is suitable for detection by photodiodes. These photon signals can be used for position sensing and/or pulse height measurement. The cross sectional view of the test setup is shown in Figure 7 and the photograph taken from top is shown in Figure 8 where the IZO electrode pattern is not seen (because of its transparency) and only the Hamamatsu S3590 10 mm x 10 mm effective area photodiode is visible through IZO S-MSGC.

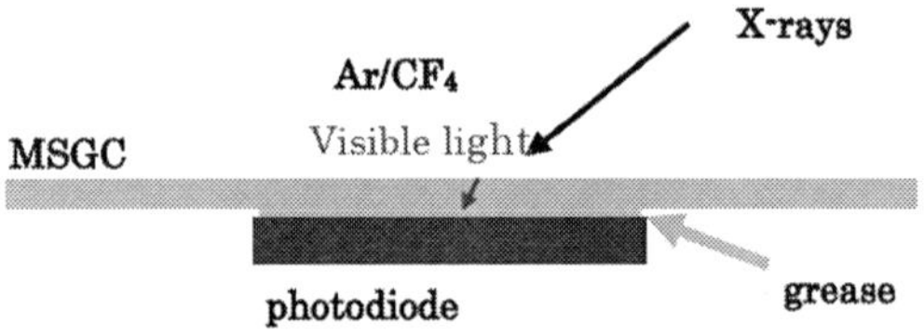

Figure 7. Cross sectional view of the test setup of an optical signal readout

Figure 8. Photograph of IZO S-MSGC with a photodiode.

The anode voltage was 508 V and the grid voltage was 239 V. The bias voltage of the PIN photodiode was 75 V. Figure 9 shows the pulse height spectra recorded from the PIN photodiode for different radiation sources. BG was obtained without the X-ray source and is considered as the noise level of the measurement system. Due to the large size of the PIN photodiode, the noise level of the preamplifier was quite large. Then, the pulse height spectra do not show photo peaks. However, the pulse height distribution shows longer tail for higher energy X-rays. This implies that larger optical signals are obtained when illuminated by higher energy X-rays, which indicates that the optical signal pulse is due to the radiation detection. This result is encouraging because the readout system is just one large PIN photodiode, whose capacitance is about 40 pF. If we utilize a small photodiode such as 150 μm x 150 μm in size, the capacitance is much smaller than that of the large PIN photodiode (10 mm x 10 mm effective area). Such photodiodes can be integrated just below the S-MSGC pixel of the same pixel size (~150 μm) and this can solve noise problems.

For further optimization of the detector, we are also thinking about the double grid version. Single grid is useful and easy to operate, however, in terms of gas gain, multi-grid design is better. For example, with a test device of two-grid structure, shown in Figure 10, much higher gain was reached. Figure 11 shows a charge spectrum of the two-grid MSGC for ^{55}Fe X-rays. The resolution is significantly improved compared to that of the S-MSGC (Figures 5 and 6). The anode pitch was 80 μm and the anode width was 2 μm in this case. Using the latest LCD process, these parameters can be realized.

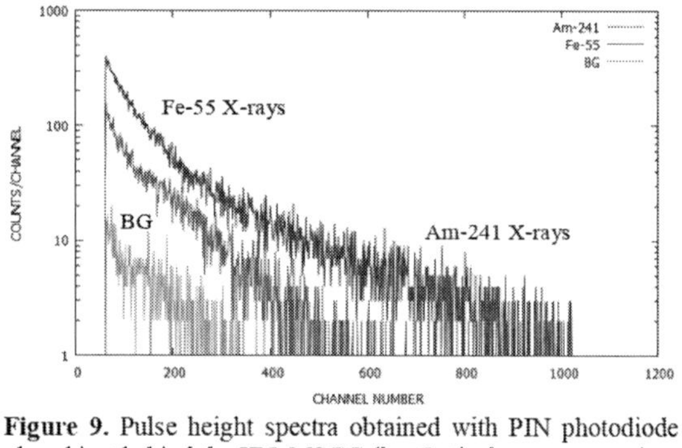

Figure 9. Pulse height spectra obtained with PIN photodiode placed just behind the IZO MSGC (i.e. Optical measurement).

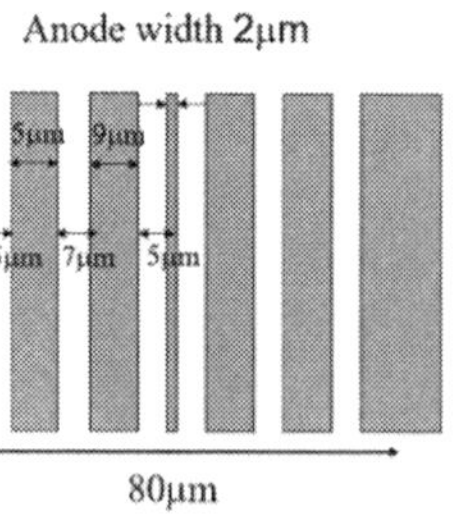

Figure 10. Electrode pattern of a double-grid-type MSGC test device.

https://doi.org/10.1051/epjconf/201817402004

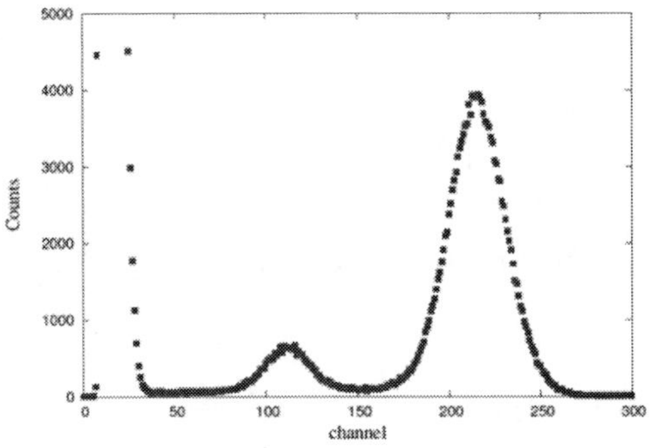

Figure 11. Pulse height spectra of the 80 μm anode pitch double-grid-type MSGC obtained with an ^{55}Fe source.

4 Integration of readout circuits

To take the full advantage of the LCD technology, integration of the readout circuit based on TFTs (Thin Film Transistors) on the same glass as the MSGC is promising. We are currently investigating the possibilities of such an integrated device and developing a circuit component using TFTs. In this approach, the use of an IGZO (indium gallium zinc oxide) is promising. Figure 12 illustrates the TFT characteristics of IGZO for this approach. It provides a very low leakage current or high resistance, which is useful for charge accumulation device. Therefore, multiplexing readout and integrator circuit will be a primary choice in this device. Also, IGZO is typically operated at tens of volts, which means the device is tolerant to the high voltage when a gaseous detector brings a giant pulse or some discharge, and is promising for the direct coupling to the gas detector signal readout.

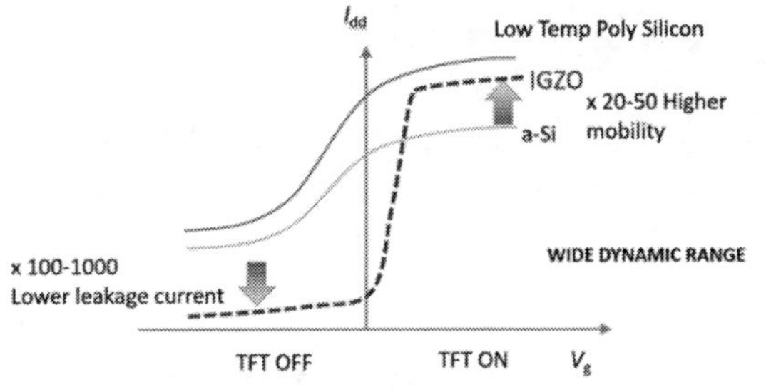

Figure 12. General characteristics of thin film transistor (TFT) available in different process technology.

4 Conclusions

We consider X-ray applications and reconsider the use of transparent multi-grid-type MSGC principle. The LCD technology is very promising for the new generation of MSGCs and the transparent IZO electrodes used in this process can provide a new readout scheme with optical photons. We have fabricated a single-grid-type MSGC (S-MSGC) using IZO. The detector has been successfully operated and we also measured optical signals by using a PIN photodiode through the glass substrate with Ar/CF_4 gas. Circuit integration on the same glass substrate is being considered. We anticipate that the highly integrated gaseous detector glass plate may find many applications not only for X-rays but also for many other areas.

References

1. A. Oed, Nucl. Instr. and Meth. A263, 351 (1988).
2. H. Takahashi, et al., Nucl. Instr. and Meth. A477, 13 (2002).
3. F.A.F. Fraga, et al., Nucl. Instr. and Meth. A 478, 357 (2002).
4. H. Takahashi, et al., Nucl. Instr. and Meth. A623, 123 (2010).
5. K. Fujita, H. Takahashi, et al., Nucl. Instr. and Meth. A580, 1027 (2007).

Study of combined THGEM/GAPD-matrix multiplier in a two-phase Cryogenic Avalanche Detector in Ar

A. Bondar[1,2], A. Buzulutskov[1,2], A. Dolgov[2], E. Shemyakina[1,2], and A. Sokolov[1,2,a]

[1] Budker Institute of Nuclear Physics SB RAS, Lavrentiev avenue 11, 630090 Novosibirsk, Russia
[2] Novosibirsk State University, Pirogov street 2, 630090 Novosibirsk, Russia

Abstract. Two-phase Cryogenic Avalanche Detectors (CRADs) with combined THGEM/GAPD-matrix multiplier have become an emerging technique in rare-event experiments such as those of coherent neutrino-nucleus scattering and dark matter search. In this work we continue to study the performance of the two-phase CRAD in Ar with THGEM/GAPD-matrix charge/optical readout. The matrix was composed of a 3x3 array of GAPDs (Geiger-mode APDs), optically recording THGEM-hole avalanches in the Near Infrared (NIR). Gain, time, amplitude and spatial resolution properties of the combined multiplier are described.

1 Introduction

The development of two-phase Cryogenic Avalanche Detectors (CRADs) based on noble liquids [1] is relevant in the field of rare-event experiments for dark matter search [2] and coherent neutrino-nucleus scattering [3, 4]. In two-phase CRADs the ionization electrons produced in the noble liquid by a particle are extracted into the gas phase and multiplied with GEM or THGEM [5] multipliers. The multiplied electrons can be recorded directly using a charge readout [6, 7] or indirectly using an optical readout of avalanche photons with either Geiger-mode APDs (GAPDs) [8] or CCD camera [9]. Gain values as large as 1000 were obtained in a two-phase CRAD operated in Ar with the double-THGEM multiplier [10]. This gain value is however not enough for effective operation with ultimate sensitivity, i.e. in single-electron counting mode and with precise 2D readout.

A solution of this problem is to use a combined THGEM/GAPD-matrix multiplier, where the GAPDs record avalanche scintillations from the THGEM holes either in the Vacuum Ultraviolet (VUV) [8] (using wavelength shifters), Near Infrared (NIR) [11, 12] or Near Ultraviolet [13] spectral range. In this work we continue to study the performance of the two-phase CRAD in Ar with THGEM/GAPD-matrix charge/optical readout; first results of the performance of such a combined multiplier were presented elsewhere [14].

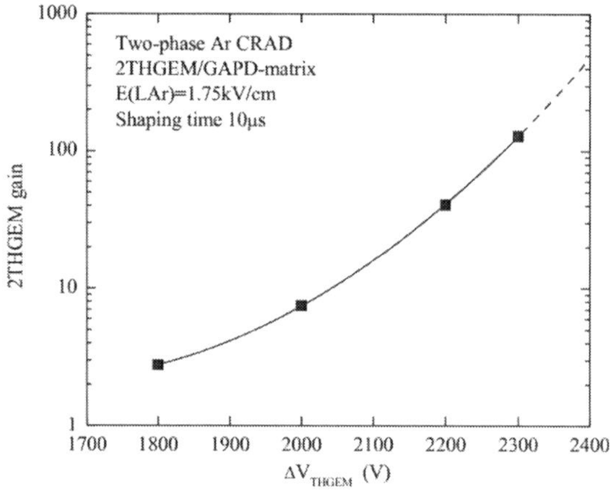

Figure 1. Charge gain of the double-THGEM multiplier as a function of the applied voltage on each THGEM, in a two-phase CRAD in Ar at 87 K.

2 Experimental setup

The experimental setup was identical to that used in our previous study of the two-phase CRAD in Ar with combined THGEM/GAPD-matrix multiplier [14]. It included a cryostat with a 9 liters volume cryogenic chamber. The chamber consisted of a cathode mesh, immersed in a ~1 cm thick liquid-Ar layer, and a double-THGEM assembly with an active area of 10×10 cm^2, placed in the gas phase above the liquid. A 3×3 matrix of GAPDs (MRS APD "CPTA 149-35") was placed in the gas phase at a distance of 6.5 mm behind the second THGEM, electrically insulated from the latter with an acrylic plate transparent in the NIR and by a wire grid at ground potential. Each GAPD had a 2.1×2.1 mm^2 active area and a Photon Detection Efficiency (PDE) of about 15% at 800 nm. The detector was operated in two-phase mode in the equilibrium state, at a saturated vapor pressure of 1.0 atm and at a temperature of 87 K. The Ar was purified by an Ox-

[a] e-mail: A.V.Sokolov@inp.nsk.su

https://doi.org/10.1051/epjconf/201817402005

isorb filter, providing an electron life-time of >13 μs in the liquid. In this study, the charge gain of the double-THGEM multiplier was kept at rather moderate level, in the range of 40-160 (see Fig. 1). The chamber has aluminum windows at the bottom for irradiation with pulsed X-rays. The detector was irradiated by 15-40 keV X-rays from a pulsed X-ray tube with Mo anode operated at 40 kV, with a pulse rate of 240 Hz, through a cylindrical collimator with a hole diameter of 2 mm. The X-ray conversion in the liquid was defined by this collimator and occurred over a small area of 2 mm in diameter. The charge signals were recorded from the last electrode of the second THGEM using a charge-sensitive amplifier. The DAQ system included an 8-channel Flash ADC CAEN V1720 (12 bits, 250 MHz): the signals from the THGEM were digitized and stored in a computer for further off-line analysis using a LabView^{TM} program.

3 GAPD signals

Fig. 2 depicts a typical signal from the GAPD. The GAPD signals are composed of a number of fast pulses (peaks), each apparently corresponding to a single photoelectron detection followed by a single-pixel discharge (or those of multi-pixels in case of cross-talks).

As seen from the upper figure in Fig. 2, the signal width exceeds few microseconds. This is due to the potential barrier at the liquid-gas interface, that delays some of the electrons for a sufficiently long time, sometimes exceeding 20 μs.

This is well illustrated in Fig. 3, where one can clearly see the fast and slow components of the signal. The fast component corresponds to the signal from the electrons, which directly penetrate through the potential barrier; the slow component corresponds to the electrons that were delayed by a potential barrier. The specific shape of the leading edge of the fast component in the distribution of Fig. 3 is due to the presence of harder and softer components in the spectrum of X-rays absorbed in the liquid Ar layer. Softer and harder X-rays are absorbed at different depths, and have, accordingly, different electron drift times. For details of the spectrum shape of pulsed X-rays absorbed in liquid Ar and their average deposited energy, amounting to 25 keV, see ref. [15].

In Fig. 2 one can see that the signals from the GAPD have rather different pulse-heights, with the tendency to smaller values. This is due to the effect of the GAPD performance degradation at lower temperatures observed recently [16, 17]. This effect manifested itself at higher photon fluxes; it resulted from the increase of the GAPD pixel recovery time due to the increase of the pixel quenching resistor at cryogenic temperatures. Some pulses were even lost below threshold, thus reducing the photon detection efficiency of the GAPD. Our recent studies [17] showed that this problem can be solved using a different type of GAPD, namely those of the MPPC type by Hamamatsu production, with reduced resistance of the pixel quenching resistor.

Since in the current measurements the GAPD pulse-height was degraded as shown above, the GAPD ampli-

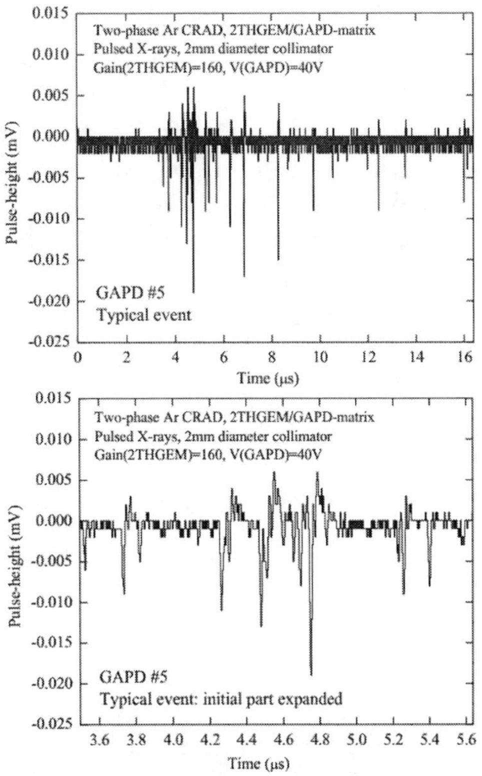

Figure 2. In the upper figure the typical signal of the GAPD in the two-phase CRAD in Ar with a THGEM/GAPD-matrix multiplier is shown. The lower figure represents the initial part of the GAPD signal. As one can see the signals have different amplitudes due to the degradation effect.

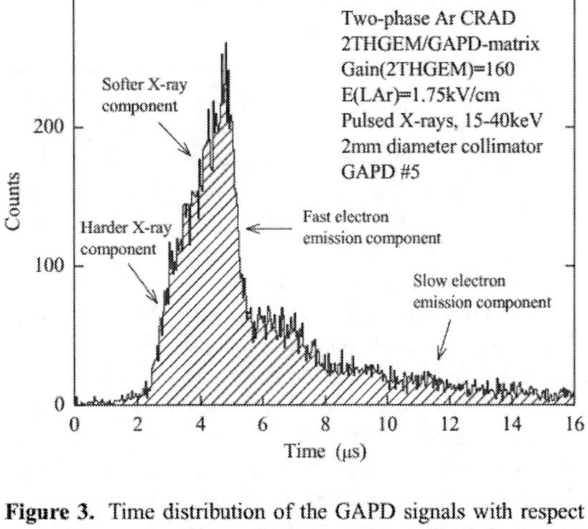

Figure 3. Time distribution of the GAPD signals with respect to the X-ray pulse trigger, in the two-phase CRAD in Ar with THGEM/GAPD-matrix multiplier, induced by pulsed X-rays. The figure clearly shows the fast and slow components of the signals.

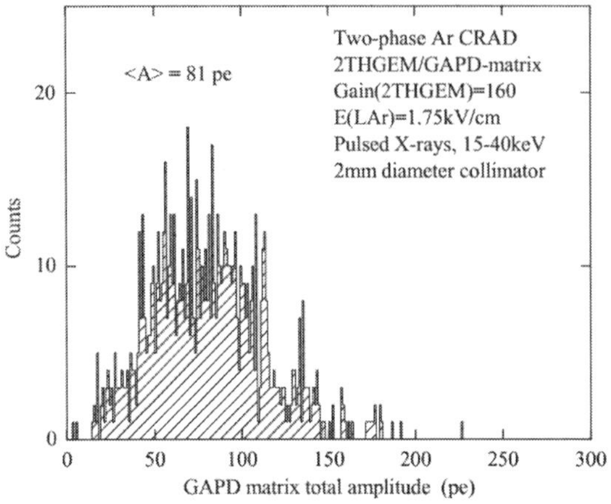

Figure 4. Distribution of GAPD-matrix total amplitude expressed in the number of photoelectrons detected, in a two-phase CRAD in Ar with THGEM/GAPD-matrix multiplier, induced by pulsed X-rays absorbed in liquid Ar with the average energy of 25 keV.

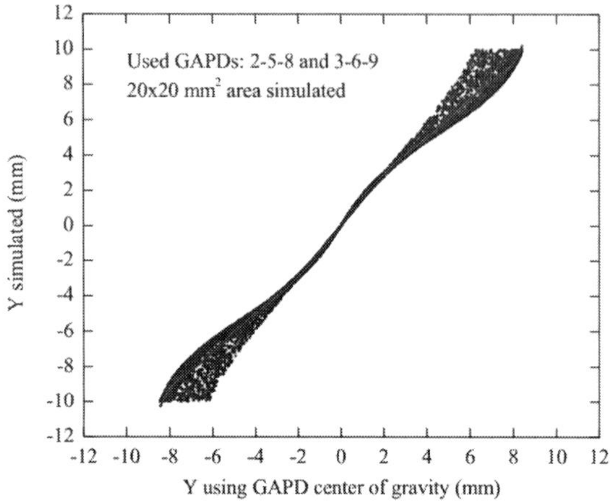

Figure 5. Simulated Y-coordinate versus that of reconstructed using 2-5-8 and 3-6-9 GAPD columns of the GAPD matrix of Fig. 6). The blue curve is the fit to the data points.

tude was calculated by just counting the number of photoelectron peaks in the signal. The result for the total GAPD-matrix amplitude is shown in Fig. 4. One can see that yield of the combined THGEM/GAPD-matrix multiplier in present conditions, namely at the double-THGEM charge gain of 160, was high enough, amounting to 81 photoelectrons per 25 keV deposited energy corresponding to the reduced multiplier yield of about 3 photoelectrons per keV.

4 Coordinate characteristics

The X-Y coordinate characteristics of the THGEM/GAPD-matrix multiplier are illustrated in Figs. 5 and 6. The X-Y coordinates were calculated from the GAPDs amplitudes using a center-of-gravity algorithm, corrected for Monte-Carlo simulation of the light propagation from the THGEM holes to a given GAPD. The result of such simulation is shown in Fig. 5. The curve in the figure, obtained by fitting to the simulated data points, was used to reconstruct the true coordinate from that of calculated using the center-of-gravity algorithm.

Fig. 6 illustrates the imaging capability of the detector: the X-Y coordinate plot shows the pulsed X-ray conversion regions defined by the 2 mm diameter collimator.

Fig. 7 shows the Y projection of the coordinate distribution. The width of distribution is 3.2 mm (FWHM). The spatial resolution of the detector can be estimated from the data by quadratic subtraction of the collimator hole diameter; it results in the value of 1 mm (a standard deviation of the distribution) for 25 keV energy deposition, at the present operation conditions.

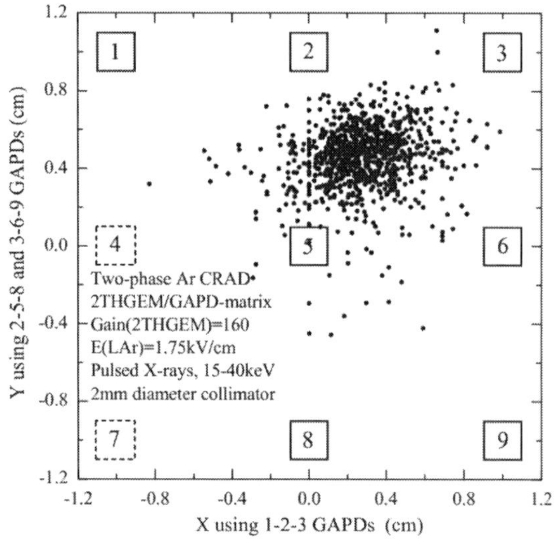

Figure 6. X-Y plots of the pulsed X-ray conversion region in the two-phase CRAD in Ar with THGEM/GAPD-matrix multiplier, defined by a 2 mm diameter collimator. Squares with numbers indicate the positioning of the active areas of the appropriate GAPDs of the matrix.

5 Conclusions

A multi-channel optical readout in a two-phase Cryogenic Avalanche Detector (CRAD) in Ar with combined THGEM/GAPD-matrix multiplier has been studied. The detector had a high yield of combined multiplier, of 3 photoelectrons per keV of deposited energy, and a good spatial resolution $\sigma = 1$ mm for 25 keV X-rays. Such a readout can be applied in low threshold rare-event experiments like those of coherent neutrino-nucleus scattering and dark matter searches.

6 Acknowledgements

We thank A. Chegodaev and R. Snopkov for technical support. This work was supported by the Russian Science Foundation (project N 14-50-00080).

References

[1] Buzulutskov A., JINST **7**, C02025, 2012.
[2] Akimov D. et al., NIM **vol 598 P 275**, 2009.
[3] Hagmann C., Bernstein A., IEEE Trans. Nucl. Sci. **vol 51 P 2151** (2004).
[4] Akimov D. et al., JINST **4**, P06010, 2009.
[5] A. Breskin et al., Nucl. Instrum. Meth. A 598 (2009) 107.
[6] A. Bondar et al., JINST **8**, P02008, 2013.
[7] A. Badertscher et al., Nucl. Instrum. Meth. A 641 (2011) 48.
[8] P.K. Lightfoot et al., JINST **4**, P04002, 2009.
[9] K. Mavrokoridis et al., JINST **10**, P10004, 2015.
[10] A. Bondar et al., JINST **8**, P02008, 2013.
[11] A. Bondar et al., JINST **5**, P08002, 2010.
[12] A. Bondar et al., JINST **7**, P06014, 2012.
[13] A. Bondar et al., EPL **112**, 19001, 2015.
[14] A. Bondar et al., Nucl. Instrum. Meth. A 732 (2013) 213.
[15] A. Bondar et al., Eprint arXiv:1505.02296, 2015.
[16] A. Bondar et al., JINST **9**, P08006, 2014.
[17] A. Bondar et al., JINST **10**, P04013, 2015.

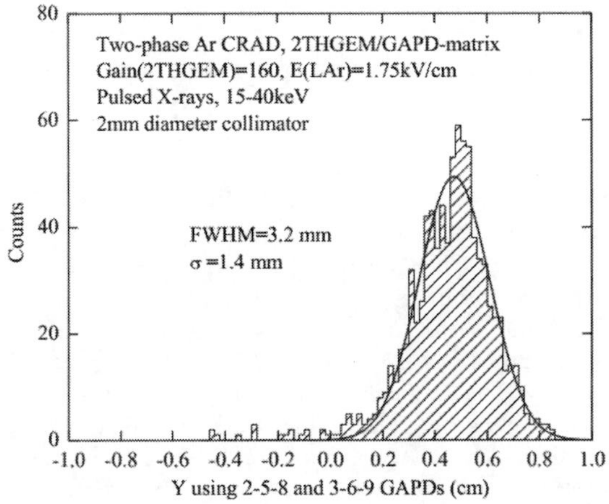

Figure 7. Y projection of 2D-coordinate distribution of the pulsed X-ray conversion region in the two-phase Ar CRAD with THGEM/GAPD-matrix optical readout, defined by 2 mm diameter collimator when irradiated with pulsed X-rays.

Study of Negative-Ion TPC Using μ-PIC for Directional Dark Matter Search

Tomonori Ikeda[1,a], Kentaro Miuch[1], Atsuhiko Ochi[1], Ryota Yakabe[1], Takashi Hashimoto[1], Ryosuke Taishaku[1], Daniel P. Snowden-Ifft[2], Jean-Luc Gauvreau[2], Toru Tanimori[3], Atsushi Takada[3], and Kiseki Nakamura[3]

[1] *Department of Physics, Kobe University, Rokkodaicho, Nada Hyogo 657-8501*
[2] *Department of Physics, Occidental College, Los Angeles, CA 90041, USA*
[3] *Department of Physics, Kyoto University, Kitashirakawaoiwakecho, Sakyo, Kyoto 606-8502, Japan*

Abstract. Negative-ion time projection chambers(TPCs) have been studied for low-rate and high-resolution applications such as dark matter search experiments. Recently, a full volume fiducialization in a self-triggering TPC was realized. This innovative technology demonstrated a significant reduction in the background with MWPC-TPCs. We studied negative-ion TPC using the μ-PIC+GEM system and obtained sufficient gas gain with CS_2 gas and SF_6 gas at low pressures. We expect an improvement in detector sensitivity and angular resolution with better electronics.

1 Introduction

Weakly interacting massive particles(WIMPs) are one of the candidates for dark matter. Among the various direct search experiments, the detection of nuclear recoil tracks is thought to provide strong evidence of dark matter detection. Gas detectors are among the possible nuclear track detection methods, and several groups have developed detectors and performed underground direction-sensitive dark matter searches[1–5].

Since the expected event rate is extremely low, a dark matter detector needs to be large(cubic-meter scale). Therefore, the electron diffusions during drift are large compared with the track lengths of the recoil nuclei(up to several mm) in conventional TPCs. Since spatial and angular resolutions are deteriorated by diffusion, the drift length is limited to several tens of cm with electron-drift detectors[2]. Transverse diffusion could be suppressed using a strong magnetic field parallel to the drift field but it would increase the cost and complexity of the system.

Negative-ion TPCs have been studied for low-rate and high-resolution applications[6]. The advantage of negative-ion TPCs is that the ion diffusions can be strongly suppressed at low pressures and without a magnetic field. The DRIFT (Directional Recoil Identification from Tracks) group has pioneered the study of negative-ion TPCs and demonstrated a strong potential for fine tracking performance[7]. Recently a discovery of "minority carriers" in CS_2 gas broadened its potential, and the measurement of absolute Z-position in self-triggering TPCs became possible. Minority carriers appeared after adding a few percent O_2 to the original gas. Although the mechanism is not completely understood, several types of negative ions are thought to drift each with slightly different velocities. Using this technique, DRIFT was able to significantly improve their detector sensitivity[8].

Electron capture, ion transport and electron release in the gas amplification process need to occur in negative ion TPC. While forming negative ions has been well studied theoretically and experimentally, the detachment process is less well understood[9]. Negative ion gases that give sufficient gas gains are very limited; carbon disulfide(CS_2) and nitromethane(CH_3NO_2) are one of the few possible choices[10]. These gases are not very safe and are not easy to use in an underground laboratory.

A safe negative-ion gas has long been sought as a TPC gas, and recently a GEM-TPC with SF_6 negative ion drift gas was reported[11]. In addition, they observed minority carriers without any additional gas. Unlike CS_2, SF_6 is non-toxic, and can be handled easily while retaining the same advantages as CS_2 gas. Fluorine has a large cross section for WIMP-nuclear spin-dependent interaction; therefore, it has good properties as a target. SF_6 is generally used as high voltage insulating gas. Its electron affinity is very large, approximately 1.1 eV. Therefore, we need strong electric fields for electron-detachment and gas multiplication to occur. Micro-pattern gaseous detectors(MPGDs) are known to have strong electric fields and would be a suitable device to use with SF_6 as a TPC gas.

NEWAGE (NEw generation WIMP-search with Advanced Gaseous tracking device Experiment) has been using a μ-PIC based TPC for direction-sensitive dark matter search with CF_4 gas. We achieved a 90% confidential level direction-sensitive spin-dependent cross section limit of 557 pb for a WIMP mass of 200 GeV/c^2[5]. The main background restricting the detector sensitivity is radioactive contamination in the μ-PIC. We initiated a new study to reduce the radioactive background using two approaches. The first approach is to develop a low back-

[a] e-mail: tomonori@stu.kobe-u.ac.jp

Figure 1. Photograph of the NEWAGE0.1c detector. The upper side of the GEM electrode is seen through the drift mesh. The μ-PIC is set below the GEM.

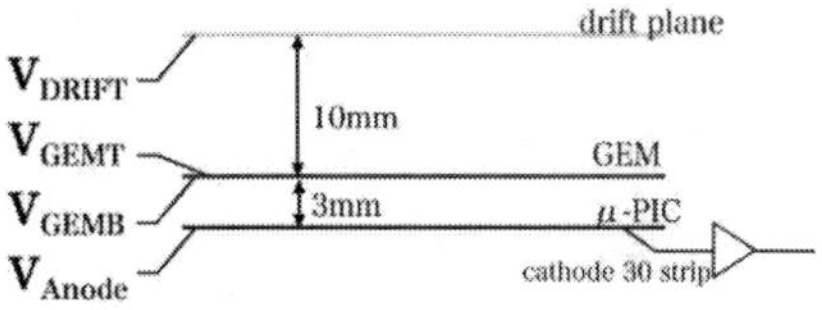

Figure 2. Schematic structure of the NEWAGE0.1c. Some cathode strips are grouped and read by one channel amplifiers(see Section 3). V_{DRIFT} is the voltage supplied at the drift plane, V_{GEMT} is the voltage at the top of the GEM, V_{GEMB} is the voltage at the bottom of the GEM, and V_{Anode} is the voltage at the anode of the μ-PIC.

ground μ-PIC. Glass fiber clothes used to reinforce the polyimide were found to contain certain amounts of U and Th which radiate α particles. We are developing a μ-PIC with a low background polyimide. The second approach is Z-fiducialization using negative ion gas. This is expected to effectively reduce the background because backgrounds from μ-PIC should be localized at low Z positions. Increased angular resolution is also expected with a negative-ion gas. The negative-ion TPC would be a good technology to construct large-volume (> 100 m^3) detectors in the future.

In this study, the first study of the μ-PIC+GEM system with the negative ion gases CS_2 and SF_6 is reported.

2 NEWAGE0.1c Detector

The NEWAGE0.1c detector which comprises of the μ-PIC and the GEM(Fig.1), was used for this work. A schematic view is shown in Fig.2. The μ-PIC has 256 × 256 pixels with a pitch of 400 μm read by 256 anode and 256 cathode orthogonally formed strips[12]. When gas multiplication occurs at an anode, the same amount of charge is read from the corresponding anode and cathode strips. For this conceptual study, the sum of a few cathode strips was used as readout. The GEM[13] was settled 3mm above the μ-PIC as a sub-amplifier with an effective area of 10 × 10 cm^2. The substrate of the GEM was a 100 μm thick liquid crystal polymer(LCP) and the hole size and pitch were 70 μm and 140 μm respectively. A stainless steel mesh was set 10mm above the GEM as a drift plane. The μ-PIC and the GEM were enclosed in a vacuum vessel.

3 Measurement Results and Discussion

3.1 CS_2 gas

We tested the μ-PIC+GEM system with CS_2 gas using ^{55}Fe at Occidental College. CS_2 gas is toxic, volatile, and inflammable; therefore, we performed the study in a dedicated gas-handling system. We supplied -1.5 kV at the drift plane. As the typical drift velocity of negative ions is about 10^{-2} cm/μs, which is two or three orders of magnitude slower than electrons, we used a slow shaping amplifier(CREMAT,CR-111). The time constant was 4 μs and the gain was 130 mV/pC. The gas pressures were 76 Torr and 38 Torr. According to SRIM simulation[14], the track length of a carbon nucleus with an energy of 100 keV is approximately 2 mm in CS_2 gas at 76 Torr which is longer than that in CF_4 gas at the same pressure.

The obtained total gas gain as a function of the voltage supplied at the anode electrodes of the μ-PIC is shown in Fig.3. The total gas gain of 10,000 at 76 Torr , was better than that achieved in CF_4 at the same pressure. Therefore, μTPC will have at least the same performance as CF_4 in low rate experiments. Total gas gains as a function of the voltage supplied to the GEM are shown in Fig.4. We confirmed that the GEM performs as a sub-amplifier at both 76 Torr and 38 Torr. This total gas gain is sufficient to reach the energy threshold we recorded in previous experiments.

Consequently, the μ-PIC+GEM system worked very well with CS_2. We should be able to observe the minority carriers by adding O_2 gas. The most remarkable point was achieving gas gains of a few thousand at 38 Torr. At this operating pressure, we can expect better angular resolution because the nuclear track is twice as long as with CF_4 at 76 Torr and track diffusion should be suppressed. We expect that μTPC with CS_2 will have high angular resolution and high background rejection capabilities.

3.2 SF_6 gas

We tested the μ-PIC+GEM system with SF_6 gas at a pressure of 20 Torr using ^{55}Fe at Kobe University. We supplied -1.8 kV at the drift plane. We used a different slow shaping amplifier(CLEAR-PULSE,CS515-1) for this experiment. The gain of CS515-1 is 1 V/pC and we coupled it with a high-pass filter that has τ_{decay} of ~100 μs.

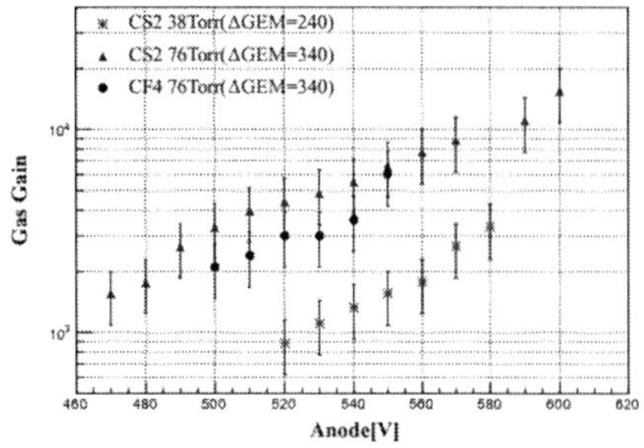

Figure 3. Total gas gains as a function of the voltage supplied at the anode electrodes of the μ-PIC. Circles, triangles and crosses show CF_4 at 76 Torr, CS_2 at 76 Torr and CS_2 at 38 Torr, respectively.

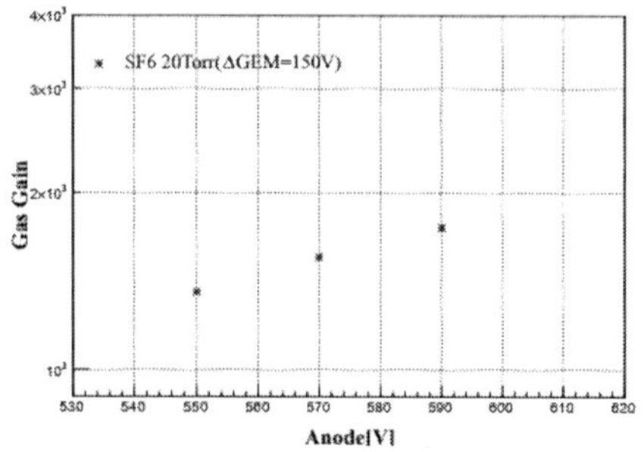

Figure 5. Total gas gains as a function of the voltage supplied at the anode electrodes of the μ-PIC. The voltages supplied at the top and bottom GEM are 840 V and 690 V, respectively.

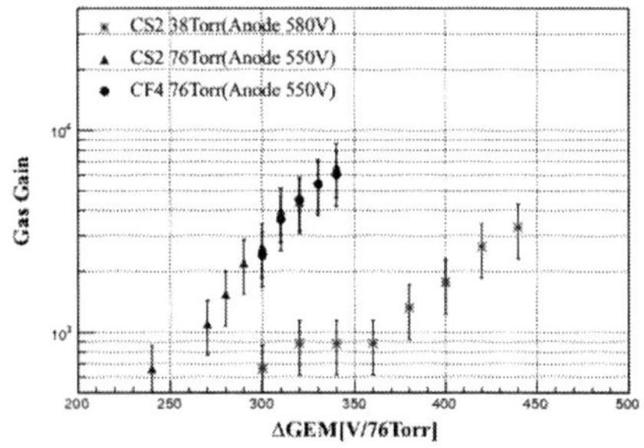

Figure 4. Total gas gains as a function of the voltage supplied between the top and bottom electrodes of the GEM. Circles, triangles and crosses show CF_4 at 76 Torr, CS_2 at 76 Torr and CS_2 at 38 Torr, respectively.

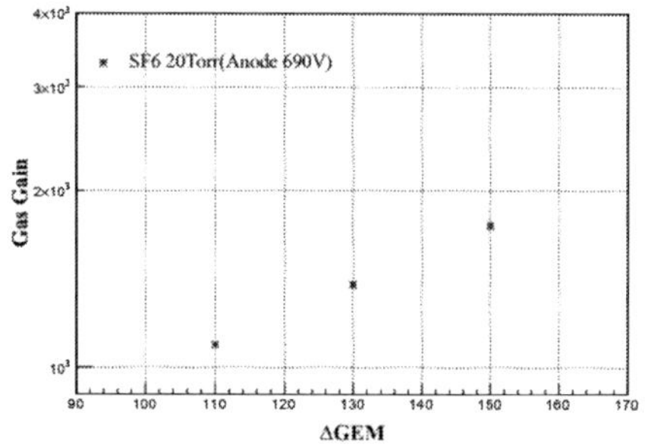

Figure 6. Total gas gains as a function of the voltage supplied between the top and bottom electrodes of the GEM. The voltage supplied at the anode is 690 V.

The obtained total gas gain as a function of the voltage supplied at the anode electrodes of the μ-PIC and the voltage supplied to the GEM are shown in Fig.5 and Fig.6, respectively. The highest gas gain was approximately 1800, while the require gas gain was $1000 \cdot \frac{76\ Torr}{P}$. This means we need to improve the S/N by factor two, which could be achieved with a dedicated amplifier.

Consequently, the μ-PIC+GEM system also worked well with SF_6. As noted in CS_2 experiments, we can expect better angular resolution because of the longer nuclear track.

4 Future work

We have confirmed that the μ-PIC+GEM system worked well with SF_6 and CS_2. For CS_2, we obtained sufficient performance; however, because of gas safety issues, we will pursue studies of negative ion μTPC with SF_6.

As our next step, we are constructing a prototype μTPC the drift length of which is 30 cm. We will attempt to observe the minority peak and to demonstrate measurement of the absolute Z position. Furthermore, we plan to develop readout electronics for negative ion μTPC. An ASIC chip with a multi-channel slow shaping amplifier is being developed for liquid argon TPC by the KEK e-sys group. This system can be adopted to negative ion TPCs with a small modification. We will then test the detector performance. We expect better spatial and angular resolution because of the smaller diffusion. A spatial resolution of ~0.2 cm along the Z axis is demonstrated by the DRIFT detector[8]; therefore we expect to reduce the background from the μ-PIC by two orders of magnitude by rejecting events from the μ-PIC plane. We expect an improvement of at least two orders of magnitude in the sensitivity because of full fiducialization.

5 Acknowledgment

This work was supported by KAKENHI Grant-in-Aids(19684005, 23684014 and 26104005) and Program for Advancing Strategic International Networks to Accelerate the Circulation of Talented Researches, JSPS, Japan(R2607).

References

[1] T. Tanimori et al., Phys. Lett. B **578** (2004)241-246

[2] D.P. Snowden-Ifft et al., Phys. Rev. D **61**(2000)101301
[3] S. Ahlen et al., Phys. Lett. B **695**(2011)124-129
[4] D. Santos et al., arxiv:1304.2255(2013)
[5] K. Nakamura et al., PTEP(2015)043F01s
[6] C.J. Martoff et al., Nucl. Instr. and Meth. A **440**(2000)355-359
[7] S. Burgos et al., Nucl. Instr. and Meth. A **600**(2009)417-423
[8] J.B.R. Battat et al., Physics of the Dark Universe 9-10(2015)1-7
[9] D.R. Nygren J.Phys.Conf.Ser **65**(2007)012003
[10] A.R. Prieskorn et al., IEEE Trans.Nucl.Sci. **58**(2001)2055-2059
[11] Nguyen Phan, Eric Lee in: Cygnus 2015 Conference in Los Angeles
[12] A. Takada et al., Nucl. Instr. Meth. A **573**(2007)195
[13] F. Sauli, Sharma, Ann. Rev. Nucl. Part Sci. **49**(1999)341
[14] J.F. Ziegler, J.P. Biersack SRIM The Stopping and Range of Ions in Matter, Code(1985)

https://doi.org/10.1051/epjconf/201817402007

Development of Gating Foils To Inhibit Ion Feedback Using FPC Production Techniques

D. Arai[1,a], K. Ikematsu[2], A. Sugiyama[2], M. Iwamura[1], A. Koto[1], K. Katsuki[1], K. Fujii[3], T. Matsuda[3]

[1] *Fujikura Ltd., 1440 Mutsuzaki, Chiba 285-8550, Japan*
[2] *Saga U., Saga 840-8502, Japan*
[3] *High Energy Accelerator Research Organization (KEK), 1-1 Oho, Tsukuba 305-0801, Japan*

Abstract. Positive ion feedback from a gas amplification device to the drift region of the Time Projection Chamber for the ILC can deteriorate the position resolution. In order to inhibit the feedback ions, MPGD-based gating foils having good electron transmission have been developed to be used instead of the conventional wire gate. The gating foil needs to control the electric field locally in opening or closing the gate. The gating foil with a GEM (gas electron multiplier)-like structure has larger holes and smaller thickness than standard GEMs for gas amplification. It is known that the foil transmits over 80 % of electrons and blocks ions almost completely. We have developed the gating foils using flexible printed circuit (FPC) production techniques including an improved single-mask process. In this paper, we report on the production technique of 335 μm pitch, 12.5 μm thick gating foil with 80 % transmittance of electrons in ILC conditions.

1 Introduction

Time Projection Chamber (TPC) with a Micro Pattern Gas Detector (MPGD) readout has been proposed as a central tracker of International Large Detector (ILD) for the International Linear Collider (ILC) [1]. GEM is one of the MPGD options to be used of the ILC-TPC gas-amplification device [2]. F. Sauli proposed the concept of GEM as a gating device [3]. Despite ion-blocking potential of GEM, a small portion of back-flow ions are accumulated during beam collisions. Ion discs are formed in the drift volume of TPC due to a very slow drift velocity of the ions. As the primary electrons pass through the ion discs on the way to the amplification module, the position resolution of TPC might be deteriorated [4]. Using a gating device to stop the ion feedback from the amplification module is effective at preventing this problem thanks to the ILC beam-train structure. This device can be used to stop ions back-flow simply by reversing the electric field in the holes of the gating foil and opening or closing the gate. Achieving a high electron transmission in a large magnetic field for an open GEM gate demands that certain conditions to be fulfilled. Simulation studies showed that only a large aperture (large hole diameter and narrow rim) and very thin foil could meet our requirements [5]. Conventional wire gating was another possibility. However, the device needs a mechanical structure to maintain wire tension, and the electron trajectories may be distorted due to E x B effects near wire. Nevertheless, the requirement of an optical aperture larger than 80 % is challenging even for FPC manufacturers who specialize in processing of fine electrodes. This report describes the requirements for gating foil for ILC-TPC in section 2, R&D of production methods of the gating foil in section 3, and finally limitations of application of the process in section 4.

[a] e-mail: daisuke.arai@jp.fujikura.com

2 Requirements for gating foil

As ILC-TPC is operated in a 3.5 T magnetic field, the diffusion of the electrons can be reduced by a gas mixture with a high $\omega\tau$. The ILC-TPC requires a good position resolution of less than 100 μm, which can be obtained even for a long drift length of up to 2.3 m. The performance of electron transmission through gating foil is also affected by the high magnetic field. It is understood that the rate is closely related to the optical aperture of the gate structure [6]. The prototype gating foil has been studied in a 1 T magnetic field and shown an electron transmission rate greater than 80 %. The results from the study above and the simulation study showed that 80 % transmission was obtained with proper potential setting of the gating foil in a 3.5 T B field. We have extended the area of the gating foil from the 100 mm x 100 mm for the test module to the 170 mm x 220 mm for the ILC-TPC module keeping the same specifications as the prototype. The 80 % optical aperture and 12.5 μm insulator thickness are needed to satisfy over 80 % electron transmission. The gating foil should have a hole diameter of less than 300 μm to keep electron position information. The details of requirements for a gating foil and an amplification GEM for ILC-TPC are summarized in Table 1.

Table 1. Requirement specs of gating foil and amplification GEM for ILC-TPC

Item	Gating foil	Amplification GEM
Optical aperture ratio	≥ 80 %	22.7 %
Hole size	≤ 300 μm	70 μm
Hole pitch	≤ 335 μm	140 μm
Rim width (Hole pitch - Hole size)	≤ 35 μm	70 μm
Insulator thickness	≤ 12.5 μm	50 μm or 100 μm
Foil size	170 mm x 220 mm	170 mm x 220 mm

3 R&D of gating foil production

Many different processes, including single mask, double mask and laser drilling, have been applied to make amplification GEMs [7]. Each processing method uses photolithography and insulator removing techniques. Those are also used in FPC production processes such as a circuit formation and a though-hole formation process. We have developed prototypes of the gating foil considering suitable conditions for each process and improvements. The following subsections briefly describe the process we have used in developing the gating foil.

3.1 Laser drilling process

Both the copper and the polyimide layer can be processed at the same time using Gaussian beam of a UV-YAG laser with a 335 nm wavelength and about a 30 μm beam size . We moved the laser beam along a circle to make 300 μm holes (see Fig.1). The gating foil has a hole pitch of 330 μm, a polyimide thickness of 25 μm, and a rim width on the front (back) side of 14 μm (28 μm) on the 10 mm x 10 mm area of the test gate. The optical aperture ratio is 75 %. We verified that the same pattern was also able to be formed on 12.5 μm thick polyimide foil. The copper electrodes on the front side are apt to come off the polyimide sheet if the rim is narrower than 10 μm. The acceptable minimum rim width is therefore about 25 μm, which allows a maximum optical aperture of 77 % with this circle hole structure. The honeycomb structure with the same rim width and pitch can provide an 85 % optical aperture in principle, however, the laser machine cannot handle different shapes other than circles because it is optimized for FPC production.

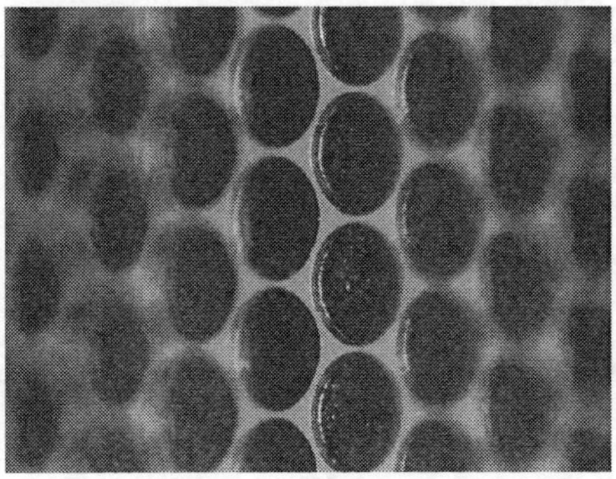

Figure 1. Gating foil produced by laser drilling process. (302 μm hole diameter, 330 μm hole pitch and 25 μm polyimide thickness on the 10 mm x 10 mm area)

3.2 Double mask process

It is necessary to produce electrode patterns on both sides of gating foil in a double mask process. The insulator is removed by a chemical process or laser etching. In this process, the copper patterns work as etching masks. The accuracy of the photomask alignment on both sides is more than 10 μm for large area processing. Creating a 30 μm rim with that process is therefore unacceptable. We conclude that the double mask process is not suitable for producing the gating foil.

3.3 Single mask process

Although mask alignment in a single mask process is not necessary, how to make an electrode pattern on the opposite side is another issue. We have approached this problem by two methods.

3.3.1 Single mask process with Ni plating

We considered using Ni plating to protect the copper electrode formed on the front side when the copper on the back side is etched later. After the circuit production on the front side, a defocused beam of a UV-YAG laser is irradiated to remove only the polyimide insulator where the copper electrode functions as a mask to stop the laser beam. Laser etching enables the creation of less tapered, uniform-quality holes over a wide area than holes created by chemical etching. A UV-YAG laser also seems to cause less damage on the surface of polyimide than a thermal process by a CO2 laser.

Figure 2. Cross-section view of the gating foil produced by single mask process with Ni plating

Figure 2 shows the cross-section of the gating foil produced by a single mask process with Ni plating. The gating foil has a polyimide thickness of 12.5 μm and a rim width on the back side of 36 μm. A gap was found between the electrodes (Ni and Copper) and the polyimide. The gap was probably caused by the entry of the etching liquid during copper etching on the back side.

3.3.2 Single mask process without Ni plating

We have invented a new single mask process, which allows the copper on the front side to be etched without protecting the copper by Ni. We prepare the sheet having a thicker copper on the front side than the back side so that enough copper is left on the front side when the copper in the hole area of the back side is etched away. After the honeycomb structure circuit on the thick copper side is formed, the polyimide is removed by a defocused beam of a UV-YAG laser. In this process, the thick copper works as a etching mask. Finally, we etched the copper from both sides, where the etching speed in the hole area was two times faster than other areas, and obtained proper patterns of the electrodes on the both sides. Figure 3 shows the schematic view of etching for the back-side copper.

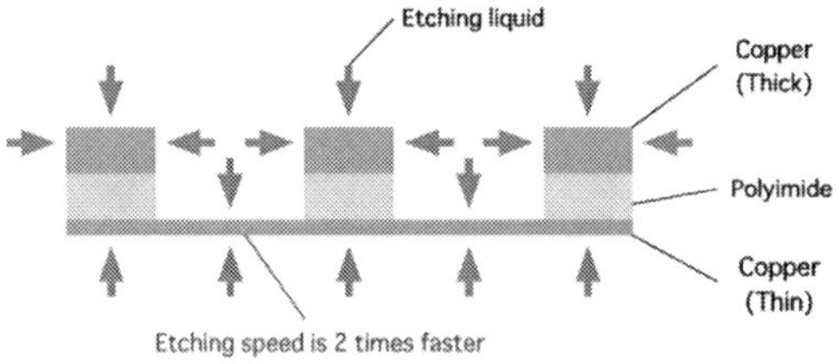

Figure 3. Schematic view of etching of back side copper (The copper of hole area of back sides is etched from both sides, so the etching speed is two times faster.)

Figure 4 and 5 show the gating foil produced by the single mask process without Ni plating. The gating foil has a hole diameter of 304 μm on the back side, a hole pitch of 335 μm, a rim width of 27 μm (31 μm) on the front (back) side and an insulator thickness of 12.5 μm on the 100 mm x 100 mm area. The optical aperture ratio is 82 %. We have produced the gating foil successfully. ILC-TPC Japan group have measured the electron transmission of the gating foil, and verified that the electron transmission was over 80 % under the magnetic field 0 T and 1 T [6].

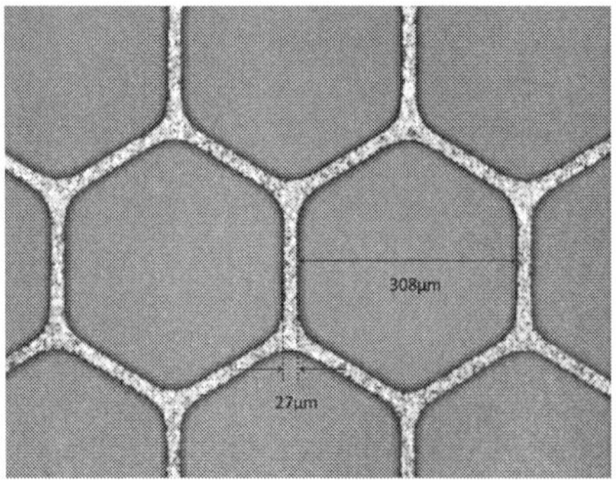

Figure 4. Micrographs of the gating foil (Front side) produced by single mask process without Ni plating

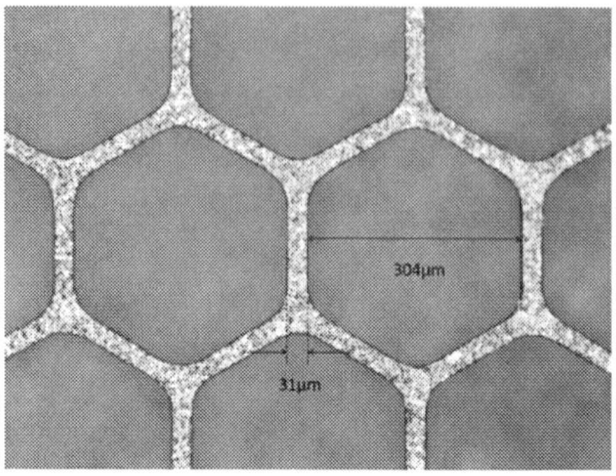

Figure 5. Micrographs of the gating foil (Back side) produced by single mask process without Ni plating

4 Limitations of gating foil processing

We have established a stable production process for the gating foil of the 100 mm x 100 mm test area. However, we still have to know the maximum area for which this method is applicable. In addition, we need to know the minimum limits of the rim width and the hole diameter (keeping the same optical aperture) for further production.

4.1 Available processing area

We have to avoid circuit breaks on the front side that lead to electrode breaks in the post etching process. The area to which this process is applicable relies on the probability of a circuit break due to contamination during photolithography. To improve the production environment and optimize the photo-resist, the gating foil with a 30 μm electrode width has been successfully produced over the 170 mm x 220 mm area of the TPC module.

4.2 Minimum rim width

Figure 6 shows the cross-section of the gating foil having a 7 μm (15 μm) rim width on the front (back) side. The narrow copper of less than 7 μm peeled off the polyimide. We conclude the minimum rim width by this process should be 15 μm.

4.3 Minimum hole size

We have tried to make gating foil with a hole diameter of 100 μm and a rim width on the back side of 20 μm to minimize the hole size. The shape of the electrodes on this foil was not a honeycomb form as shown in Fig.7. The corner of that small hexagonal holes could not be formed using a common chemical process for circuit formation. A diameter of 150 μm seems to be the minimum limit of the process.

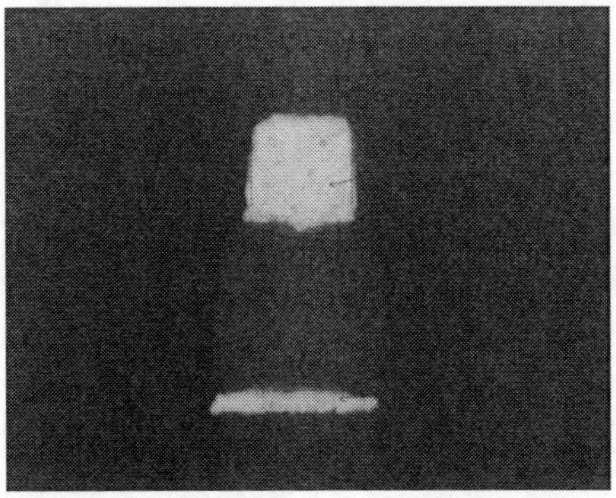

Figure 6. Cross-section view of the gating foil which has a rim width on the front (back) side of 7 μm (15 μm)

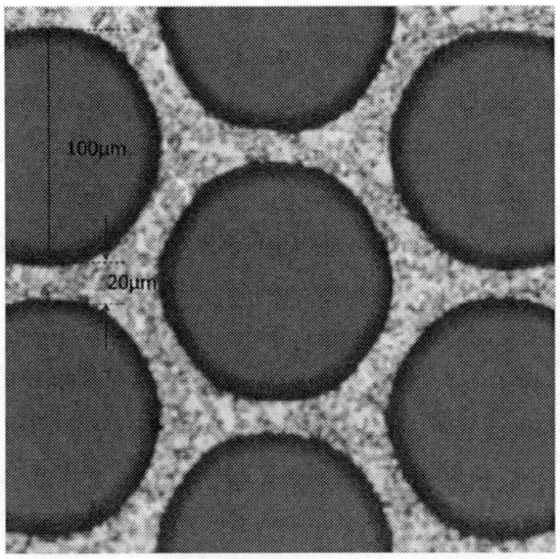

Figure 7. Micrographs of gating foil which has a hole diameter on the front side of 100 μm

5 Conclusion

We have developed the gating foil that has a GEM-like structure and can provide an 82 % optical aperture over the 100 mm x 100 mm test area and the 170 mm x 230 mm module area. This will be a viable gating device to block ion-backflow for ILC-TPC. The gating foil mounted on the ILC-TPC prototype module has been produced and the electron transmission will be measured soon.

References

[1] *The ILC Technical Design Report (TDR) : Volume 4 - Detectors, http://www.linearcollider.org/ILC/Publications/ Technical-Design-Report.*

[2] R. Yonamine et al. JINST. **7**, C06011 (2012)

[3] F. Sauli et al. NIM **A560**, 269 (2006)

[4] D. Arai, *Master thesis (2012), Tokyo University of Agriculture and Technology (TUAT)*

[5] P. Gros et al.JINST **8**, C11023 (2013)

[6] K. Ikematsu on behalf of the LCTPC-Japan Collaboration, *IEEE Nuclear Science Symposium and Medical Imaging Conference (NSS/MIC) 2014 Conference Record (N44-7).*

[7] T. Tamagawa et al. NIM **A560**, 418-424 (2006)

Operation of an InGrid based X-ray detector at the CAST experiment

Christoph Krieger[1,a], Klaus Desch[1], Jochen Kaminski[1], and Michael Lupberger[1]

[1] *Physikalisches Institut, University of Bonn, Nußallee 12, 53115 Bonn, Germany*

Abstract. The CERN Axion Solar Telescope (CAST) is searching for axions and other particles which could be candidates for Dark Matter and even Dark Energy. These particles could be produced in the Sun and detected by a conversion into soft X-ray photons inside a strong magnetic field. In order to increase the sensitivity for physics beyond the Standard Model, detectors with a threshold below 1 keV as well as efficient background rejection methods are required to compensate for low energies and weak couplings resulting in very low detection rates.

Those criteria are fulfilled by a detector utilizing the combination of a pixelized readout chip with an integrated Micromegas stage. These InGrid (Integrated Grid) devices can be build by photolithographic postprocessing techniques, resulting in a close to perfect match of grid and pixels facilitating the detection of single electrons on the chip surface. The high spatial resolution allows for energy determination by simple electron counting as well as for an event-shape based analysis as background rejection method. Tests at an X-ray generator revealed the energy threshold of an InGrid based X-ray detector to be well below the carbon K_α line at 277 eV. After the successful demonstration of the detectors key features, the detector was mounted at one of CAST's four detector stations behind an X-ray telescope in 2014. After several months of successful operation without any detector related interruptions, the InGrid based X-ray detector continues data taking at CAST in 2015.

During operation at the experiment, background rates in the order of $10^{-5}\,\mathrm{keV}^{-1}\,\mathrm{cm}^{-2}\,\mathrm{s}^{-1}$ have been achieved by application of a likelihood based method discriminating the non-photon background originating mostly from cosmic rays. For continued operation in 2016, an upgraded InGrid based detector is to be installed among other improvements including decoupling and sampling of the signal induced on the grid as well as a veto scintillator to further lower the observed background rates and improving sensitivity.

1 The CAST experiment

The CERN Axion Solar Telescope (CAST) utilizes a decommissioned LHC prototype dipole magnet to search for weakly interacting particles emerging from the Sun [1]. When the experiment started its search in 2002, the main focus was on the search for axions. Those particles arise from a solution to the strong CP problem via the Peccei-Quinn mechanism. As axions provide only weak coupling to ordinary matter as well as a low mass and long lifetime, they are a candidate for at least part of what is called Dark Matter. Axions are produced in the Sun's core through the particle's coupling to photons in the presence of strong electromagnetic fields. To detect axions on the earth, the inverse effect is utilized: Axions can convert back to photons inside a strong magnetic field. As axions originate from the Sun's core, the resulting photons are in the soft X-ray regime of a few keV.

As CAST is able to track the Sun for 2×1.5 h during sunrise and sunset, four detector stations are equipped with X-ray detectors all providing high sensitivity and low background rates. One of these stations provides an X-ray telescope [2] (a spare from the ABRIXAS space mission, in the following mentioned as the MPE-XRT). Until 2013 a pn-CCD detector was mounted behind the X-ray telescope which then needed replacement. Together with the other detectors, mostly Micromegas, it was possible to set the most stringent limits on solar axions up to now [3–5].

When the axion search at CAST was going to finish, CAST started to extend its field of search to other, even more exotic particles with solar origin. Among these is the chameleon [6] which comes along with one of the yet few explanations for what is observed as Dark Energy. The production and detection mechanisms for chameleons are quite similar to those for axions. Chameleons however would not be produced in the Sun's core but in the tachocline which is a region located at 0.7 solar radii and features high differential rotation causing strong magnetic fields. According to the lower temperature in the tachocline compared to the solar core chameleons are produced at much lower energies than axions, resulting in a spectrum peaking below 1 keV. Therefore the chameleon search requires detectors with a very low energy threshold in addition to the low background rates. A first glance at CAST's potential in chameleon search has been taken at the end of 2013 using an off the shelf silicon drift detector which was able to set an upper limit on the chameleon photon coupling [7].

[a] e-mail: krieger@physik.uni-bonn.de

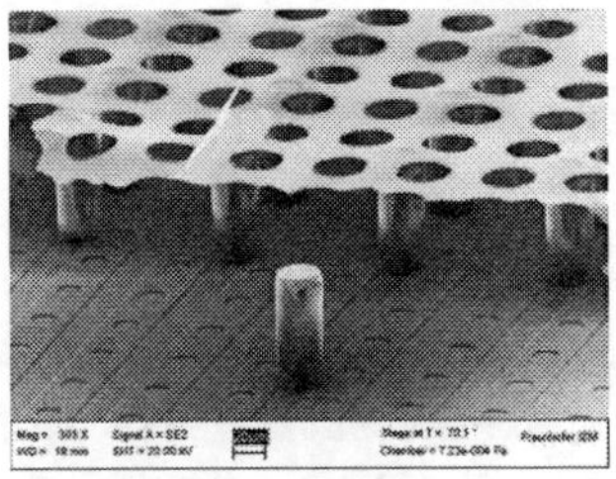

Figure 1: SEM image of an InGrid on top of a Timepix ASIC. Part of the metallic grid has been removed, to show the underlying pixels and pillars. Taken from [13].

2 Integrated Micromegas

Among the Micro-Pattern Gaseous Detectors (MPGDs) especially the Micro-MEsh Gaseous Structure detectors (Micromegas) have proven their performance and reliability in many particle physics experiments. Usually, in a Micromegas detector a highly granular gas amplification stage is combined with a rather coarse readout plane, thus, loosing the initial high spatial resolution. As for pad or strip based readout schemes, a granularity in the order of the gas amplification stage's granularity is unfeasible due to the high channel density and necessity to have the electronic readout chain routed to each pad or strip, the only way is to combine a Micromegas-like gas amplification stage with highly integrated readout electronics as it can be found in pixelized readout chips.

To avoid misalignment and the resulting Moire patterns, it is critical to have a close to perfect match between pixels and holes in the Micromegas' mesh. This is achieved by application of modern photolithographic postprocessing techniques, which allow to build up a Micromegas-like structure directly on top of a pixelized readout chip. A reasonable choice for the pixelized readout chip is the Timepix ASIC [8], which provides many useful features for the application in an MPGD, e.g. a Time Projection Chamber (TPC). The Timepix ASIC offers 256 × 256 pixels on $2\,cm^2$ of active surface with a pixel pitch of 55 µm. Each pixel contains its own preamplifier, discriminator and counting logic allowing for either measurement of the arriving charge or the time the discriminator threshold was surpassed.

The production of these integrated Micromegas (also called InGrids, derived from Integrated Grid) devices was first pioneered at the University of Twente [9, 10] and is nowadays done at the IZM in Berlin [11] allowing for the processing of full eight inch Timepix wafers and thus, fulfilling the demands of R&D regarding the number of InGrids available at a time. Figure 1 shows an SEM image of an InGrid structure made at the IZM in Berlin. To protect the pixels' electronics from discharges, a thin (typically 4 or 8 µm) resistive layer made from silicon nitride [12] is deposited on the chip's surface before building up the Micromegas structure. This protection layer does not prevent discharges but spreads the arriving charge over many pixels in case of spark happening.

3 The InGrid based X-ray detector

An X-ray detector based on the InGrid technology, treated in the previous section, has been build. After successful tests with a first prototype [14] the detector version to be installed at the CAST experiment was designed basing on the CAST (sunset) Micromegas detectors [15]. An exploded view of the detector can be found in figure 2. The Timepix ASIC with its InGrid stage on top is the central element of the detector housed in the exchangeable readout module which is read out by an FPGA based readout system developed at Bonn [13]. To reduce field distortions at the chip's borders, it is covered by a field shaping electrode with a cutout just the size of the chip's active area to which the electric potential according to its position within the drift field is applied. The drift region of the detector is defined by a ring made of acrylic glass closed by the cathode frame made from copper covered with a film of 2 µm metalized Mylar.

To allow very low energetic X-ray photons to enter the detector volume, the copper is removed on a 16 by $16\,mm^2$ area, small ribs have been kept to provide a strongback for the thin window below. At CAST, the detector needs to be connected to the vacuum of the MPE-XRT, thus, requiring the window to withstand a pressure difference of 1050 mbar and not leaking too much of the detectors gas filling into the vacuum. The 2 µm Mylar film has been chosen as a good compromise between robustness and transparency. Similarly, the gas used to operate the detector, a mixture of Argon and isobutane (Ar/iC_4H_{10} 97.7/2.3), was chosen as a compromise between gas amplification properties and the wish to avoid the usage of gas mixtures regarded as flammable.

An X-ray photon entering the detector will likely hit a gas atom and produce a bunch of primary electrons through ionization. The number of primary electrons depends strongly on the energy of the initial photon. Due to the electric field between cathode and InGrid, the electrons

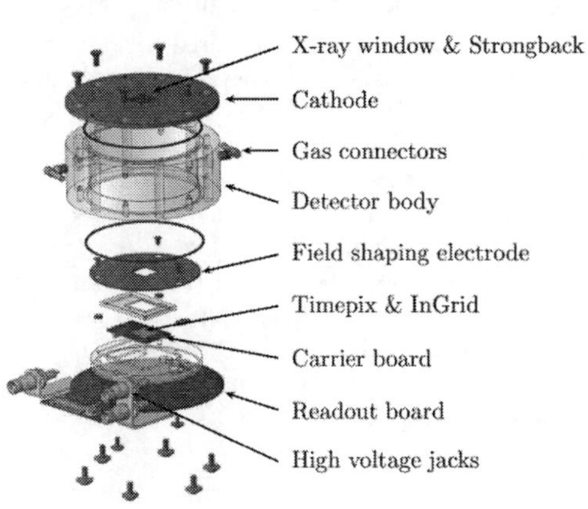

Figure 2: An exploded view of the InGrid based X-ray detector including labeling of main parts. Taken from [16].

https://doi.org/10.1051/epjconf/201817402008

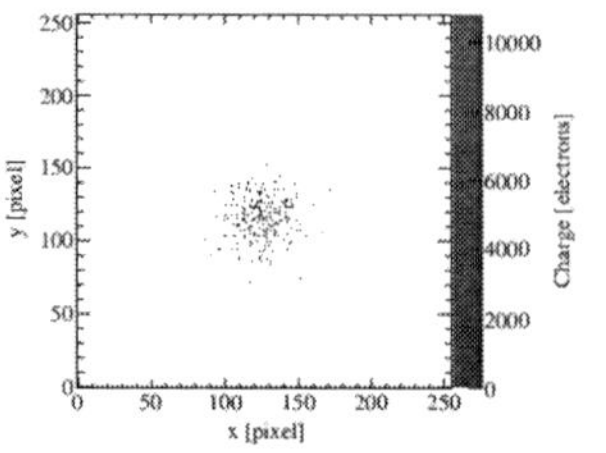

Figure 3: Cloud of primary electrons originating from an 8 keV X-ray photon, recorded with an InGrid based X-ray detector. Depicted area is the complete active area of a Timepix ASIC. Charge resulting from gas amplification above each pixel is color coded on the z-axis.

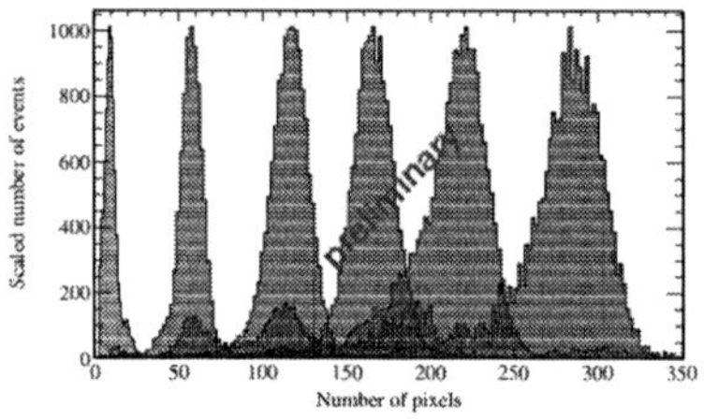

Figure 4: Characteristic X-ray lines recorded during tests at an X-ray generator, each line was recorded separately, all spectra have been normalized to the same height. From left to right: 277 eV, 1.5 keV, 3 keV, 4.5 keV, 5.9 keV and 8 keV.

will drift towards the gas amplification structure, diffusing thereby. This results in an electron cloud of almost circular shape being detected on the readout chip. Due to the high granularity of Timepix ASIC and InGrid primary electrons can be identified as pixels hit, if they are seperated by more than the pixel pitch of 55 µm at the level of the grid. This allows for energy measurement based on simple pixel counting as well as for an event-shape-based discrimination of non-photon events. A sample X-ray event is shown in figure 3. The energy threshold of this kind of detector is given by the energy needed to create one primary electron, typically about 25 eV. To verify the ability to detect very low energetic X-ray photons (below 1 keV), the detector was tested at an X- ray generator at CERN [16] prior to the installation at the CAST experiment. These tests successfully demonstrated the ability to detect X-ray photons down to energies as low as the carbon K_α line at 277 eV, see figure 4 for a set of characteristic X-ray lines recorded during the tests.

4 Installation behind the MPE-XRT

To install the InGrid based X-ray detector behind the MPE-XRT at the CAST experiment, a small vacuum system had to be set up connecting the gas filled detector to

(a) (b)

Figure 5: Pictures of the setup for the InGrid based X-ray detector at the CAST experiment: (a) shows the MPE-XRT, its vacuum system including the part allowing to connect the detector which is housed in a lead shielding on the right; (b) shows the detector within its partially assembled lead shielding.

the good vacuum of the telescope. To be able to reach vacuum pressures low enough so the gate valves to the telescope can be opened a differential pumping scheme was applied using a thin differential window made from 0.9 µm Mylar. This vacuum system along with the infrastructure for the detector itself was installed at CAST in 2014. To allow for calibration of the detector while taking data at the experiment, an ^{55}Fe source can be moved in front of the detector inside the vacuum. The vacuum system of the InGrid based detector and of the MPE-XRT is illustrated in figure 5a. From October 2014 until November 2014, the detector took first data while tracking the Sun. Data taking continued in 2015 for several months until the detector was dismounted end of November 2015.

To reduce background events induced by cosmic rays, the detector is shielded within a small enclosure made from low activity lead provided by the University of Zaragoza. Figure 5b shows the detector inside the lead shielding with only the top layer of lead bricks missing.

5 Background reduction & measured rates

The Timepix ASIC of the InGrid based X-ray detector is operated untriggered and is read out framewise, each frame is taken with an exposure time of 0.98 s, which is reasonable as the expected event rate is low. About 90 % of the frames recorded during a background run are empty. Reconstruction of the events is performed utilizing a modified clustering algorithm. From the resulting reconstructed events, only those which were most likely caused by an X-ray photon are of interest. To separate those from the non-photon-like events, a likelihood based method is applied. The likelihood is calculated from the distributions of three event-shape variables taking advantage of the Timepix ASIC's and InGrid's high granularity. The variables are excentricity, length divided by transverse rms and fraction of pixels within a radius of one transverse rms. These variables have been chosen in a way so they do not depend on gas parameters like temperature. The reference distributions are taken from the tests at the X-ray

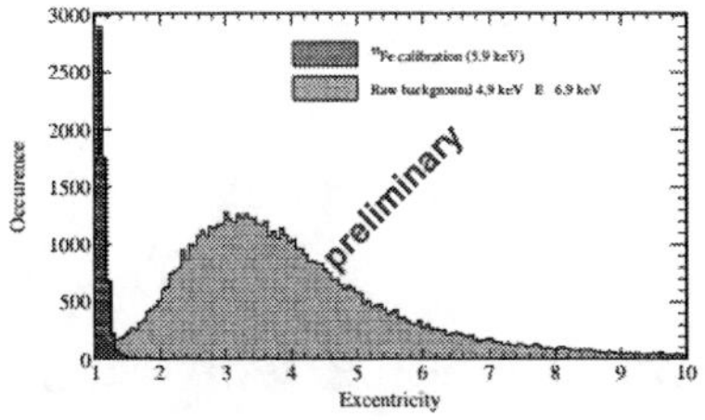

Figure 6: Excentricity distribution for 5.9 keV photons from an ^{55}Fe source and for raw background events of comparable energy. A clear separation is visible, only few of the raw background events look photon-like.

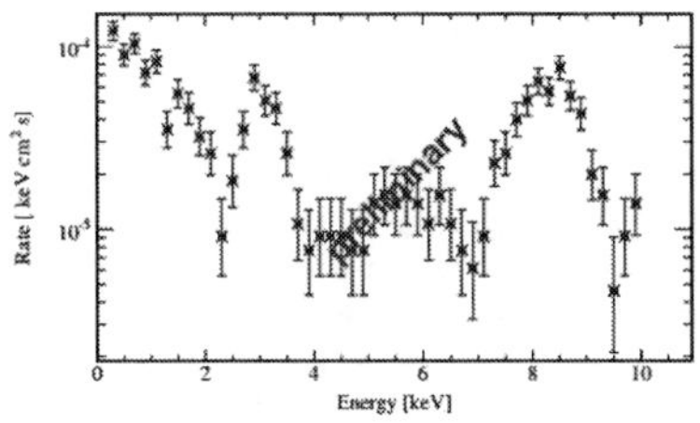

Figure 7: Background spectrum obtained with the InGrid based X-ray detector operated at the CAST experiment. Rate is calculated from events within the innermost 5 by 5 mm^2 area of the chip.

generator, the reference dataset to be used is selected according to the energy of the reconstructed events. The cut value for the likelihood is chosen so that for all energy ranges a software efficiency of 80 % is maintained. Figure 6 shows the distribution of the excentricity for raw background events and a calibration dataset taken under usage of an ^{55}Fe source. Only a small fraction of the raw background events appears to be photon-like.

When applying the likelihood based method to reduce the background by only selecting photon-like events the background spectrum shown in figure 7 is obtained which shows the rate within the innermost 5 by 5 mm^2 area of the chip, where lowest background rates are achieved. For energies above 2 keV, a background rate of a few $10^{-5}\,\mathrm{keV}^{-1}\,\mathrm{cm}^{-2}\,\mathrm{s}^{-1}$ is achieved except for two peaks at 3 keV and 8 keV. While the latter one originates from florescence photons created by cosmic rays traversing the copper cathode, the other one is caused by florescence photons from the Argon gas in the detector. Additionally, cosmic tracks perpendicular to the chip's surface can mimic X-ray-like events, which also contribute to the peak at about 8 keV. For energies below 2 keV, the background rate rises as the likelihood method's efficiency decreases due to blurring of the reference distributions. This is caused by the reduced number of activated pixels for low energy events, increasing the spread for the event-shape variables. Still, the background rate below 2 keV with at maximum $10^{-4}\,\mathrm{keV}^{-1}\,\mathrm{cm}^{-2}\,\mathrm{s}^{-1}$ is one order of magnitude lower than for the silicon drift detector used for CAST's first look at chameleons [7] and should therefore allow for an improved limit on the chameleon couplings in case of absence of a signal.

6 Summary & Outlook

An X-ray detector based on integrated Micromegas technology utilizing a pixelized readout chip has been successfully build, tested and installed at the CAST experiment. It was operated from October 2014 until end of November 2015 and was able to achieve background rates down to a few $10^{-5}\,\mathrm{keV}^{-1}\,\mathrm{cm}^{-2}\,\mathrm{s}^{-1}$ above 2 keV and at maximum $10^{-4}\,\mathrm{keV}^{-1}\,\mathrm{cm}^{-2}\,\mathrm{s}^{-1}$ below.

For 2016, an upgraded version of the InGrid based X-ray detector is to be installed and operated at the CAST experiment. Upgrades will include decoupling and sampling of the signal induced on the InGrid's grid, integration of one or more veto scintillators, both aiming to further reduce the background levels by providing a better handle to discriminate events caused by cosmic rays. Additionally, an ultra thin (few hundred nm) detector window made of silicon nitride on a silicon and ceramic strongback is foreseen to improve the detector's sensitivity by increasing X-ray transmission of the entrance window.

References

[1] K. Zioutas et al., Nucl. Instr. Meth. Phys. Res. A **425**, 480 (1999)

[2] M. Kuster et al., New J. Phys. **9**, 169 (2007)

[3] E. Arik et al., J. Cosmol. Astropart. Phys. **2009**, 008 (2009)

[4] M. Arik et al., Phys. Rev. Lett. **112**, 091302 (2014)

[5] M. Arik et al., Phys. Rev. D **92**, 021101 (2015)

[6] P. Brax et al., Phys. Rev. D **85**, 043014 (2012)

[7] V. Anastassopoulo et al., Phys. Lett. B **749**, 172 (2015)

[8] X. Llopart et al., Nucl. Instr. Meth. Phys. Res. A **581**, 485 (2007)

[9] M. Chefdeville et al., Nucl. Instr. Meth. Phys. Res. A **556**, 490 (2006)

[10] H. van der Graaf, Nucl. Instr. Meth. Phys. Res. A **580**, 1023 (2007)

[11] M. Lupberger et al., *InGrid: Pixelated Micromegas detectors for a Pixel-TPC*, in [17]

[12] Y. Bilevych et al., Nucl. Instr. Meth. Phys. Res. A **629**, 66 (2011)

[13] M. Lupberger, J. Instrum. **5**, C01033 (2014)

[14] C. Krieger et al., Nucl. Instr. Meth. Phys. Res. A **729**, 905 (2013)

[15] J. Galán et al., J. Instrum. **5**, P01009 (2010)

[16] C. Krieger et al., *An InGrid based Low Energy X-ray Detector for the CAST Experiment*, in [17]

[17] *Proceedings of the 3rd International Conference on Technolgy and Instrumentation in Particle Physics* (PoS, 2014)

Imaging Demonstration of a Glass Gas Electron Multiplier with Electronic Charge Readout

Yuki Mitsuya[1,a], Patrik Thuiner[2,4], Eraldo Oliveri[4], Filippo Resnati[4], Miranda van Stenis[4], Takeshi Fujiwara[3], Hiroyuki Takahashi[1], and Leszek Ropelewski[4]

[1] *The University of Tokyo*
[2] *Vienna University of Technology*
[3] *National Institute of Advanced Industrial Science and Technology*
[4] *European Organization for Nuclear Research (CERN)*

Abstract. We have developed a Glass Gas Electron Multiplier (Glass GEM, G-GEM), which is composed of two copper electrodes separated by a photosensitive etchable glass substrate having holes arranged in a hexagonal pattern. In this paper, we report the result of imaging using a G-GEM combined with a 2D electronic charge readout. We used a crystallized photosensitive etchable glass as the G-GEM substrate. A precise X-ray image of a small mammal was successfully obtained with position resolutions of approximately 110 to 140 μm in RMS.

1 Introduction

The gas electron multiplier is one of the most commonly used Micro Pattern Gas Detectors [1]. It has a large-area imaging capability with high position resolution because of its numerous microholes. Its high degree of freedom – detector design – is also attractive; since it only multiplies electrons, it can be combined with a variety of readout systems, or other types of MPGDs as their preamplifier.

Recently, we developed a new type of MPGD, Glass Gas Electron Multiplier (G-GEM) [2]. The G-GEM's substrate is composed of the photosensitive etchable glass, PEG3 (HOYA Corporation, Japan). Holes of the G-GEM are uniformly fabricated using a photolithography technique over the entire sensitive area. The G-GEM is appropriate for sealed-type detectors since it is free of outgassing. The G-GEM has the sufficiently thick glass substrate to support itself over the sensitive area. However, if we use the thick substrate, the charge-up of avalanche electrons on the hole surface becomes serious because the large part of the insulator is exposed to the operational gas. This problem is resolved by using the low volume resistivity glass for the substrate. The resistivity of PEG3 (8.5×10^{12} Ω·cm) is sufficiently low to reduce charge-up.

The large-area imaging device with a gas detector will be attractive for various fields: imaging of heavy charged particles, which normally cause severe radiation damages to solid state detectors; large-area neutron imaging combined with a solid state converter like ^{10}B; and imaging of low energy X-rays (several keV), which cannot be observed clearly with solid state detectors because of the low signal levels against them. In addition, if the detector does not outgas and can be operated stably under sealed condition, all the gas circulating or purifying systems can be removed; It will greatly improve the portability of the imaging device. The portability is necessary for the on-site use of the detector: for example, at the scene of radiotherapy in the hospitals. The sealed-type detector can be also used for gaseous photomultiplier applications. The sealed-type detector is also compatible with the rare and expensive gases like Xenon, which have higher detection efficiency against X-rays. Realizing this kind of portable gas-sealed type imaging device is made possible by G-GEM's non-outgassing property. In addition, the high gain capability of G-GEM with single stage and self-supporting structure, make an attractive imaging device.

In this study, we demonstrate the imaging capability of G-GEM using a two-dimensional electronic charge readout system. We report on the results, and discuss the position resolution obtained with the setup.

[a]e-mail: yukimitsuya@sophie.q.t.u-tokyo.ac.jp

2 Experimental setup

In this experiment, we used a single G-GEM made of crystallized glass, which is a crystallized version of PEG3, called PEG3C (HOYA Corporation, Japan) for measurements. PEG3C has higher resistivity (4.5×10^{14} Ω·cm) than that of PEG3 (8.5×10^{12} Ω·cm). However, the mechanical strength of PEG3C is improved compared with PEG3 glass, which means that the PEG3C can be used with a wider variety of designs including those with thinner structure [3]. The sensitive area of the G-GEM is 100 $\times$ 100 mm^2. The thickness, hole pitch, and hole diameter are 680 μm, 280 μm, and 170 μm respectively.

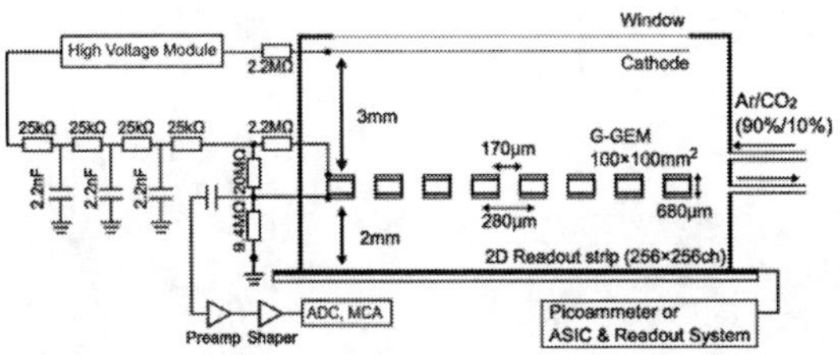

Figure 1. Setup of experiments

Figure 1 is the experimental setup. We installed a drift cathode 3 mm above and a readout 2 mm under the G-GEM. The G-GEM was powered with a voltage divider composed of high voltage resistors. The drift field was 1 kV/cm, and the induction field was 3 kV/cm. A charge-sensitive preamplifier was connected to the bottom of G-GEM with an AC coupling capacitor, and the preamplifier output was connected to a shaping amplifier and an MCA for capturing pulse height spectra. The detector was flushed with an Ar/CO_2 90/10 gas mixture at 6 L/h of flow rate.

The 2D readout anode consists of two planes of 256 parallel copper strips, with the upper one (x-axis) arranged perpendicular to and separated from the lower one (y-axis) by Kapton ridges of 50 μm thickness. The strips of both axes have a pitch of 400 μm, with a strip width of 80 μm and 340 μm for the x-axis and y-axis, respectively, to allow equal charge sharing between both planes. For gain calibration, all the strips of the anode were shorted and read out with a Keithley 6487 picoammeter. For gain uniformity measurement and imaging demonstration, we installed four APV25s and a Scalable Readout System (SRS) to read analogue signals from each strip independently [4].

3 Gain and spectrum measurement

Effective gain of PEG3C G-GEM was calibrated with ^{55}Fe source. For gain measurement, a picoammeter was connected to the readout board. Effective gain G_{eff} was calibrated with the following equation.

$$I = n \cdot e \cdot f \cdot G_{eff} \tag{1}$$

Where I is the readout current, e is the elementary charge, f is the interaction rate measured with MCA and also with the scaler, which counted the rate of pulses from the bottom of G-GEM.

The maximum effective gain reached 2.5 × 10^4 (figure 2). High gain was obtained with a single PEG3C G-GEM. The pulse height spectrum was also measured with the signals from the bottom of G-GEM (figure 3). The energy resolution was 25.7 % at full-width half maximum (FWHM) of the Gaussian-fitted result at the gain of 1.07 × 10^4. We also observed a polarization effect after powering the G-GEM. It caused approximately 20 % of gain decrease in the timescale of 40 min; therefore, the detector was kept powered on for more than several hours before the imaging demonstration in order to exclude the polarization effect.

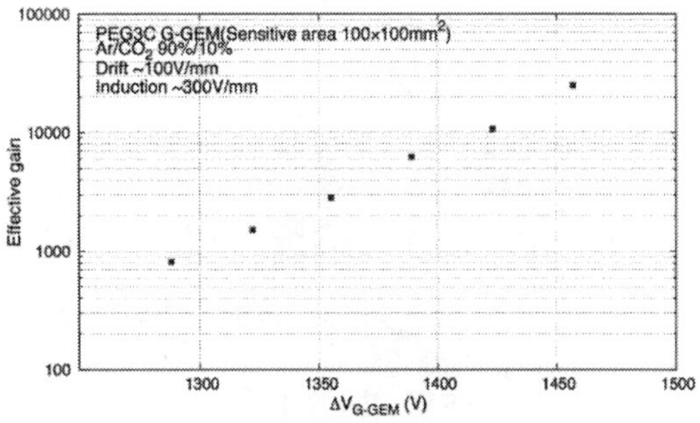

Figure 2. Gain curve of PEG3C G-GEM

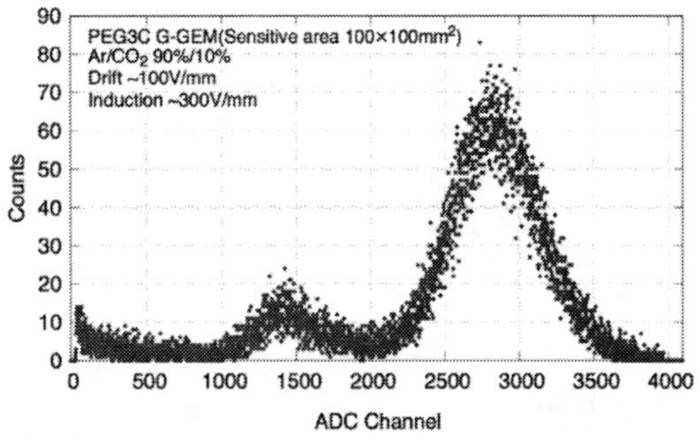

Figure 3. Pulse height spectrum obtained from the bottom of the G-GEM

4 Gain uniformity

Before imaging, the global gain uniformity was measured for the PEG3C G-GEM. If there are large differences in gains, by point, over the sensitive area, some parts of the sensitive area could not be biased sufficiently for imaging. In this case, signal-to-noise ratios become worse at the parts where the gains are not enough, or signal levels become incompatible with the following electronics.

For uniformity measurement, the gain of G-GEM was set at 2500. First, we measured the total collected charges by ^{55}Fe irradiation over the 100 × 100 mm^2 sensitive area. We also measured the total hit counts at the same time. Then the charge map and hit count map were binned into 10 × 10 bins. Finally, the charge map was divided by the hit count map, which provides information equivalent to the gain map. Figure 4 shows the gain map, which was normalized at an arbitrary point.

The edge 36 bins were excluded for a uniformity evaluation because they are normally affected by field distortions. Some strips having abnormally less hit counts in the longitudinal direction of the middle of the sensitive area were also excluded from evaluation due to statistical inconsistency: The reason of less hit counts in this region is considered attributable to strip or channel defects in the ASIC. As a result, the maximum normalized gain became 1.09, and the minimum became 0.87. Good uniformity

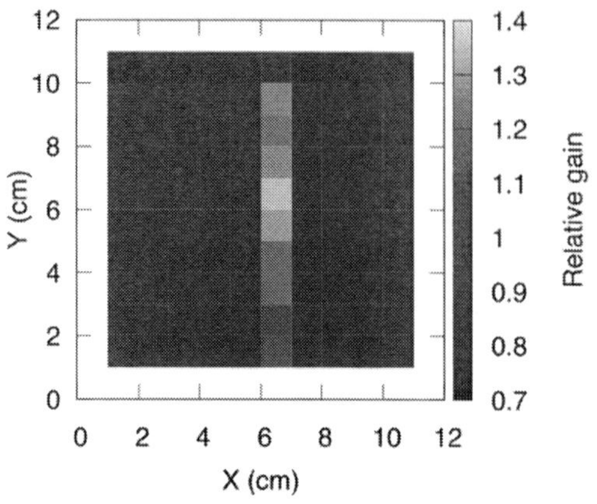

Figure 4. Normalized gain map of 100 × 100 mm² sensitive area. The high gain parts for the longitudinal bins in the middle are considered to be due to strip or channel defects.

over the sensitive area was obtained, which proved a sufficient level for the imaging applications.

5 X-ray imaging

We demonstrated imaging of a small mammal (bat) and evaluated the total position resolution. The bat has been used for the imaging demonstration of the standard GEM in the past [5]. We took its X-ray image again with the G-GEM. The bat was placed onto the detector window together with an aluminum block, and Kapton strips of 200 μm width and 1 mm pitch – with the last two being used for position resolution estimation. The ^{55}Fe source was placed approximately 60 cm over the detector to ensure sufficiently parallel beams. The gain was set to approximately 2500. Figure 5 is a photograph of the bat and the Kapton strip before being placed on the detector. Figure 6 and 7 is the imaging result of 100 × 100 mm^2 area with four million hit events in total. The skeletal structure is clearly visible, although a large number of the photons were stopped by the bones because of the low energy source. Additionally, each of the Kapton 200 μm strips can clearly be discriminated. In order to estimate the position resolutions in detail, the edge profiles (arrays of hit count bins) were extracted for both X and Y directions from the edges of aluminum blocks. The edge profiles were then fitted with the Error function as shown below.

$$f(x) = a \cdot erf\left(\frac{x-\mu}{\sqrt{2}\sigma}\right) + c \quad (2)$$

where x is the position, μ is the expectation of edge position, σ is the standard deviation, a is the scaling factor and c is the offset. Figure 8 and 9 show edge profiles and fitted curves. The position resolutions were $\sigma = 137$ μm in the X direction, and 114 μm in the Y direction (RMS). It should be noted that the resolutions were not guaranteed sufficiently due to the rather small data counts for each bin.

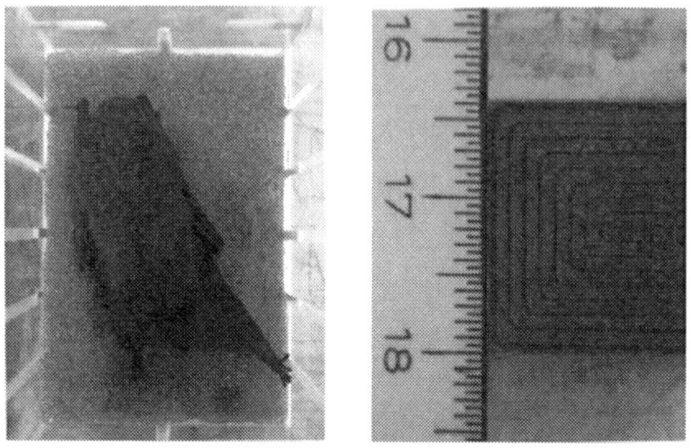

Figure 5. Photographs of the bat and kapton strip

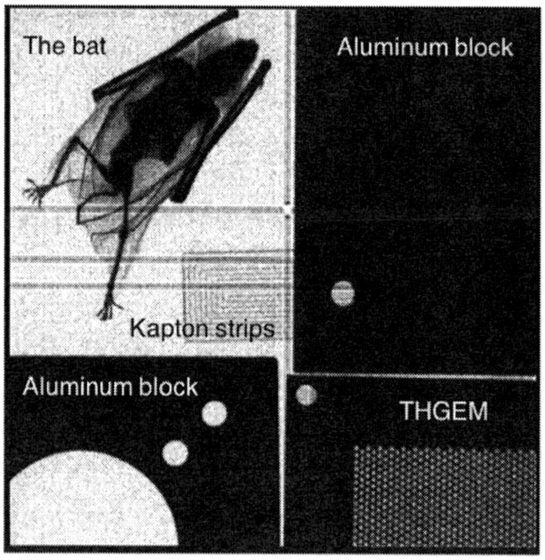

Figure 6. X-ray image of the whole sensitive area

6 Conclusion

In this study, the imaging demonstration of a G-GEM with electronic charge readout was successfully conducted. We used a G-GEM made of crystalized photosensitive etchable glass, PEG3C. The G-GEM showed high gain of 2.5 × 10^4 at maximum with single stage. The global gain uniformity was approximately within ± 10 % over the sensitive area. We obtained a precise X-ray image of a small mammal with the G-GEM combined with the electronic charge readout system and achieved high position resolutions of $\sigma = 137$ μm in the X direction and $\sigma = 114$ μm in the Y direction (RMS). As these results illustrate, the G-GEM is prospective as an imaging detector, and the future development of a sealed-type imaging device is expected.

References

[1] F. Sauli, Nucl. Instr. and Meth. Phys. Res. A **386**, 531(1997)

https://doi.org/10.1051/epjconf/201817402009

[2] H. Takahashi, Y. Mitsuya, T. Fujiwara, et al., Nucl. Instr. and Meth. Phys. Res. A **724**, 1 (2013)
[3] F. Tokanai, et al., Nucl. Instr. and Meth. Phys. Res. A **732** 273 (2013)
[4] S. Martoiu, et al., IEEE NSS-MIC Conference record, N43-5, 2036 (2011)
[5] S. Bachmann, L. Ropelewski, F. Sauli, et al, Nucl. Instr. and Meth. Phys. Res. A, **478** 104 (2002)

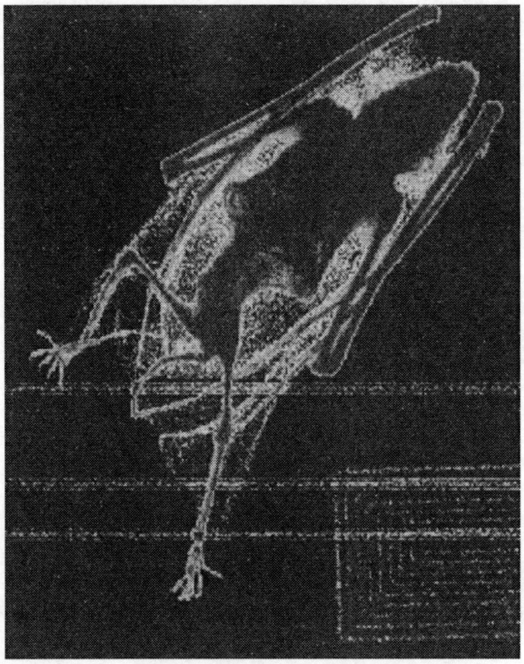

Figure 7. X-ray image of the bat

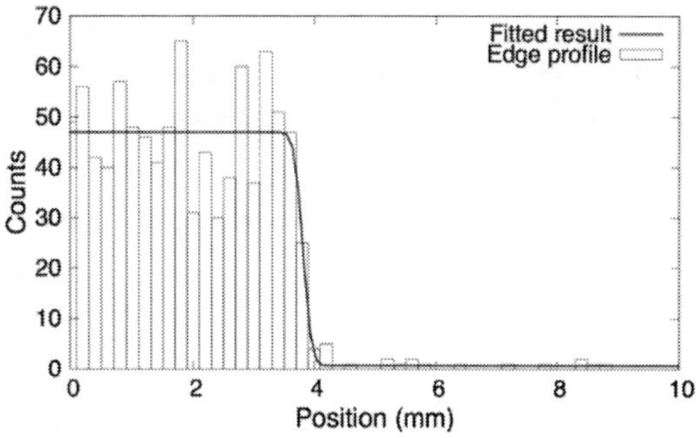

Figure 8. Fitted result of an edge in X direction using the Error function

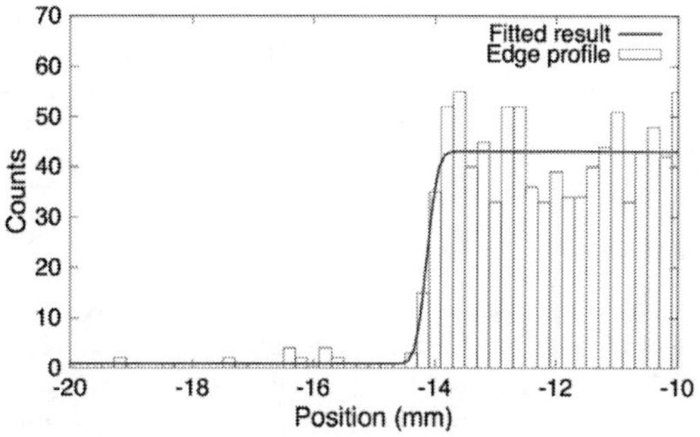

Figure 9. Fitted result of an edge in Y direction using the Error function

Development of the micro pixel chamber based on MEMS technology

T. Takemura[1,a], A. Takada[1], T. Kishimoto[1], S. Komura[1], H. Kubo[1], Y. Matsuoka[1], K. Miuchi[2], S. Miyamoto[1], T. Mizumoto[1], Y. Mizumura[1], T. Motomura[3], Y. Nakamasu[1], K. Nakamura[1], M. Oda[1], K. Ohta[3], J. D. Parker[4], T. Sawano[5], S. Sonoda[1], T. Tanimori[1], D. Tomono[1], and K. Yoshikawa[1]

[1] *Kyoto University, Department of Physics, Japan*
[2] *Kobe University, Department of Physics, Japan*
[3] *Dai Nippon Printing Co., Ltd.*
[4] *CROSS-Tokai, Neutron R&D Division, Japan*
[5] *Kanazawa University, School of Mathematics and Physics, Japan*

Abstract. Micro pixel chambers (μ-PIC) are gaseous two-dimensional imaging detectors originally manufactured using printed circuit board (PCB) technology. They are used in MeV gamma-ray astronomy, medical-imaging, neutron imaging, the search for dark matter, and dose monitoring. The position resolution of the present μ-PIC is approximately 120 μm (RMS), however some applications require a fine position resolution of less than 100 μm. To this end, we have started to develop a μ-PIC based on micro electro mechanical system (MEMS) technology, which provides better manufacturing accuracy than PCB technology. Our simulation predicted the gains of MEMS μ-PICs to be twice those of PCB μ-PICs at the same anode voltage. We manufactured two MEMS μ-PICs and tested them to study their behavior. In these experiments, we successfully operated the fabricated MEMS μ-PICs and we achieved a maximum gain of approximately 7×10^3 and collected their energy spectra under irradiation of X-rays from ^{55}Fe. However, the measured gains of the MEMS μ-PICs were less than half of the values predicted in the simulations. We postulated that the gains of the MEMS μ-PICs are diminished by the effect of the silicon used as a semiconducting substrate.
KEYWORDS: gaseous detector, Micro-pattern detector, Micro Electro Mechanical System

1 Introduction

A micro-pixel chamber (μ-PIC) [1] is a gaseous two-dimensional imaging detector (Figure 1). Each pixel of the μ-PIC, consisting of an anode via with a diameter of 60 μm and a cathode electrode with a hole diameter of 250 μm, is positioned with a pitch of 400 μm. A gas avalanche occurs near the anode via because of the strong electric field between these electrodes; therefore, each pixel functions as a proportional counter. The anode vias are connected to an anode strip electrode on the back of the substrate, and the cathodes are formed in strip electrodes on the front surface. A two-dimensional readout is available using anode and cathode strips arranged orthogonally. μ-PICs are manufactured by printed circuit board (PCB) technology and are composed of copper electrodes and 100-μm-thick polyimide substrate. PCB technology enables the low-cost manufacturing of large-area μ-PICs (10×10 cm^2 [2] or 30×30 cm^2[3]). PCB μ-PICs feature a fine position resolution (RMS $\approx$ 120 μm) [4], a high gas gain (nominal 6000, maximum 15000) [5] , and good gain uniformity (RMS $\approx$ 5%). PCB μ-PICs have been used in a wide range of applications including MeV gamma-ray astronomy [6], medical imaging [7], the search for dark matter [8], neutron imaging [9], and dose monitoring [10]. However, neutron imaging applications require a position resolution of less than 100 μm and MeV gamma-ray astronomy requires precise tracking of charged particles with an angular resolution of less than 5°. Satisfying these requirements ne-

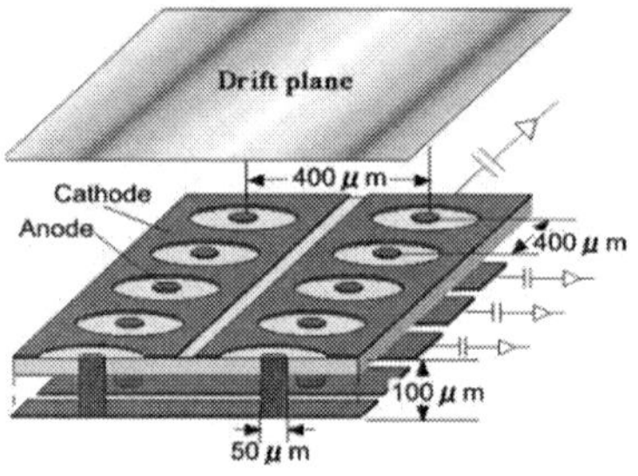

Figure 1. Schematic of a μ-PIC based on PCB technology [1]. The μ-PIC is composed of copper electrodes and a 100-μm polyimide substrate. Each pixel of the μ-PIC has an anode via with a diameter of 60 μm and a cathode electrode with a hole diameter of 250 μm and is fabricated with a pitch of 400 μm. A gas avalanche occurs near the anode via because of the strong electric field between these electrodes; therefore, each pixel functions as a proportional counter.

[a] e-mail: takemura@cr.scphys.kyoto-u.ac.jp

 https://doi.org/10.1051/epjconf/201817402010

Table 1. Thickness and relative permittivity values used in simulations of each layer in the μ-PICs.

material	relative permittivity	MEMS [μm]	PCB [μm]
copper electrode	10^{10}	15	15
polyimide	3.2	4	~100
silicon	11	400	—
SiO_2	4.5	1 or 10	—

cessitates the development of a μ-PIC with a fine position resolution.

Because of the limitation of the manufacturing accuracy of PCB technology, the diameter of μ-PIC anode vias is limited to approximately 50 μm. Therefore, a more advanced technology is needed for manufacturing fine-pitch μ-PICs. Micro electro mechanical system (MEMS) technology is one such capable technology. MEMS technology is well established, and MEMS manufacturing costs have been reduced to level similar to those of PCB technology. In addition, MEMS technology enables the use of thicker substrates compared with those for PCB technology. In the case of μ-PICs, the gain depends on the thickness of the substrate: a two fold thicker substrate improves the gain by a factor of two [11]. Thus, a MEMS μ-PIC is expected to exhibit a fine position resolution and a high gas gain. However, the substrates used in MEMS technology comprise silicon, which is a semiconductor; thus, the effect of a semiconducting substrate being substituted for the insulating substrate of μ-PICs should be investigated. In this work, we manufactured two MEMS μ-PICs to investigate their behavior. Here, we report the results of both the simulations and experiments.

2 Simulation study of MEMS μ-PIC

We fabricated a MEMS μ-PIC with a structure similar to that of a PCB μ-PIC to focus on the effect using a semiconducting instead of an insulating substrate. Figure 2 shows a photo and a schematic of the cross-section of the manufactured MEMS μ-PIC. The differences between a PCB μ-PIC and a MEMS μ-PIC lie in the substrate material and its thickness, which are listed in Table 1. For simulation purpose, the MEMS μ-PIC substrate was assumed to comprise a 400-μm-thick silicon layer, a thin SiO_2 layer, and a thin polyimide layer; by contrast, the PCB μ-PIC substrate was assumed to be composed of a polyimide layer with a thickness of 100 μm. The thin SiO_2 layer and polyimide layer are insulated on MEMS μ-PICs to enhance their discharge tolerance. We varied the thickness of the SiO_2 layer in the manufactured MEMS μ-PICs to study its effect on device performance. Other parameters (such as the thickness of the strips, pitches of pixels, diameter of the cathode hole, diameter of the anode via, and width of a cathode strip) were approximately the same.

We evaluated the geometries of the PCB and MEMS μ-PICs to study their gains. We fabricated μ-PIC pixel electrodes using the Gmsh program [12] and calculated the electric field around the μ-PICs using the Elmer program [13]. We simulated the avalanche and obtained gas gains using the Garfield++ program with a three-dimensional mesh [14]. We used a gas of Ar/C_2H_6 (pressure ratio 90:10) at 1 atm and adopted a penning-effect rate of 0.31 [15]. The drift electric field and cathode voltage were fixed at 1.0 kV/cm and 0 V, respectively. Figure 3 shows the electric fields of the cross sections of the PCB μ-PIC and MEMS μ-PIC for the case of an anode voltage of 460 V. The electric field at 1 μm above the edge of the anode via was 1.7×10^5 V cm^{-1} for the PCB μ-PIC and 2.2×10^5 V cm^{-1} for the MEMS μ-PIC. The thick substrate of the MEMS μ-PIC concentrates the electric field lines to an anode via, because the thick substrate obstructs the electric field lines between the cathode and anode strips. Therefore, the MEMS μ-PIC exhibits a stronger electric

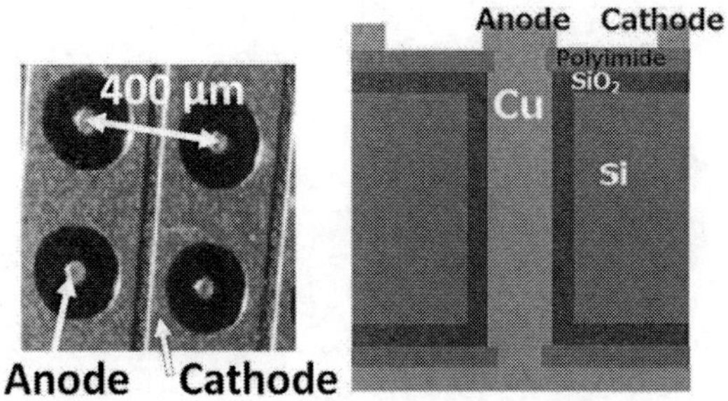

Figure 2. Photograph (left) and schematic of the cross-section (right) of a MEMS μ-PIC. The substrate of the MEMS μ-PIC comprises to a silicon layer with a thickness of 400 μm, a thin SiO_2 layer, and a thin polyimide layer. The thin SiO_2 and polyimide layers are insulated on the MEMS μ-PIC to enhance the discharge tolerance.

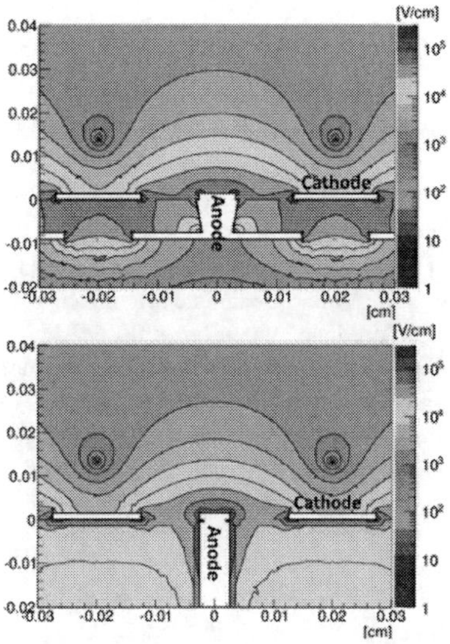

Figure 3. Electric fields of the cross section of a PCB μ-PIC and a MEMS μ-PIC calculated using the program Elmer for the case of an anode voltage of 460 V. The drift electric field and cathode voltage were fixed at 1.0 kV/cm and 0 V, respectively.

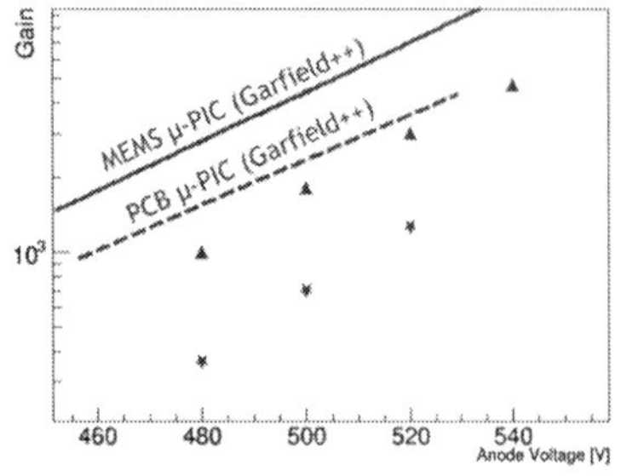

Figure 4. The simulated and measured gains of PCB and MEMS μ-PICs. The dashed and solid lines represent the simulated gains of a PCB μ-PIC and a MEMS μ-PIC, respectively. The simulation predicted the gains of the MEMS μ-PIC are to be twice those of the PCB μ-PIC at the same anode voltage. Stars and triangles represent the measured gains of the MEMS μ-PIC with 1-μm and 10-μm SiO_2 layers, respectively. The gains were measured using the setup shown in Figure 5.

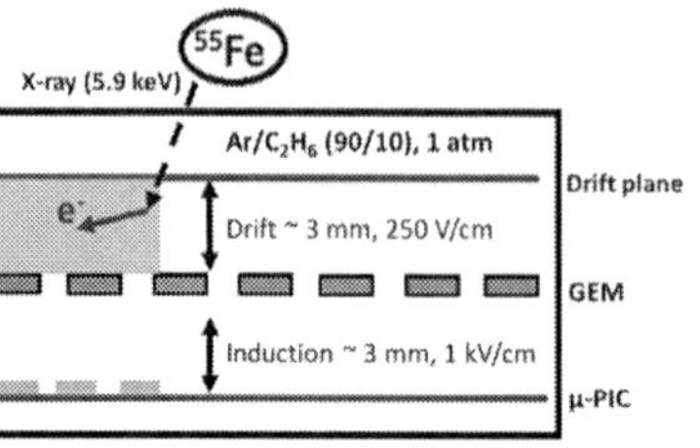

Figure 5. Schematic of the setup of the test operation. The details are described in the text.

field. The dependence of the simulated avalanche size on the anode voltage is shown in Figure 4. In this figure, the dashed and solid lines represent the simulated gains of the PCB [16] and MEMS μ-PICs, respectively. The measured gains of the PCB μ-PICs agree well the simulated gains [16]; we therefore considered Garfield++ to be a reliable simulator of an avalanche around a μ-PIC anode. The simulation predicted the gains of the MEMS μ-PIC to be twice those of the PCB μ-PIC at the same anode voltage. In addition, we studied the effect of the SiO_2 layer in the simulation, and observed that the gains of the MEMS μ-PIC were independent of the thickness of the SiO_2 layer when the layer thickness was less than 15 μm.

3 Measurements of the gas gains of MEMS μ-PICs

To confirm the simulation results, we operated the manufactured MEMS μ-PICs using the setup shown in Figure 5. The active area of the MEMS μ-PIC was 5 × 10 mm^2, and it comprised 6 cathode and 10 anode strips. The MEMS μ-PIC was placed in a sealed aluminum vessel filled with Ar/C_2H_6 gas (in a pressure ratio of 90:10) at atmospheric pressure. We positioned a gas electron multiplier (GEM) [17] 3 mm above the μ-PIC. The GEM has a polyimide substrate with a thickness of 100 μm, and its hole diameter is 70 μm with a pitch of 140 μm; and the active area of the GEM is 10 × 10 cm^2. We operated the GEM at a gain of approximately 20; the electric field of the induction field was 1 kV/cm. The drift plane was 3 mm above the GEM, and the electric field of 250 V/cm was applied in the drift space. Therefore, the volume of the detection space was 5 × 10 × 3 mm^3; this space is indicated by the hatched area in Figure 5. The signals from anode and cathode strips were read out by a CMOS ASIC chip [18]. This ASIC has 16 analog input channels, 16 digital output channels and an analog-sum signal. The analog-sum signal is fed to an 8-bit FADC operated at 25 MHz. The details of the data acquisition system used in these experiments have been described elsewhere [18].

We first operated the MEMS μ-PIC with a 10-μm SiO_2 layer under irradiation of 5.9 keV X-rays from ^{55}Fe. Figure 6 shows the signal when anode voltage was 460 V. The MEMS μ-PIC was stably operated at a discharge rate of less than 0.03 Hz and at the anode voltage of 560 V. In the case of the MEMS μ-PIC with 1-μm-thick SiO_2 layer, the leak current was approximately 160 nA under the applied anode voltage; the leak current remained greater than 20 nA for longer than 4 h. The discharge rate of the MEMS μ-PIC with a 1-μm-thick SiO_2 layer was greater than 0.5 Hz at an anode voltage of 530 V. Therefore, SiO_2 layer is essential for its stable operation.

Figure 7-(a) and (b) show the ^{55}Fe energy spectra of PCB (25 × 25 mm^2) and MEMS μ-PICs, respectively. The energy resolution of the PCB μ-PIC at 5.9 keV was 40%, with a gain of 1.1 × 10^3; by contrast the energy resolution of the MEMS μ-PIC (1-μm-thick SiO_2 layer) was 5.9 keV was 57% with a gain of 1.2 × 10^3. The energy resolution of the MEMS μ-PIC (10-μm-thick SiO_2 layer) was nearly equal to that of the MEMS μ-PIC (1-μm-thick SiO_2 layer). We considered that the poor energy resolution observed for the MEMS μ-PICs as compared to the PCB μ-PIC is due to the detection area, which was 10% smaller than that of the PCB μ-PIC. The electrons escaped from the smaller detection area more easily, depositing only partial energy and, thus, leading to worsened energy resolution. In future experiments, we plan to obtain the energy spectrum with imaging for the rejection of escape electrons, after we achieve stable operation with a gain of 10^4.

In Figure 4, stars and triangles represent the measured gains of MEMS μ-PIC with 1-μm- and 10-μm-thick SiO_2 layers, respectively. The maximum gain of the MEMS μ-PIC reached approximately 7 × 10^3. The gains of the MEMS μ-PICs of 1-μm- and 10-μm-thick SiO_2 layers were 16% and 40% that of the simulated gains at an anode voltage of 500 V, respectively. Because semiconductors contain charge carriers, the silicon near the anodes and cathodes became negatively and positively charged, respectively under an applied anode voltage. However, our simulations with the Elmer and Garfield++ programs did not account for such an effect. The electric field around the anode via would be weaker than that predicted by the simulations because of the charge-up effect of silicon; there-

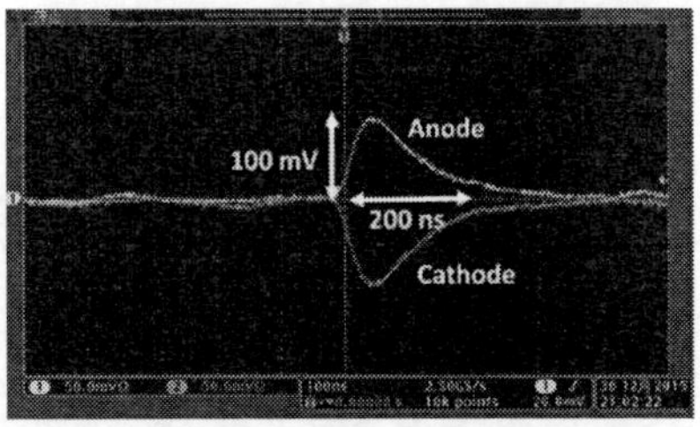

Figure 6. The signals of MEMS μ-PIC with SiO_2 layer of 10 μm by the 5.9 keV X-ray of ^{55}Fe source. Positive and negative signals are obtained from cathodes and anodes, respectively. Voltage between GEM top and bottom is 320 V, and anode voltage is 460 V. The filled gas is with a normal-pressure Ar/C_2H_6 (pressure ratio 90:10) gas mixture.

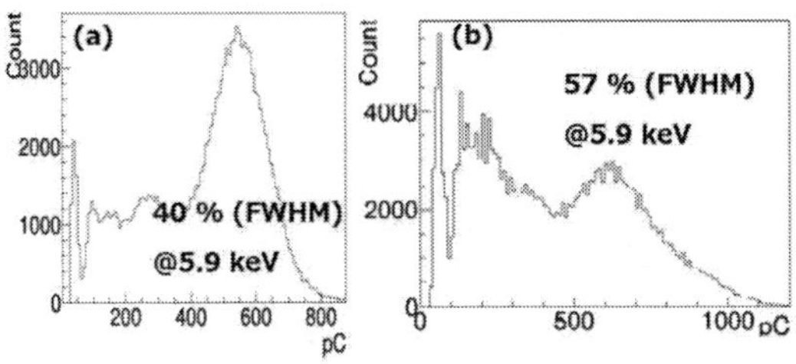

Figure 7. The energy spectrum of the 5.9 keV X-ray of ^{55}Fe source (a)PCB μ-PIC and (b)MEMS μ-PIC, respectively.

fore, the gains of the MEMS μ-PIC were reduced. In addition, the measured gas gain depends on the thickness of the SiO_2 layer, which served as an insulator positioned between the electrodes and the silicon substrate. As described in section 2, the simulated gain is independent of the thickness of the SiO_2 layer; however, thinner SiO_2 layer results in a greater charge-up effect. The leak current during startup is additional evidence of this charge-up effect.

To achieve both stable operation and a high gain with a MEMS μ-PIC, we need to reduce the charge-up effect of the silicon. We considered two solutions to this issue. One is to fabricate a MEMS μ-PIC using a glass substrate that contains no semiconductor. We can manufacture a MEMS μ-PIC on a glass substrate, but the cost will be expensive. The other solution is to fabricate a MEMS μ-PIC with a thick SiO_2 layer. We believe that we can fabricate a MEMS μ-PIC with a SiO_2 layer with 15-30-μm-thick, and we plan to manufacture them in the future. We expect that the resulting MEMS μ-PICs will be useful in various applications.

4 Summary

The applications of μ-PICs require a finer position resolution than that of current PCB μ-PICs. To satisfy this requirement, we fabricated and studied MEMS μ-PICs. Specifically, we fabricated MEMS μ-PICs with a structure similar to that of a PCB μ-PIC to focus on the effects of using semiconducting instead of an insulating substrate. Simulations predicted the gains of the MEMS μ-PICs to be twice those of PCB μ-PICs at the same anode voltage. The MEMS μ-PIC with a 1-μm-thick SiO_2 layer was unstable, demonstrating that a SiO_2 layer is essential to the stable operation of MEMS μ-PICs. For the first time, we successfully obtained a signal and energy spectra using MEMS μ-PICs under irradiation of X rays from ^{55}Fe. The maximum gain of the MEMS μ-PICs reached approximately 7×10^3. However, the gains measured at an anode voltage of 500 V were 16-40% of that of the simulated values. We considered that the gains of the MEMS μ-PICs were diminished by the effect of silicon as a semiconductor. To minimize the effect of the semiconductor, we plan to manufacture MEMS μ-PICs with thicker SiO_2 layers or glass substrates.

Acknowledgements

This work was supported by the "UCHUGAKU" project of the Unit of Synergetic Studies for Space, Kyoto University.

References

[1] A. Ochi, et al., Nuclear Instruments and Methods in Physics Research Section A **471**, 264-276 (2001).
[2] H. Kubo, et al., Nuclear Instruments and Methods in Physics Research Section A **513**, 94-98 (2003).
[3] A. Takada, et al., Nuclear Instruments and Methods in Physics Research Section A **573**, 195-199 (2007).
[4] T. Nagayoshi, et al., Nuclear Instruments and Methods in Physics Research Section A **525**, 20-27 (2004).
[5] K. Miuchi, et al., IEEE Transactions on Nuclear Science **50**, 825-830 (2003).
[6] T. Tanimori, et al., The Astrophysical Journal **810**, 28 (2015).
[7] S. Kabuki, et al., Nuclear Instruments and Methods in Physics Research Section A **623**, 606-607 (2010).
[8] K. Nakamura, et al., Progress of Theosetical and Experimental Physics , 043F01 (2015).
[9] J. D. Parker, et al., Nuclear Instruments and Methods in Physics Research Section A **697**, 23-31 (2013).
[10] T. Mizumoto, et al., Journal of Instrumentation **10**, C06003 (2015).
[11] T. Nagayoshi, et al., Nuclear Instruments and Methods in Physics Research Section A **546**, 457-465 (2005).
[12] http://geuz.org/gmsh.
[13] http://www.csc.fi/english/pages/elmer.
[14] http://garfieldpp.web.cern.ch/garfieldpp.
[15] Ö. Şahin, et al., Journal of Instrumentation **5**, P05002 (2010).
[16] A. Takada, et al., Journal of Instrumentation **8**, C10023 (2013).
[17] F. Sauli, Nuclear Instruments and Methods in Physics Research Section A **386**, 531-534 (1997).
[18] T. Mizumoto, et al., Nuclear Instruments and Methods in Physics Research Section A **800**, 40-50 (2015).

MPGD2015: Low-energy electron source to characterize Micromegas/InGrid and study of dE/dx for low energy electrons

David Attie[2], Sergey Barsuk[1], Oleg Bezshyyko[3], Leonid Burmistrov[1], Andrii Chaus[2], Paul Colas[2], Oleksii Fedorchuk[3], Larisa Golinka-Bezshyyko[3], Mykyta Haranko[3], Vladyslav Krylov[1,3], Viacheslav Kubytskyi[1], Roberto Lopez[4], Hugues Monard[1], Daniil Sukhonos[3], Maxim Titov[2], Davide Tomassini[4], Alessandro Variola[1], and Volodymyr Rodin[3]

[1] *Laboratoire de l'Accélérateur Linéaire (LAL, IN2P3/CNRS and PSud University), Orsay, France*
[2] *Commissariat à l'énergie atomique, Institut de Recherche sur les lois Fondamentales de l'Univers (CEA IRFU), Saclay, France*
[3] *Taras Shevchenko National University of Kyiv, Kyiv, Ukraine*
[4] *CERN, Geneva 23, CH-1211, Switzerland*

Abstract. Insert your english abstract here.A new versatile facility LEETECH for detector R&D, tests and calibration is designed and constructed. It uses electrons produced by the photoinjector PHIL at LAL, Orsay and provides a powerful tool for wide range R&D studies of different detector concepts delivering "mono-chromatic" samples of low energy electrons with adjustable energy and intensity. Among other innovative instrumentation techniques, LEETECH will be used for testing various gaseous tracking detectors and studying new Micromegas/InGrid concept which has very promising characteristics of spatial resolution and can be a good candidate for particle tracking and identification. In this paper the importance and expected characteristics of such facility based on detailed simulation studies are addressed.

1 Introduction

Development of new high-energy physics collider experiments calls for a rapid evolution of already established and development of new innovative detector techniques. To characterize new types of detector systems and ensure quality of the already developed instruments high accuracy tests need to be performed. Test beam facilities play a key role in such tests.

In this article we describe the design and simulation of versatile source of electrons with adjustable bunch intensity and particles energy, LEETECH (Low Energy Electron TECHnique). The facility is designed to deliver electron samples with energy spread of about 2% and sub-nanosecond time spread within a sample. A LEETECH like facilities are particularly required for a wide range of tests where the energy needs to be changed continuously and detector response to low intensity particle flows (down to 1 particle) needs to be studied. LEETECH will be particularly useful for the tests of Micromegas/InGrid technique [1] – the micro-pattern gaseous detector with pixel readout electronics, which provides 3D coordinate measurement and has a performance approaching that of silicon detectors. The counting of individual primary ionization clusters, using InGrid detector with 55x55 μm^2 pixels, could allow to achieve an ultimate energy loss (dE/dx) measurement [2].

The LEETECH source uses electrons from the PHIL facility [3] at LAL, Orsay. The PHIL is a photoinjector and a linear accelerator delivering electron bunches with energy up to 5 MeV with 5 Hz repetition rate and intensity of 10^8 to 10^{10} particles per bunch.

Test beams are usually obtained from big and expensive accelerator facilities with high energies between tens of MeV and 800 GeV [4–8]. Operational cost of these test platforms are quite high and user time is deficient. Such an important combination of beam characteristics as small time (tens of picoseconds) and energy spread (down to 1%), possibility to precisely adjust energy and reduced operational costs is highly required. Precise testing of timing characteristics of modern detectors requires test beams with short bunches in picoseconds range. Therefore LEETECH facility delivering electron samples of low and adjustable energy, variable intensity in the sample, with precise timing and small energy spread, accessible and having reduced operational costs, finds its place as a test facility for large variety of detector techniques.

2 Photoinjector PHIL

The PHIL (PHotoInjector at LAL) is an electron beam accelerator at LAL. This accelerator is dedicated to tests and characterization of the electron photoguns and high frequency structures for future accelerator projects (lepton colliders of next generation — CLIC, ILC). This machine has been designed to produce electron bunches of low energy (E < 10 MeV), small emittance ($\epsilon = 10\pi \cdot mm \cdot$ rad), high current (charge nearly 2 nC/bunch) electron bunch at

https://doi.org/10.1051/epjconf/201817402011

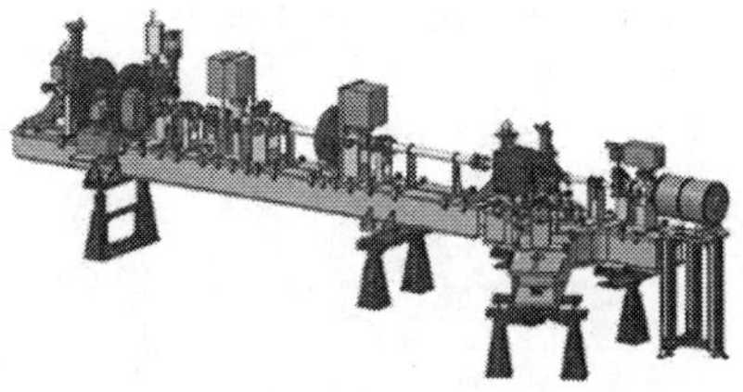

Figure 1. General view of PHIL photoinjector.

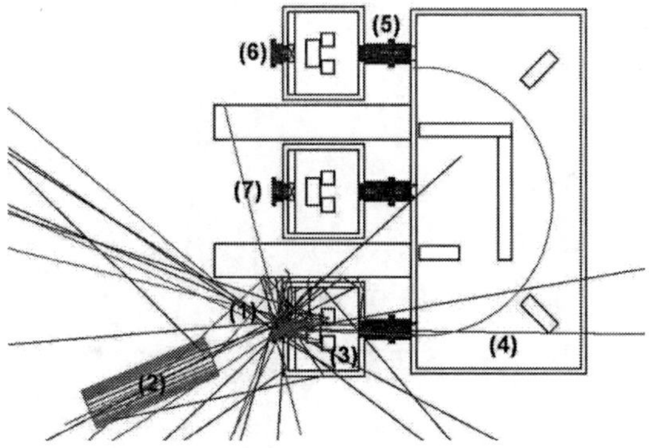

Figure 2. Geometry implemented into full Geant4 simulation. End of the PHIL beam pipe (2), target (1), vacuum chamber with magnetic field (4), entrance and exit collimator systems (3, 5) and lead shielding are shown.

low repetition frequency (<10 Hz) [3]. At the end of the accelerator, the normalized emittance is about $4\pi \cdot mm \cdot$ rad. Bunch length is short and determined by laser pulses having FWHM duration of 5 ps. PHIL is currently the 6-meters long accelerator with two extracted beam lines (see Fig. 1).

The direct beam line is mainly dedicated to the 2D transverse emittance and bunch length measurement. The deviated line is devoted to the measurements of the beam parameters (mean and dispersion energy measurement of the beam). The injection in the deviated line is performed by the Tesla Test Facility injector dipole. The direct beam line is equipped with:

- 2 Beam Position Monitors.
- 1 phosphorescent transverse beam profile monitor.
- 1 Faraday Cup.

The beam profile monitor is the phosphorescent screen oriented at 45° relative to the beam axis. The screen is a ceriumdoped yttrium:aluminum:garnet (YAG:Ce) crystal scintillator of 300 μm thickness and 40 mm of diameter.

In 2010, three other phosphorescent YAG:Ce screen monitors were installed on PHIL. The first one is mounted at the entrance of the dipole. It provides important information on the beam behavior just before the dipole magnet, which is used to adjust the beam for the mean and energy spread measurement on the deviate line. Each phosphorescent screen is complemented with a versatile optical system (made of one or more achromatic lens) and a Gigaethernet CCD camera (2 with 1/3" sensor format with 7.4 μm pixel size and 2 with 1/2" sensor format with 4.65 μm pixel size). The CCD dynamic range is 8 bit. In order to avoid pixel saturation during the measurement a remote control optical density wheel is mounted in front of each camera.

One very important feature of the PHIL electron bunches that duration of one pulse is a few of picoseconds and signals from different bunches are very well separated in time which enables us to obtain good time resolution for investigating of detector timing characteristics. A two meters extension of the direct beam line has been constructed to accommodate the LEETECH facility.

3 Principle of the setup

The LEETECH platform uses electrons from PHIL to provide electron samples with adjustable energy and intensity. Schematic view of the setup is shown in Fig. 2:

The principle of setup is described below. An Al attenuator (1) is installed at the exit window of PHIL beam pipe (2). After the attenuator the electrons form a wide energy spectrum and solid angle distribution. The entrance collimators system (3) selects a direction of electrons sent to the spectrometer also adjusting the intensity. Thus obtained narrow secondary beam passes the magnetic field region (4) inside the vacuum chamber. At the exit (5) the electrons are again filtered by exit collimators system and through the thin exit window impinged in the tested detector (6). The second exit window (7) with collimators system is dedicated to the training and monitoring purposes.

The electrons of a desired energy are selected by choosing the magnetic field controlled by the current in the dipole magnet. Output intensity can be complementarily adjusted by changing the attenuator thickness and magnetic field. Remote control and monitoring of the collimator systems and dipole current is provided via user friendly interface from the PHIL control room.

4 Simulations

The full Geant4 simulation of LEETECH platform was performed [9]. Geometry of the setup used in the simulation is shown on Fig. 2. A dedicated shielding in the vacuum chamber suppresses contribution from reflected particles and increases signal to noise ratio.

For all particles entering this volume information about their type, position, momentum and time is stored to the ROOT [10] file. If required this information is used as an input to further simulation. For example this technique is used to simulate the response of diamond detector [12] installed at the exit of the LEETECH spectrometer.

A typical simulation spectrum of electron energy at the exit of LEETECH is shown in Fig. 3. A small collimator

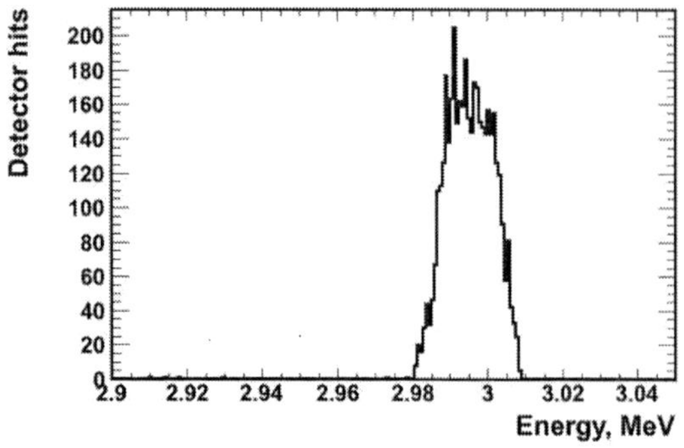

Figure 3. Spectrum of electrons at the exit of LEETECH facility.

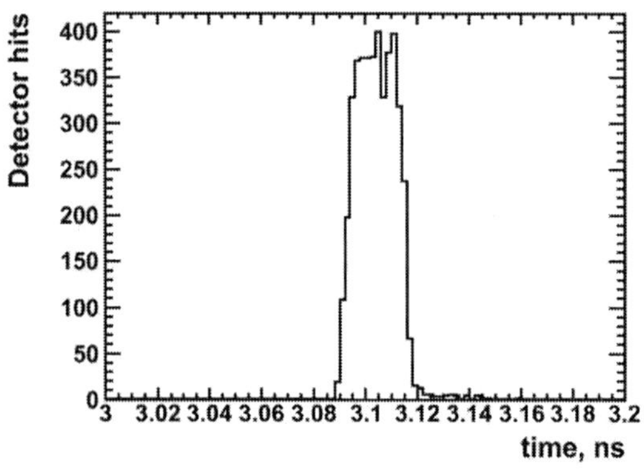

Figure 4. LEETECH timing characteristics for "narrow" mode of collimator system.

window was used to study how good energy resolution can we obtain in experiments.

The timing distribution obtained with the simulation is shown in Fig. 4. Both distributions correspond to the "narrow" collimators mode – collimators opening is 0.5x0.5 mm at the entrance and 1x1 mm at the exit. As we can see from Fig. 3 and Fig. 4 the energy resolution of 1% and time resolution of 40 ps can be achieved.

5 Attenuator thickness role

Dependence of angular and momentum spectra on the thickness of attenuator was studied. This simulation was performed with the detector positioned behind the attenuator. Momentum spectra of electrons after passing the Aluminum attenuators of different thickness are shown on Fig. 5. The obtained spectra show that different attenuator thickness favours different energy samples, so that attenuators of different thickness are foreseen to be produced.

The angular distribution of electrons' flight direction after the attenuator is shown on Fig. 6. In agreement to naive expectations, for thin attenuators the flight direction is smeared around the original flight direction. For thicker attenuators wider distribution of deviation angles

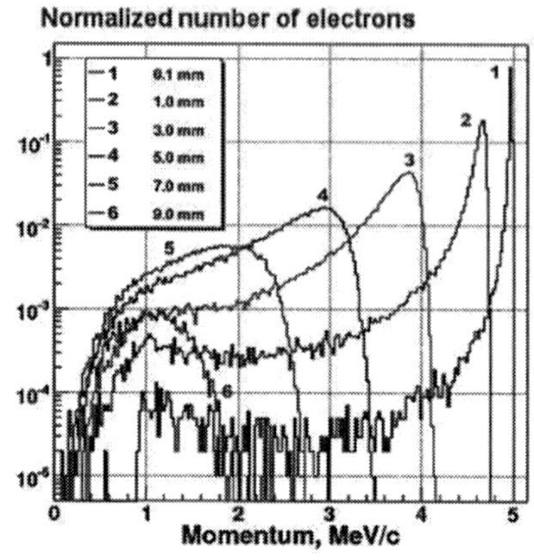

Figure 5. Momentum spectra of electrons after passing the Aluminum attenuators of different thickness. Momentum of primary electrons is 5 MeV/c

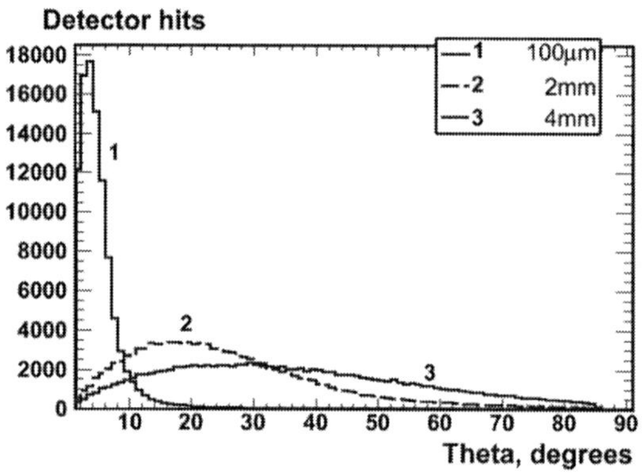

Figure 6. Angular distribution of electrons' direction after the attenuator.

is observed. This fact has to also be taken into account for calculations of the intensity of the delivered samples.

6 Variation of magnetic field

In this section we study the intensity of the delivered samples as a function of applied magnetic field. This dependence for different thickness of the attenuator is particularly important since it can be directly measured in the experiment and confronted to the simulation results. Simulation was performed using primary electrons of 3.5 MeV/c momenta for aluminum targets of 100 μm, 2 mm and 4 mm. The obtained intensities of delivered sample depending on the applied magnetic field are shown on Fig. 7.

These distributions are in agreement with the spectra shown on Fig. 5. Note that the magnetic field shown on the horizontal axis of Fig. 7 is proportional to the electron momentum shown on the horizontal axis of Fig. 5. For each attenuator, a maximum corresponding to the maximum energy of electrons after the attenuator and slow ris-

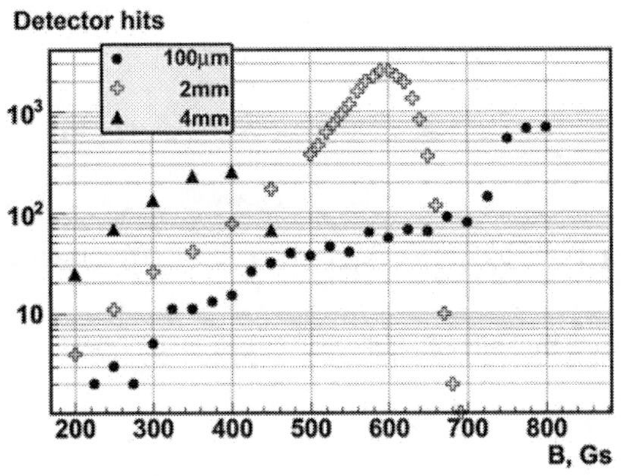

Figure 7. Magnetic field scans for different attenuator thickness.

ing edge, are observed. However, maximum intensity for the attenuator of 100 μm on Fig. 7 is lower than that for the attenuator of 2 mm. This can be explained by the inclination of the beam pipe with respect to the vacuum chamber by angle of 23°. In case of the 100 μm attenuator the beam solid angle after the attenuator is narrow and the main part of the beam goes aside and hits the collimators. Therefore the number of electrons accessing the vacuum chamber becomes significantly smaller than the total number of electrons after passing the attenuator. In case of the 4 mm attenuator the beam solid angle is wider so bigger part of electrons accesses the chamber.

7 Summary

A new facility LEETECH delivering low-energy electron samples with adjustable energy and intensity of the delivered samples was designed. The R&D, calibration and quality control programs with LEETECH are identified for a wide range of instrumentation techniques. The intensity of delivered samples ranges between few and 10^3 electrons. Bunches with a duration of hundreds (tens for further studies) of picoseconds are delivered at a frequency of 5 Hz. Samples energy is chosen in the range between hundreds of keV and the maximum PHIL energy of 5 MeV, with resolution of a few percents (down to 1% for further studies). Remote control of the entire setup has been developed.

The full Geant4 simulation of the facility has been performed. Desired parameters of the simulated samples were obtained by adjusting the dipole current and collimators position and by choosing the attenuator thickness. Time of the simulation was significantly reduced using tools for parallel computing. Further improvement of timing characteristics of the delivered samples will be studied.

Research was conducted in the scope of the IDEATE International Associated Laboratory (LIA).

References

[1] M. Chefdeville et al. An electron-multiplying *Micromegas grid made in silicon wafer post-processing technology* Nuclear Instruments and Methods. – 2006. – Vol. A556. – P. 490–494.

[2] M. Hauschild *2D and 3D Cluster Counting with GEMs and Small Pads: the Digital TPC* Proceedings of the International Workshop on Linear Colliders. – 2002. – P. 464-469.

[3] R. Roux, S. Cavalier, M. Bernard et al. *PHIL: a test beamline at LAL* Proceedings of EPAC08, Genoa, Italy. – 2008. – WEPP078. – P. 2698–2700.

[4] Gregor I-M. *Test beam for Detector Qualification In HEP* INFIERI2013 School talk. – 2013. – P. 8-42.

[5] Matz W. et al. *ROBL - a CRG Beamline for Radiochemistry and Materials Research at the ESRF* Journal of Synchrotron Radiation. – 1999. – Vol. 6. – P. 1076-1085.

[6] Quintieri L. et al. *Photo-Neutron Source by High Energy Electrons on Target: Comparison between Monte Carlo Predicitons and Experimental Measurements* 2010 IEEE Nuclear Science Symposium Conference Record. – 2010. – Vol. 31, No. 1. – P. 915–919.

[7] Buonomo B. et al. *A wide range electrons, photons, neutrons beam facility* Proceedings of EPAC08, Genoa, Italy. – 2008. – THPC143. – P. 3321–3323.

[8] Tajudin S.M., Namito Y., Sanami T., Hirayama H. *Quasi-monoenergetic 200 keV photon field using a radioactive source with backscatter layout* Japanese Journal of Applied Physics. – Vol. 53, No. 11. – P. 15–23.

[9] Agostinelli S. et al. *Geant4 - A Simulation Toolkit* Nuclear Instruments and Methods. – Vol. A506. – 2003. – P. 250–303.

[10] Brun R., Rademakers F. *ROOT – An Object Oriented Data Analysis Framework* // Nuclear Instruments and Methods. – 1997. – Vol. A389. – P. 81

[11] Barsuk S. et al. *New Versatile Platform Based on the Low Energy Electron Beam Delivered by the PHIL Photoinjector* WDS'13 Proceedings of Contributed Papers. – 2013. – Vol. 3. – P. 33–38.

[12] Mathes M. et al., *Characterization of a Single Crystal Diamond Pixel Detector in a High Energy Particle Beam* Journal of Instrumentation. – 2008 – Vol. 3. – P. 1–12.

EPJ Web of Conferences **174**, 03001 (2018)
MPGD 2015

https://doi.org/10.1051/epjconf/201817403001

Development of large area resistive electrodes for ATLAS NSW Micromegas

Atsuhiko Ochi[1,a], on behalf of the ATLAS Muon Collaboration

[1] *Kobe University, Kobe 657-8501, Japan*

Abstract. Micromegas with resistive anodes will be used for the NSW upgrades of the ATLAS experiment at LHC. Resistive electrodes are used in MPGD devices to prevent sparks in high-rate operation. Large-area resistive electrodes for Micromegas have been developed using two different technologies: screen printing and carbon sputtering. The maximum resistive foil size is 45 × 220 cm with a printed pattern of 425-μm pitch strips. These technologies are also suitable for mass production. Prototypes of a production model series have been successfully produced. In this paper, we report the development, the production status, and the test results of resistive Micromegas.

1 Requirements for ATLAS NSW Micromegas and resistive strips

The ATLAS detector will be upgraded in view of the luminosity increase of LHC. For endcap muon systems, a New Small Wheel (NSW) will be introduced instead of the current inner endcap muon detectors. A NSW consists of two types of gaseous detectors: sTGC and Micromegas. In this NSW, Micromegas are mainly designed for high resolution tracking, and to provide auxiliary trigger signals [1]. The NSW covers $1.3 < \eta < 2.7$ (η is pseudo rapidity) and will be placed 7 m from the interaction point. NSW installation is planned for the next LHC long shutdown (2019-2020).

For increased luminosity, we require an endcap muon detector with both high position resolution and tolerance to a high rate of Highly Ionizing Particles (HIP). The position resolution requirements for the η direction are less than 100 μm, and they are a few mm for the ϕ direction (azimuth angle around the beamline). At the NSW, the neutron and γ ray irradiation rate will rise up to 15 kHz/cm^2. Such incident particles might cause sparks in the detector because the deposit energy inside it could exceed the Raether limit [2–4]. To stably operate the muon detector to meet such requirements, ATLAS Micromegas is designed with resistive strips as anode electrodes [1].

The schematic structure of the NSW Micromegas is shown in Fig. 1. The avalanche charges on the resistive strips induce a signal on the readout strips. The readout strips are made from Printed Circuit Boards (PCBs). Resistive strips are formed on thin (50 μm) polyimide foils and attached to the readout PCBs.

A NSW consists of 16 sectors (eight large and eight small types), as shown in Fig. 2. The dimensions of the small and large sectors are shown in Fig. 3. The sectors consist of two modules, which have 3 + 5 readout boards. The shape of the resistive strip foils is almost identical as the readout PCBs. The maximum foil size is 45 × 222 cm with a printed pattern of 425-450 μm pitch strips. An example of a magnified resistive strip pattern is shown in Fig. 4. The resistivity requirements of the anode strip are 20 MΩ/cm.

Before full production of the NSW Micromegas modules, test chambers are produced for checking the technical availability. Those prototype modules are called Module-0s [5]. All of the chamber types in the large and small

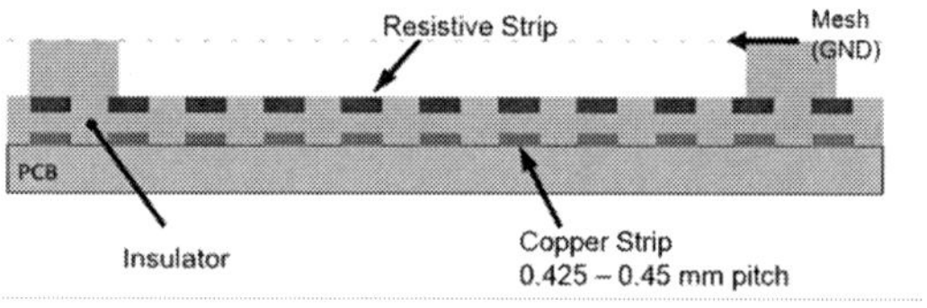

Figure 1. Schematic structure of Micromegas with resistive anodes designed for ATLAS NSW

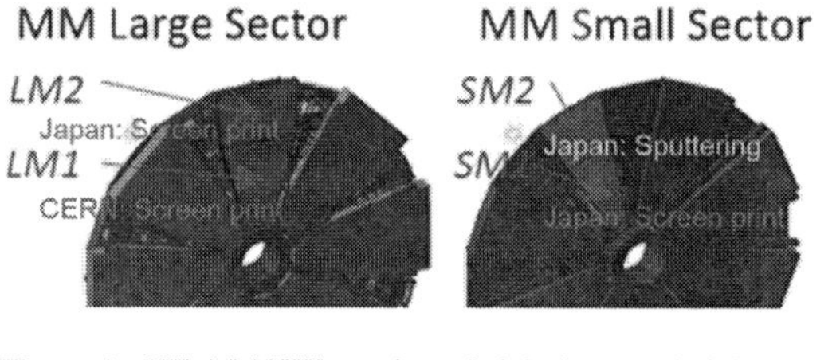

Figure 2. ATLAS NSW consists of eight large and eight small sectors. There are two Micromegas modules for each sector: LM1 and LM2 (large sectors) and SM1 and SM2 (small sectors). Resistive foils for module-0 for LM1 are made by screen printing in CERN, those for LM2 and SM1 are made by screen printing in Japan, and those for SM2 are made by carbon sputtering in Japan.

[a] e-mail: ochi@kobe-u.ac.jp

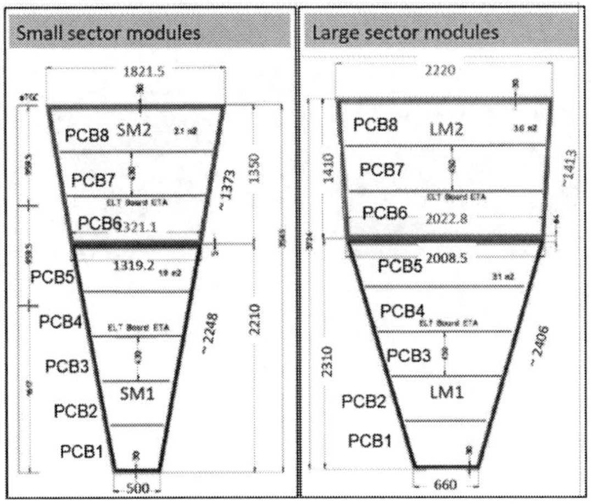

Figure 3. Size parameters of Micromegas modules for both small and large sectors

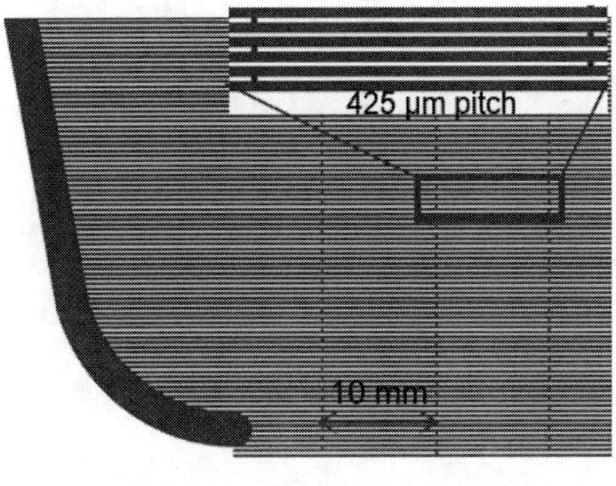

Figure 4. Example of resistive strip pattern. The thick line on left is HV supplier line. The pitch of the strips is 425 μm with interconnecting lines with 1-cm spacing between strips.

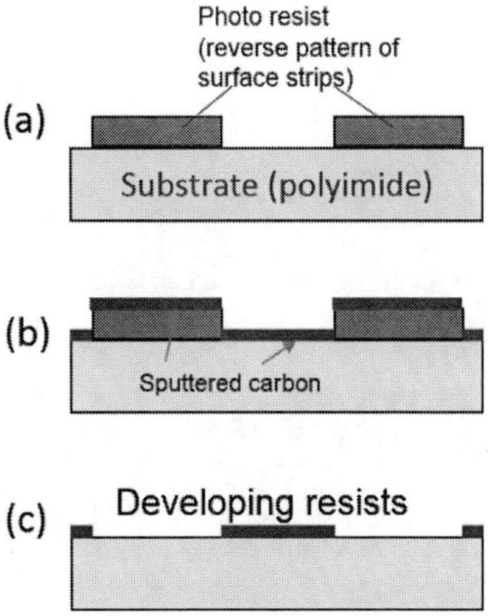

Figure 5. Resistive patterning process using carbon sputtering and liftoff process: (a) photoresist on polyimide substrate is etched by photolithography; (b) carbon is sputtered on surface; (c) remaining resist is removed by strongly alkaline water. The alkaline attacks to only photo-resists from the side of them.

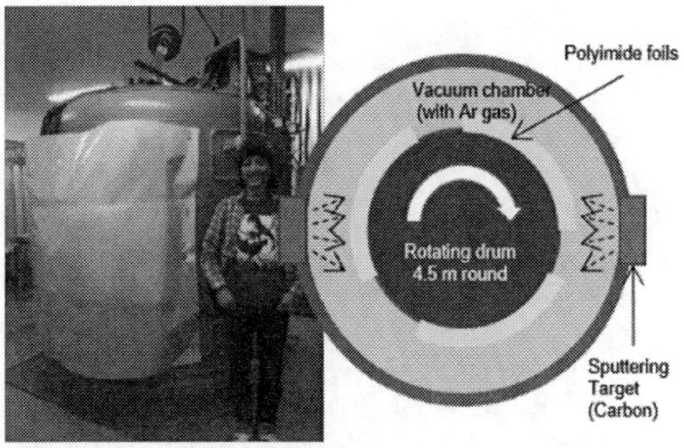

Figure 6. Large-size sputter chamber (Be-Sputter Co., Ltd.) and a schematic diagram of sputtering machine

sectors were produced for the Module-0 prototype. There are 32 types of resistive strip patterns.

Two different types of resistive strips foils have been developed for NSW Micromegas. One is carbon sputtering with a liftoff method, and the other is the screen printing of carbon paste. These two production processes have been studied for large-size resistive strip foils in module-0 production. The foils for the SM2 type (Fig. 3) were made using carbon sputtering, and those for the SM1/LM2 types were made using screen printing.

2 Carbon sputtering and liftoff

Carbon sputtering is an excellent method for making resistive electrodes for MPGDs [6]. The fine structure of electrodes can be formed by the liftoff method (Fig. 5). In the first step, (a) the photoresist on the polyimide substrate is etched by photolithography. Second, (b) carbon is sputtered on the surface. Finally, (c) the remaining resist is removed by strongly alkaline water. Making large size patterns using the photolithography is a common technology for producing PCBs. Large-size (1 × 4.5 m maximum, see Fig. 6) sputtering equipment is also available from Be-Sputter Co., Ltd. Resistive pattern masking and liftoff process were made by Raytech Inc. The sputtered carbon's resistivity can be controlled by its thickness and nitrogen doping, which has been newly developed in our studies. The target resistivity of NSW resistive foils is about 0.8 MΩ/sq., which corresponds to the strip resistivity of 20MΩ/cm with 0.4mm pitch. This value is easily available for thin (70-nm thick) carbon thickness with appropriate nitrogen doping [6].

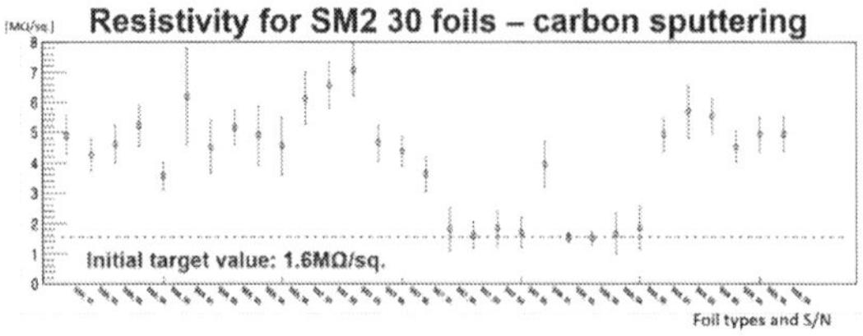

Figure 7. Surface resistivity for 30 resistive foils produced using carbon sputtering for SM2 type modules

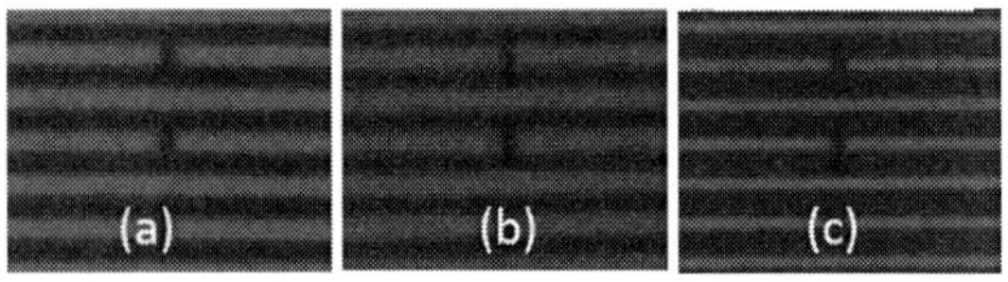

Figure 8. Microscopic image of sputtered electrodes. Differences in line width can be seen for different foils.

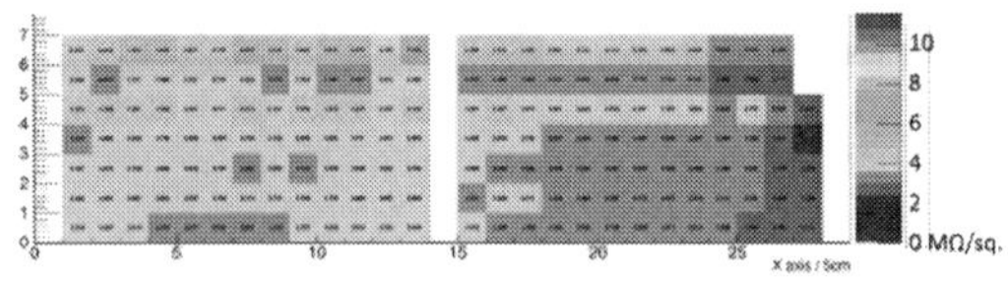

Figure 9. Surface resistivity map of one sputtered foil. Colors correspond to surface resistivity for each 5 × 5cm area.

For the module-0 prototype, we produced 30 foils using the carbon sputtering method. In the first large area production, we found some large variabilities of the resistivity (Fig. 7), caused by the differences of the width of the resistive strips due to mistakes in the initial mask pattern (Fig. 8). But the uniformity of the resistivity within each foil is very good (less than 50% resistivity fluctuations as shown in Fig. 9).

3 Screen printing

The resistive strip patterns for NSW Micromegas can also be printed on substrate by a screen printing technique. The carbon loaded epoxy, RS12115 (from ESL ElectroScience), was used as a printing paste. The printing process was done by Matsuda-Screen Inc., where a large (2.6 × 1.3 m) screen printing machine is available in a clean room (Fig. 10). After printing the paste on the substrate, the foils were cured by a large-size oven at 170°C for two hours. The quality of the patterns and the resistivity is controlled by the line/space (L/S) width on the screen plate. For printing a 400-μm pitch of strip patterns, five L/S patterns were tested: L/S = 150 μm/250 μm, 180 μm/220 μm, 200 μm/200 μm, 220 μm/180 μm, and 250 μm/150 μm. Fig. 11 shows microscopic pictures of the printed patterns and a resistivity table for each pattern. From this test, we conclude that a 150-μm line width was too narrow to make

Figure 10. Pictures of screen printing for resistive foils (Matsuda-Screen Inc.). Large printing machine was used in a clean room.

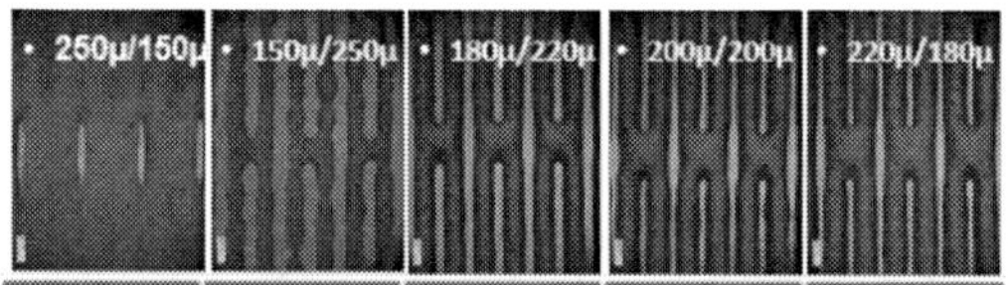

Line on screen	150μm	180μm	200μm	220μm
Resistivity	>2MΩ	570kΩ	360kΩ	270kΩ
Resistivity / line	>50MΩ	14MΩ	9.1MΩ	6.9MΩ

Figure 11. Microscopic pictures of screen-printed resistive strips and measured resistivities

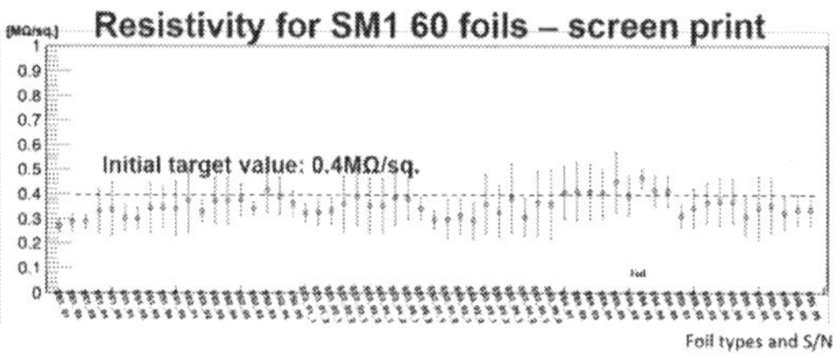

Figure 12. Surface resistivity for 60 resistive foils produced using screen printing. Those foils were made for SM1 type module.

strips, and a 150-μm space width was too narrow to separate each strip. We chose L/S values of 180 μm and 220 μm, which show the closest resistivity of 20 MΩ/cm.

For the module-0 prototype, we produced 60 foils using the screen-printing method. Based on our previous experiment, the resistive value will double after the foils are attached to the readout PCB board. Then the initial target resistivity was set to 0.4 MΩ/sq. Fig. 12 shows the results of the resistivity measurements for each screen-printed foil. The resistivity of all the foils satisfies our requirement.

https://doi.org/10.1051/epjconf/201817403001

Figure 13. Successfully produced module-0 foils. This picture is one large foil made by sputtering method.

4 Conclusion: Comparing the two technologies

We used both carbon sputtering and screen printing to produce Micromegas resistive foils for module-0s. Fig. 13 shows a large resistive foil for the NSW module-0. The pattern precision of carbon sputtering greatly surpasses that of screen printing (Figs. 8 and 11). However, the availabilities and qualities of both are adequate for NSW series production. From the viewpoint of production cost, screen printing is suitable for mass production, and we need almost 2000 foils for NSW construction. A rough cost estimation suggests that full production cost using screen printing is about three times lower than carbon sputtering. Therefore, we chose screen printing technology as a first candidate for resistive foil production technology for the ATLAS NSW Micromegas.

This work was supported by JSPS KAKENHI Grant Numbers 23340072, 26104707, and 15H02092.

References

[1] ATLAS Collaboration, "New Small Wheel Technical Design Report", *CERN-LHCC-2013-006* (2013)

[2] H. Raether, *Electron Avalanches and Breakdown in Gases.* London, UK: Butterworth, (1964)

[3] A. Bressan, et al., *Nucl. Instr. Meth.* **A 424** (1999) 321

[4] S. Bachmann, et al., *Nucl. Instr. Meth.* **A 479** (2002) 294

[5] J. Bortfeldt et al., in this volume of proceedings.

[6] A. Ochi et al., *Proc. of Sci.* (TIPP2014) 35

A novel application of Fiber Bragg Grating (FBG) sensors in MPGD

D. Abbaneo[18], M. Abbas[18], M. Abbrescia[2], M. Abi Akl[13], O. Aboamer[8], D. Acosta[16], A. Ahmad[20], W. Ahmed[20], A. Aleksandrov[30], P. Altieri[2], C. Asawatangtrakuldee[3], P. Aspell[18], Y. Assran[8], I. Awan[20], S. Bally[18], Y. Ban[3], S. Banerjee[20], V. Barashko[16], P. Barria[5], G. Bencze[7], N. Beni[10], L. Benussi[15a], V. Bhopatkar[24], S. Bianco[15], J. Bos[18], O. Bouhali[13], A. Braghieri[27], S. Braibant[4], S. Buontempo[26], C. Calabria[2], M. Caponero[14,15], C. Caputo[2], F. Cassese[26], A. Castaneda[13], S. Cauwenbergh[19], F.R. Cavallo[4], A. Celik[9], M. Choi[34], S. Choi[32], J. Christiansen[18], A. Cimmino[19], S. Colafranceschi[18], A. Colaleo[2], A. Conde Garcia[18], S. Czellar[10], M.M. Dabrowski[18], G. De Lentdecker[5], R. De Oliveira[18], G. de Robertis[2], S. Dildick[9,19], B. Dorney[18], G. Endroczi[7], F. Errico[2], A. Fenyvesi[10], M.Ferrini[15,28], S. Ferry[18], I. Furic[16], P. Giacomelli[4], J. Gilmore[9], V. Golovtsov[17], L. Guiducci[20], F. Guilloux[29], A. Gutierrez[12], R.M. Hadjiiska[30], J. Hauser[23], K. Hoepfner[1], M. Hohlmann[24], H. Hoorani[20], P. Iaydjiev[30], Y.G. Jeng[34], T. Kamon[9], P. Karchin[12], A. Korytov[16], S. Krutelyov[9], A. Kumar[11], H. Kim[34], A.Lalli[15,28], J. Lee[34], T. Lenzi[5], L. Litov[31], F. Loddo[2], A. Madorsky[16], T. Maerschalk[5], M. Maggi[2], A. Magnani[27], P.K. Mal[6], K. Mandal[6], A. Marchioro[18], A. Marinov[18], N. Majumdar[21], J.A. Merlin[18,35], G. Mitselmakher[16], A.K. Mohanty[25], A. Mohapatra[24], J. Molnar[10], S. Muhammad[20], S. Mukhopadhyay[21], M. Naimuddin[11], S. Nuzzo[2], E. Oliveri[18], L.M. Pant[25], P. Paolucci[26], I. Park[34], L. Passamonti[15], G. Passeggio[26], B. Pavlov[31], B. Philipps[1], D. Piccolo[15], D. Pierluigi[15], H. Postema[18], F. Primavera[15], A. Puig Baranac[18], A. Radi[8], R. Radogna[2], G. Raffone[15], A. Ranieri[2], G. Rashevski[30], C. Riccardi[27], M. Rodozov[30], A. Rodrigues[18], L. Ropelewski[18], S. RoyChowdhury[21], A. Russo[15], G. Ryu[34], M.S. Ryu[34], A. Safonov[9], S. Salva[19], G. Saviano[15,28], A. Sharma[2], A. Sharma[18], R. Sharma[11], A.H. Shah[11], M. Shopova[30], J. Sturdy[12], G. Sultanov[30], S.K. Swain[6], Z. Szillasi[10], J. Talvitie[22], A. Tatarinov[9], T. Tuuva[22], M. Tytgat[19], M. Valente[28], I. Vai[27], M. Van Stenis[18], R. Venditti[2], E. Verhagen[5], P. Verwilligen[2], P. Vitulo[27], S. Volkov[17], A. Vorobyev[17], D. Wang[3], M. Wang[3], U. Yang[33], Y. Yang[5], R. Yonamine[5], N. Zaganidis[19], F. Zenoni[5], and A. Zhang[24]

[1] *RWTH Aachen University, III Physikalisches Institut A, Aachen, Germany*
[2] *INFN Bari and University of Bari, Bari, Italy*
[3] *Peking University, Beijing, China*
[4] *INFN Bologna and University of Bologna, Bologna, Italy*
[5] *Universite Libre de Bruxelles, Brussels, Belgium*
[6] *National Institute of Science Education and Research, Bhubaneswar*
[7] *Institute for Particle and Nuclear Physics, Wigner Research Centre for Physics, Hungarian Academy of Sciences, Budapest, Hungary*
[8] *Academy of Scientific Research and Technology - Egyptian Network of High Energy Physics, ASRT-ENHEP, Cairo, Egypt*
[9] *Texas A&M University, College Station, U.S.A.*
[10] *Institute for Nuclear Research of the Hungarian Academy of Sciences (ATOMKI), Debrecen, Hungary*
[11] *University of Delhi, Delhi, India*
[12] *Wayne State University, Detroit, U.S.A*
[13] *Texas A&M University at Qatar, Doha, Qatar*
[14] *ENEA - Frascati Research Centre, Frascati RM- Italy*
[15] *Laboratori Nazionali di Frascati - INFN, Frascati, Italy*
[16] *University of Florida, Gainesville, U.S.A.*
[17] *Petersburg Nuclear Physics Institute, Gatchina, Russia*
[18] *CERN, Geneva, Switzerland*
[19] *Ghent University, Dept. of Physics and Astronomy, Ghent, Belgium*
[20] *National Center for Physics, Quaid-i-Azam University Campus, Islamabad, Pakistan*
[21] *Saha Institute of Nuclear Physics, Kolkata, India*
[22] *Lappeenranta University of Technology, Lappeenranta, Finland*
[23] *University of California, Los Angeles, U.S.A.*
[24] *Florida Institute of Technology, Melbourne, U.S.A.*
[25] *Bhabha Atomic Research Centre, Mumbai, India*
[26] *INFN Napoli, Napoli, Italy*
[27] *INFN Pavia and University of Pavia, Pavia, Italy*
[28] *University of Rome "La Sapienza" (IT) - Facoltà di Ingegneria, Ingegneria Chimica Materiali ed Ambiente, Roma, Italy*
[29] *IRFU CEA-Saclay, Saclay, France*
[30] *Institute for Nuclear Research and Nuclear Energy, Sofia, Bulgaria*

[31] *Sofia University, Sofia, Bulgaria*
[32] *Korea University, Seoul, Korea*
[33] *Seoul National University, Seoul, Korea*
[34] *University of Seoul, Seoul, Korea*
[35] *Institut Pluridisciplinaire - Hubert Curien (IPHC), Strasbourg, France*

Abstract. We present a novel application of Fiber Bragg Grating (FBG) sensors in the construction and characterisation of Micro Pattern Gaseous Detector (MPGD), with particular attention to the realisation of the largest triple (Gas electron Multiplier) GEM chambers so far operated, the GE1/1 chambers of the CMS experiment at LHC. The GE1/1 CMS project consists of 144 GEM chambers of about 0.5 m^2 active area each, employing three GEM foils per chamber, to be installed in the forward region of the CMS endcap during the long shutdown of LHC in 2108-2019. The large active area of each GE1/1 chamber consists of GEM foils that are mechanically stretched in order to secure their flatness and the consequent uniform performance of the GE1/1 chamber across its whole active surface. So far FBGs have been used in high energy physics mainly as high precision positioning and re-positioning sensors and as low cost, easy to mount, low space consuming temperature sensors. FBGs are also commonly used for very precise strain measurements in material studies. In this work we present a novel use of FBGs as flatness and mechanical tensioning sensors applied to the wide GEM foils of the GE1/1 chambers. A network of FBG sensors have been used to determine the optimal mechanical tension applied and to characterise the mechanical tension that should be applied to the foils. We discuss the results of the test done on a full-sized GE1/1 final prototype, the studies done to fully characterise the GEM material, how this information was used to define a standard assembly procedure and possible future developments.

1 Introduction

To upgrade the Compact Muon Solenoid (CMS[1]) muon system 144 GEM chambers will be installed in the high pseudo-rapidity region of CMS during Long Shutdown 2 (LS2) of the Large Hadron Collider [2]. The GEMs can provide extra leverage on precision studies of standard model physics, as well as open up a window to explore exotic signatures with muons in the high pseudorapidity region [3]. The GEM chambers will installed, as shown in fig. 1, very close to the beam pipe where a high flux of low Pt muons is expected. The GEM chambers can easily handle this rate due to their high rate capability of 100 MHz/cm^2. The large active area of each GE1/1 (GEM Endcap) chamber, approximately 0.4 m^2 [4], consists of a triple-GEM foil stack stretched by means of screws placed around the stack frame (fig.2). These foils need to be stretched simultaneously in order to secure the planarity and consequent uniform performance of the GE1/1 chamber [5]. The GE1/1 detector technology used for CMS is described in detail somewherelse [6].

The FBG sensors act as low cost precision spatial and temperature sensing tools and they are commonly used for strain measurements [7] [8] [9]. These sensors are commercially available and have relative low cost. In this work FBG sensors are used to measure the planarity and mechanical tension of the GEM foils in the GE1/1 chambers. The GE1/1 assembly procedure employs a mechanical stretching procedure to apply tension to the GEM foils by means of a series of lateral screws inserted into the internal GE1/1 frame. This technology allows mechanical assembly of the GEM chamber without the use of internal spacers or glue.

[a]Corresponding author; email address: Luigi.Benussi@lnf.infn.it

Figure 1: GEM detector installation position in CMS [4]

2 FBG sensors as a strain measurement

A FBG is a type of distributed Bragg reflector, constructed in a short segment of optical fiber that reflects particular wavelengths of light and transmits all others. The sensitivity of FBG in terms of strain, defined as relative elongation w.r.t. the initial position is of the order of 0.1 μ. This is achieved by creating a periodic variation in the refractive index of the fiber core, which generates a wavelength-specific dielectric mirror. Therefore it can be used as a strain measurement tool since variation of the FBG translates into different light frequency response.

In order to validate the mechanical stretching technique a network of FBG sensors is affixed on the triple-GEM stack as shown in figure 3. Each sensor is glued on the GEM foil using a very thin layer of epoxy glue.

Figure 2: Lateral screws for stretching the GEM foils stack.

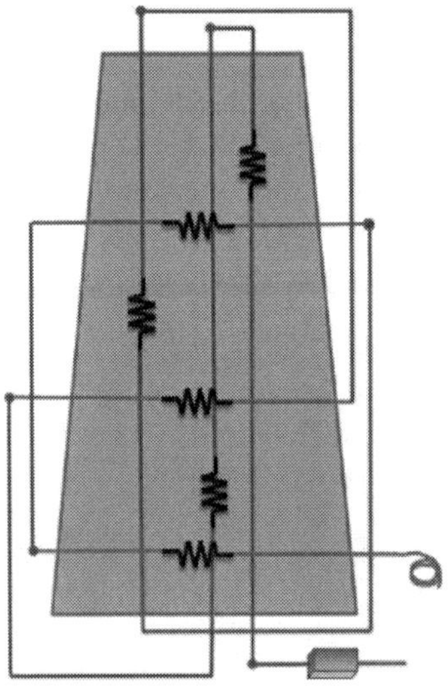

Figure 3: Conceptual schema of the FBG sensors connection on one GEM foil.

Due to the high temperature sensitivity of the FBG the thermal dilatation of the materials used are corrected by using a separate FBG sensor on a 10×10 cm^2 GEM detector. The test is performed by modifying the stretching conditions of the GEM foils stack with real time monitoring and recording of the FBG sensors data. The test starts with the chamber normally assembled with the GEM stack mechanically stretched to the nominal tensile load. After some time while steady in the starting condition, the mechanical stretching of the GEMs is released and kept in such condition for several hours. Finally the GEMs are stretched again up to the nominal tensile load. The trends of the FBG sensors are shown in figure 4a.

The steep variations of the strain evident in figure 4a correspond to the actions of un-screwing and screwing the mechanical stretchers during the test. The initial stretch value is assumed as reference condition with strain = 0. When stretchers are un-screwed the strain goes to the lower value, different strain values apply to different foils as they fold quasi-free and assume unequal conditions. After the stretchers are screwed back, the strain value is similar for all foils, showing that they all experience similar stretching, about the original value of the reference condition. Thus it can be inferred that at the predetermined tensile load all foils reach a similar stretched level although they started from different values. From the plot it can be seen that all the sensors of the network react at the same moment. These results allow us to validate the mechanical stretching assembly technique for GE1/1 chambers. Further tests are ongoing to confirm other important parameters such as the optimal tensile load to be applied to the GEMs and the maximum planarity obtainable for the GEMs without applying a load beyond the Young's region for GEM foils.

3 Another FBG application: the load gauge

Another important application of FBG in GEM chamber construction is the possibility to be used as load gauge for precise measurement of the tensile load applied to the foils, of the different layers, in the same moment. This is extremely important in the case of the GE1/1 chambers since, during their assembly procedure, the foils are stretched by means of screwing nuts with dynamometric screwdriver. It is thus very important to know precisely how much to pull the foils in order to avoid to stretch them too much, reaching the mechanical load with will cause the GEM foils to be operated outside their elastic range (Young region). In order to demonstrate the idea of FBG as load gauge we have done the following test which results in terms of FBG response are shown in figure 4b. After having properly stretched the GE1/1 GEM stack, we have removed a single stretching screw from the chamber and replaced it with an eyelet screw on which we have fixed a stainless steel wire used to attach different weights. The idea was to add weights in different steps till they applied load reproduce in the FBG facing the eyelet the same response when the GEMs are stretched by means of the original screw.

In this test few sample of weights (lead bricks) are used. We took 4 points corresponding to the following weights A=2.86 kg, B=5.65 kg, C=8.5 kg, D=11.3 kg. Fig 4b shows the response of the FBG during this test. The RED and BLUE lines correspond to the two sensors closest to the point where the load is applied (RED sensor parallel to the line of the stainless steel wire, BLUE orthogonal to the wire direction). In the plot are visible different regions corresponding to the different actions performed to the foils. The first flat region correspond to the initial position of the foils stack, meaning the GEMs properly stretched with the original screw. Then the screw is removed and the two sensors immediately record the effect on the foils tension. There are some movements in the following flat regions due to the handling during the substitution of the screw. The first weight A applied shows its effect on the sensors. Then weight A is removed to load weight B (it is visible the short foils relaxation during this action). The third step was to add the weight A to B and

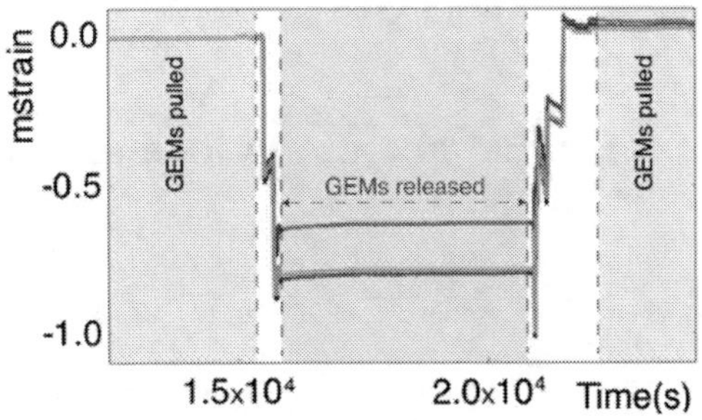

(a)

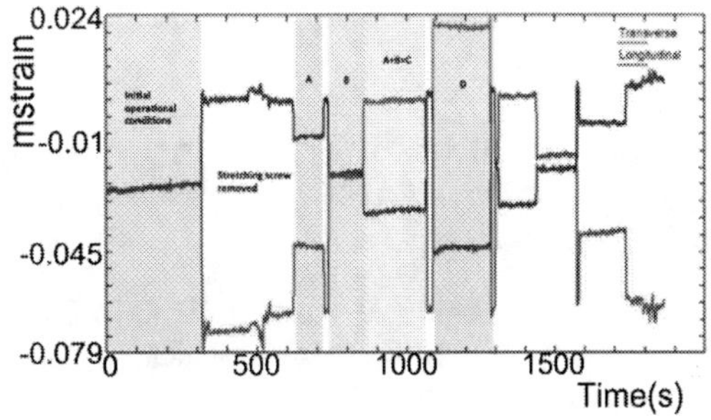

(b)

Figure 4: (a) Three regions corresponding to the mechanical stretched, loose and again stretched triple GEM foils stack respectively. (b) The response of FBG sensors during the test as load gauge

after removing A+B (weight C) we add weight E. Finally we redo the same steps backwards as shown in the plot, removing the weights in steps. By analysing the plots shown in fig. 4b we were able to determine the load applied on a single screw once the GEM stack is stretched to its operational value. This value is of the order 5.65 kg and is very close to the expected value of 3 kg/cm considering that the crimping length of one screw is 2 cm. This result is important since other measurements presented in this conference shown that the GEM foils Young region ends around 8 kg/cm. This value ensures that the mechanical tension applied to GEM foils is still in the elastic regime by a wide margin.

4 Conclusion

With the use of FBG sensors we successfully demonstrated that the novel glue-less technique adopted to assemble the GE1/1 chambers for the LS2 update of CMS is reliable and guarantees the correct tensioning of the three GEM foils. By applying the correct tension across the GEM stack the uniform gaps spacing is obtained, which is extremely important to get the required performance of the detector. We have also demonstrated the FBG are excellent load gauges that can be used to determine very precisely the applied stretching force to GEM foils which is a critical point for the correct assembly of the GE1/1 chambers and to secure their homogeneity. Other tests are ongoing by using the same FBG sensors to optimise the tensile load in order to avoid damage and guarantee planarity of the GEM foils.

Acknowledgments

We gratefully acknowledge the support of FRS-FNRS (Belgium), FWO-Flanders (Belgium), BSF-MES (Bulgaria), BMBF (Germany), DAE (India), DST (India), INFN (Italy), NRF (Korea), QNRF (Qatar), and DOE (USA). This project has received funding from the European Union's Horizon 2020 Research and Innovation programme under Grant Agreement no. 654168.

References

[1] S. Chatrchyan *et al.* [CMS Collaboration], JINST **3**, S08004 (2008). doi:10.1088/1748-0221/3/08/S08004

[2] D. Abbaneo et al., Performance of a Large-Area GEM Detector Prototype for the Upgrade of the CMS Muon Endcap System, arXiv:1412.0228v2, 8 Dec 2014.

[3] D. Abbaneo et al., Upgrade of the CMS muon system with triple-GEM detectors, JINST 9 (2014) C10036, http://iopscience.iop.org/1748-0221/9/10/C10036/.

[4] A. Colaleo et al., CMS TECHNICAL DESIGN REPORT FOR THE MUON ENDCAP GEM UPGRADE, CERN-LHCC-2015-012. CMS-TDR-013, https://cds.cern.ch/record/2021453.

[5] D. Abbaneo et al.,Status of the Triple-GEM project for the upgrade of the CMS Muon System, DOI: 10.1088/1748-0221/8/12/C12031.

[6] Gilles De Lentdecker for the CMS GEM collaboration"Status Report of the Upgrade of the CMS muon system with triple-GEM detectors". Elba 2015 confernce proceeding under publication.

[7] L. Benussi et al.,The Omega-like: A novel device using FBG sensors to position vertex detectors with micrometric precision, Nucl.Phys.Proc.Suppl. 172.

[8] M. Caponero et al., Use of fiber optic technology for relative humidity monitoring in RPC detectors, Published in PoS RPC2012 (2012) 073.

[9] L. Benussi et al., A Novel Temperature Monitoring Sensor for Gas-Based Detectors in Large HEP Experiments, doi = "http://dx.doi.org/10.1016/j.phpro.2012.02.400"

Quality control for the first large areas of triple-GEM chambers for the CMS endcaps

D. Abbaneo[17], M. Abbas[17], M. Abbrescia[2], M. Abi Akl[13], O. Aboamer[8], D. Acosta[15], A. Ahmad[19], W. Ahmed[19], A. Aleksandrov[28], P. Altieri[2], C. Asawatangtrakuldee[3], P. Aspell[17], Y. Assran[8], I. Awan[19], S. Bally[17], Y. Ban[3], S. Banerjee[20], V. Barashko[15], P. Barria[5], G. Bencze[7], N. Beni[10], L. Benussi[14], V. Bhopatkar[23], S. Bianco[14], J. Bos[17], O. Bouhali[13], A. Braghieri[26], S. Braibant[4], S. Buontempo[25], C. Calabria[2], M. Caponero[14], C. Caputo[2], F. Cassese[25], A. Castaneda[13], S. Cauwenbergh[18], F.R. Cavallo[4], A. Celik[9], M. Choi[32], S. Choi[30], J. Christiansen[17], A. Cimmino[18], S. Colafranceschi[17], A. Colaleo[2], A. Conde Garcia[17], S. Czellar[10], M.M. Dabrowski[17], G. De Lentdecker[5], R. De Oliveira[17], G. de Robertis[2], S. Dildick[9,18], B. Dorney[17], G. Endroczi[7], F. Errico[2], A. Fenyvesi[10], S. Ferry[17], I. Furic[15], P. Giacomelli[4], J. Gilmore[9], V. Golovtsov[16], L. Guiducci[4], F. Guilloux[27], A. Gutierrez[12], R.M. Hadjiiska[28], J. Hauser[22], K. Hoepfner[1], M. Hohlmann[23], H. Hoorani[19], P. Iaydjiev[28], Y.G. Jeng[32], T. Kamon[9], P. Karchin[12], A. Korytov[15], S. Krutelyov[9], A. Kumar[11], H. Kim[32], J. Lee[32], T. Lenzi[5], L. Litov[29], F. Loddo[2], A. Madorsky[15], T. Maerschalk[5], M. Maggi[2], A. Magnani[26], P.K. Mal[6], K. Mandal[6], A. Marchioro[17], A. Marinov[17], N. Majumdar[20], J.A. Merlin[17,33], G. Mitselmakher[15], A.K. Mohanty[24], A. Mohapatra[23], J. Molnar[10], S. Muhammad[19], S. Mukhopadhyay[20], M. Naimuddin[11], S. Nuzzo[2], E. Oliveri[17], L.M. Pant[24], P. Paolucci[25], I. Park[32], G. Passeggio[25], B. Pavlov[29], B. Philipps[1], D. Piccolo[14], H. Postema[17], A. Puig Baranac[17], A. Radi[8], R. Radogna[2], G. Raffone[14], A. Ranieri[2], G. Rashevski[28], C. Riccardi[26], M. Rodozov[28], A. Rodrigues[17], L. Ropelewski[17], S. RoyChowdhury[20], G. Ryu[32], M.S. Ryu[32], A. Safonov[9], S. Salva[18], G. Saviano[14], A. Sharma[2], A. Sharma[17], R. Sharma[11], A.H. Shah[11], M. Shopova[28], J. Sturdy[12], G. Sultanov[28], S.K. Swain[6], Z. Szillasi[10], J. Talvitie[21], A. Tatarinov[9], T. Tuuva[21], M. Tytgat[18,a], I. Vai[26], M. Van Stenis[17], R. Venditti[2], E. Verhagen[5], P. Verwilligen[2], P. Vitulo[26], S. Volkov[16], A. Vorobyev[16], D. Wang[3], M. Wang[3], U. Yang[31], Y. Yang[5], R. Yonamine[5], N. Zaganidis[18], F. Zenoni[5], and A. Zhang[23]

[1]*RWTH Aachen University, III Physikalisches Institut A, Aachen, Germany*
[2]*INFN Bari and University of Bari, Bari, Italy*
[3]*Peking University, Beijing, China*
[4]*INFN Bologna and University of Bologna, Bologna, Italy*
[5]*Universite Libre de Bruxelles, Brussels, Belgium*
[6]*National Institute of Science Education and Research, Bhubaneswar*
[7]*Institute for Particle and Nuclear Physics, Wigner Research Centre for Physics, Hungarian Academy of Sciences, Budapest, Hungary*
[8]*Academy of Scientific Research and Technology - Egyptian Network of High Energy Physics, ASRT-ENHEP, Cairo, Egypt*
[9]*Texas A&M University, College Station, U.S.A.*
[10]*Institute for Nuclear Research of the Hungarian Academy of Sciences (ATOMKI), Debrecen, Hungary*
[11]*University of Delhi, Delhi, India*
[12]*Wayne State University, Detroit, U.S.A*
[13]*Texas A&M University at Qatar, Doha, Qatar*
[14]*Laboratori Nazionali di Frascati - INFN, Frascati, Italy*
[15]*University of Florida, Gainesville, U.S.A.*
[16]*Petersburg Nuclear Physics Institute, Gatchina, Russia*
[17]*CERN, Geneva, Switzerland*
[18]*Ghent University, Dept. of Physics and Astronomy, Ghent, Belgium*
[19]*National Center for Physics, Quaid-i-Azam University Campus, Islamabad, Pakistan*
[20]*Saha Institute of Nuclear Physics, Kolkata, India*
[21]*Lappeenranta University of Technology, Lappeenranta, Finland*
[22]*University of California, Los Angeles, U.S.A.*
[23]*Florida Institute of Technology, Melbourne, U.S.A.*
[24]*Bhabha Atomic Research Centre, Mumbai, India*
[25]*INFN Napoli, Napoli, Italy*
[26]*INFN Pavia and University of Pavia, Pavia, Italy*
[27]*IRFU CEA-Saclay, Saclay, France*
[28]*Institute for Nuclear Research and Nuclear Energy, Sofia, Bulgaria*
[29]*Sofia University, Sofia, Bulgaria*
[30]*Korea University, Seoul, Korea*

 https://doi.org/10.1051/epjconf/201817403003

[31] *Seoul National University, Seoul, Korea*
[32] *University of Seoul, Seoul, Korea*
[33] *Institut Pluridisciplinaire - Hubert Curien (IPHC), Strasbourg, France*

Abstract. The CMS Collaboration plans to equip the very forward muon system with triple-GEM detectors that can withstand the environment of the High-Luminosity LHC. This project is at the final stages of R&D and moving to production. An unprecedented large area of several 100 m^2 are to be instrumented with GEM detectors which will be produced in six different sites around the world. A common construction and quality control procedure is required to ensure the performance of each detector. The quality control steps will include optical inspection, cleaning and baking of all materials and parts used to build the detector, leakage current tests of the GEM foils, high voltage tests, gas leak tests of the chambers and monitoring pressure drop vs. time, gain calibration to know the optimal operation region of the detector, gain uniformity tests, and studying the efficiency, noise and tracking performance of the detectors in a cosmic stand using scintillators.

1 Introduction

The CMS Collaboration is planning several muon system upgrades in order to maintain its high level of performance achieved during LHC Run 1 also during the future High Luminosity LHC [1]. A quadrant of the CMS muon system is shown in Figure 1, where both the present detectors and the proposed extensions are displayed. In the $1.5 < |\eta| < 2.4$ forward endcap region, currently only Cathode Strip Chambers are installed. To increase redundancy and enhance the muon trigger and reconstruction capabilities in that region, additional muon detectors are foreseen. In particular, two new stations, called GE1/1 and GE2/1, would be equipped with GEM detectors. Furthermore, in order to increase the coverage for muon detection up to $|\eta| \approx 3$, another GEM station called ME0 is proposed to be installed behind new endcap calorimeters.

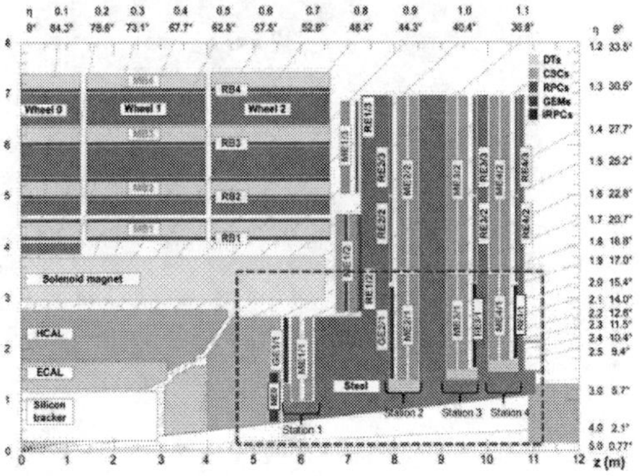

Figure 1. Quadrant of the CMS muon system showing present detectors, i.e. Drift Tubes (DTs), Resistive Plate Chambers (RPCs) and Cathode Strip Chambers (CSCs), and the locations of the proposed GEM detectors, i.e. the ME0, GE1/1 and GE2/1 stations.

Preparations for a demonstrator consisting of four GE1/1 superchambers covering a 40° sector are well advanced; their installation is foreseen during the LHC 2016 Year End Technical Stop. The full GE1/1 station is scheduled for installation in CMS during LHC Long Shutdown 2 (2018-2020). The installation of GE2/1 and ME0 chambers is planned for Long Shutdown 3 (2024-2026).

2 The GE1/1 Station

For the muon detectors to be installed in the CMS $1.5 < |\eta| < 2.2$ endcap region in the face of the high luminosity, GEMs are an excellent choice due to their thin profile and the ability to operate well at the high particle fluxes expected there under HL-LHC conditions. The GE1/1 station will be equipped with triple-GEM chambers that were developed over the years starting in 2009 [2]. The detector front-end electronics will be based on the custom designed VFAT3 chip, providing both fast input to the CMS level-1 muon trigger and full granularity information for the offline muon reconstruction. A full description of the GE1/1 station can be found in [3].

2.1 Chambers and Superchambers

The GE1/1 stations to be installed in each of the two CMS endcaps consist of 36 superchambers, each made of a double layer of trapezoidal triple-GEM chambers that cover about 10° in ϕ. The superchambers alternate in ϕ between long ($1.5<|\eta|<2.18$) and short ($1.6<|\eta|<2.18$) versions to maximize the coverage of the available space. The total thickness of a superchamber is 88 mm.

The 144 GE1/1 chambers needed to produce the 72 superchambers in total, have a trapezoidal shape with a 10.15° opening angle and are 3.5 cm thick. They contain a triple-GEM detector with a 3/1/2/1 mm drift/transfer-1/transfer-2/induction field gap configuration. The active readout area is 0.345/0.409 m^2 for the short/long chambers. The foils for the short/long chambers are segmented in 40/47 HV sectors and in 3×8 readout sectors in $\phi \times \eta$, each with 128 readout strips, yielding a total of 3072 readout channels per chamber. The chambers will be operated with an Ar/CO_2 70:30 gas mixture, at a nominal detector gain of $1\text{-}2\times10^4$.

2.2 Detector Production and Quality Control

The GEM foils are produced at the CERN Surface Treatment Workshop using a single-mask technique. A new

[a] Corresponding author; e-mail: Michael.Tytgat@UGent.be

https://doi.org/10.1051/epjconf/201817403003

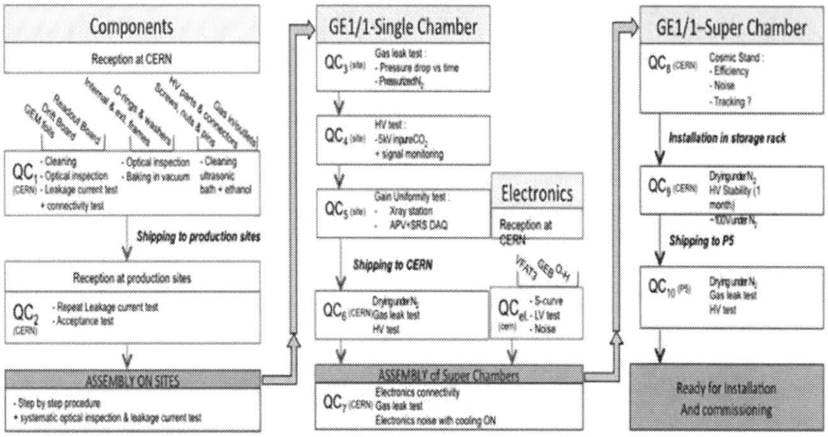

Figure 2. GE1/1 chamber and superchamber assembly and quality control workflow.

Figure 3. Leakage current measurement station for the GE1/1 GEM foils.

technique in which the three GEM foils of a chamber are simultaneously stretched in a mechanical way, allows to assemble a full chamber in a few hours time [4].

As depicted in Figure 2, the general GE1/1 chamber and superchamber assembly and quality control workflow consists of 3 main parts:

- Component production and quality control (QC1-2)
- Assembly and commissioning of GE1/1 chambers at production sites (QC3-5)
- Assembly and commissioning of superchambers at CERN before installation in CMS (QC6-10)

After cleaning of all parts using ultrasonic baths, baking and sand blast procedures, the inspection of the main chamber components includes an I-V curve measurement of the voltage divider, an optical and electrical inspection of the drift board, and a verification of the readout board checking for possible shorts between readout strips. The GEM foils are optically inspected for defects and then submitted to a leakage current test (see Figure 3) in which the foils should not draw a current higher than 30nA when a potential difference of 500V is applied between the GEM metal sides.

Upon completion of the assembly, every chamber is checked for gas leaks by pressurizing it up to 20 mbar

Figure 4. The chamber gain uniformity measurements, including pictures of the setup at CERN (top) and an example graph showing the relative gain of the chamber as function the position along its surface.

Figure 5. Cosmic stand at CERN for the final quality control of GE1/1 superchambers.

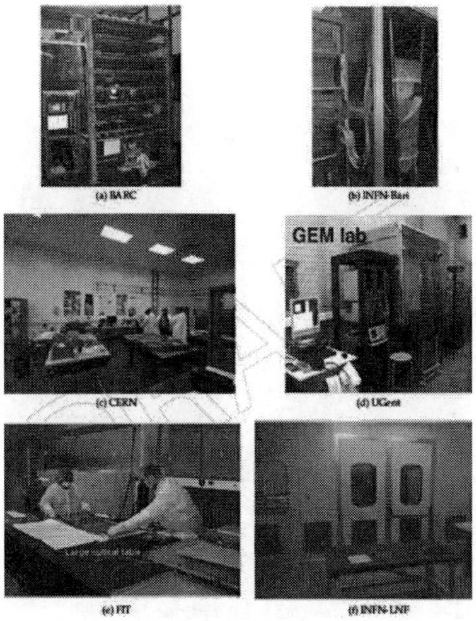

Figure 6. The following assembly sites for GE1/1 production are being considered by CMS: (a) Bhabha Atomic Research Center (BARC), India; (b) INFN Sezione di Bari, Italy; (c) CERN, Switzerland; (d) Ghent University, Belgium; (e) Florida Institute of Technology (FIT), USA; (f) INFN Laboratori Nazionali di Frascati (LNF), Italy.

with dry nitrogen. Next, the output signal of the chambers is monitored after flushing for several hours with Ar/CO_2 and turning them on at moderate HV. Furthermore, the uniformity of the gain of each chamber across its surface is verified in a dedicated setup inside a Copper protective box (see Figure 4), with an SRS/APV25 hybrid based data-acquisition [5] and an X-ray source placed at a distance of about 1 m, illuminating the full chamber at once.

Superchambers are fabricated by coupling together two GE1/1 single chambers. Inside a cosmic stand, a high voltage scan is performed on each chamber and the relevant parameters such as detector gain, noise levels and cluster size are measured with the final detector electronics. Figure 5 shows the present cosmic stand at CERN, which allows to test up to 15 superchambers at once and features fully automated HV scans, measurements of cosmic muon tracks and a DAQ system comparable to the final setup to be used inside the CMS experiment.

Throughout the entire detector assembly and quality control procedure, all detailed information and test results will be stored in an Oracle based database for future reference.

2.3 Assembly Sites

At present, six assembly sites are being considered for GE1/1 mass production. All sites will follow the same assembly and quality control protocols and will include at least the following setups:

- A certified cleanroom of class 1000 minimum to assemble full GE1/1 chambers; the cleanroom must be equipped with clean and dry nitrogen gas lines used to blow chamber parts during the assembly
- A nitrogen-flushed box to measure GEM foil leakage currents at voltages up to 500V
- An X-ray setup for gain uniformity measurements across the chamber surface
- A gas leak measurement station for verify gas tightness of the chambers

Pictures of different assembly site candidates are shown in Figure 6.

3 Summary

Facing the High-Luminosity LHC, the CMS Collaboration is planning several muon system upgrades in order to maintain its high level performance in terms of muon triggering and reconstruction. CMS recently approved the installation of the so-called GE1/1 station: during LHC Long Shutdown 2, the $1.5<|\eta|<2.2$ region of the first endcap disks will be equipped with a total of 144 new triple-GEM detectors, arranged in 72 super-chambers. The GE1/1 chamber assembly will be done at several locations in- and outside CERN; a detailed chamber assembly and quality control workflow is being worked out.

Acknowledgements

We gratefully acknowledge support from FRS-FNRS (Belgium), FWO-Flanders (Belgium), BSF-MES (Bulgaria), BMBF (Germany), DAE (India), DST (India), INFN (Italy), NRF (Korea), QNRF (Qatar), and DOE (USA).

References

[1] The CMS Collaboration, *Technical Proposal for the Phase-II Upgrade of the CMS Detector*, CERN-LHCC-2015-010, LHCC-P-008

[2] D. Abbaneo *et al.*, JINST **9** (2014) 10, C10036

[3] The CMS Collaboration, *CMS Technical Design Report for the Muon Endcap GEM Upgrade*, CERN-LHCC-2015-012, CMS-TDR-013

[4] D. Abbaneo *et al.*, Nucl. Instrum. Meth. A**732**, 203-207 (2013)

[5] S. Martoiu *et al.*, 2011 IEEE Nucl. Sci. Symp. Conf. Rec., 2036-2038 (2011)

GEM Foil Quality Assurance For The ALICE TPC Upgrade

Erik Brücken[1,a] and Timo Hildén[1] for the ALICE TPC Collaboration

[1] Helsinki Institute of Physics, P.O. Box 64, FIN-00014 University of Helsinki, Finland

Abstract. The ALICE (A Large Ion Collider Experiment) experiment at the Large Hadron Collider (LHC) at CERN is dedicated to heavy ion physics to explore the structure of strongly interacting matter. The Time Projection Chamber (TPC) of ALICE is a tracking detector located in the central region of the experiment. It offers excellent tracking capabilities as well as particle identification. After the second long shutdown (LS2) the LHC will run at substantially higher luminosities. To be able to increase the data acquisition rate by a factor of 100, the ALICE TPC experiment has to replace the Multi-Wire Proportional Chamber (MWPC) –based readout chambers. The MWPC are operated with gating grid that limits the rate to O (kHz). The new ReadOut Chamber (ROC) design is based on Gas Electron Multiplier (GEM) technology operating in continuous mode. The current GEM productions scheme foresees the production of more than 800 GEM foils of different types. To fulfill the requirements on the performance of the GEM TPC readout, necessitates thorough Quality Assurance (QA) measures. The QA scheme, developed by the ALICE collaboration, will be presented in detail.

1 Introduction

ALICE at the LHC/CERN is dedicated to heavy ion physics, to explore the structure of strongly interacting matter.

The LHC will be upgraded during the next long shutdown LS2 to run with higher collision rates. The ALICE experiment that currently runs with a recording event rate of 500 Hz will be upgraded to cope with an event rate of 50 kHz. To accomplish this, the ReadOut Chambers (ROC) of the present TPC, based on Multi-Wire Proportioanl Chambers (MWPC) [1] will be replaced by new ROCs utilizing Gas Electron Multiplier technologies (GEM) [2, 3].

The MWPC come with a huge ion back flow (IBF) that creates space charge distortions in the drift area. To avoid this a gating grid was introduced to eliminate the IBF with the disadvantage of having a limited readout rate of O (kHz). GEM based readout chambers allow running in continuous mode while suppressing the IBF substantially. The remaining space charge distortions can be effectively corrected.

GEM detectors are a versatile and robust type of micropattern gaseous radiation detectors [2]. The active component is a thin (50 μm) polyimide foil coated with copper from both sides. Microscopic holes with typical sizes around 70 μm are etched chemically through the foil in a hexagonal pattern. When high voltage (HV) is applied over the copper electrodes, a high electrostatic field inside the holes, around 50 kV/cm, allows gas multiplication of electrons

[a] e-mail: erik.brucken@iki.fi

Each hole in a GEM foil acts as an independent multiplication channel, enabling high resolution 2d detection of charged particles. Fig. 1 shows a magnified section of a GEM foil including a simulated electron avalanche by using `garfield++` [4]. A drifting electron entering from the top and the electron avalanche are shown in white. The back drifting positive ions are shown in yellow.

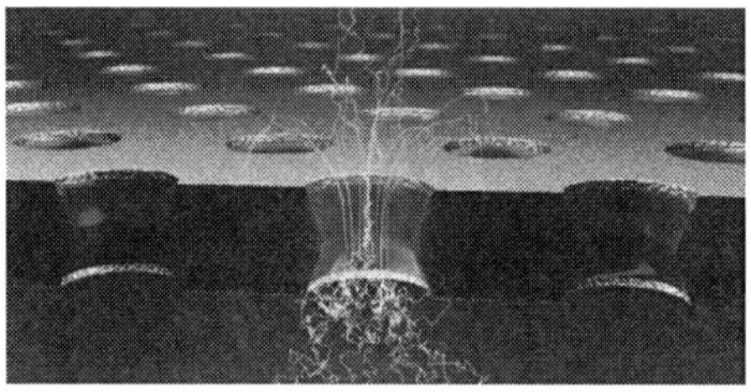

Figure 1. Garfield simulation of an electron avalanche in a GEM foil.

The ROC will be constructed with stack of 4 GEM foils that allow for continuous readout. The configuration can be tuned such to fulfill the design criteria, especially to maintain the present energy resolution and to have low ion feedback without the need for a gating grid. This could not be achieved with a stack of 3 GEM foils. More detailed information can be found in proceedings by Andreas Mathis [5].

To fulfill the requirement of excellent detector performance, low IBF and stable operation, thorough quality assurance (QA) measures are of utmost importance. In this paper we focus solely on the QA of individual GEM foils but not on the QA of the ROCs.

2 The ALICE TPC GEM foils

The GEM foils for the ALICE TPC are processed with single mask technique by the Micro-Pattern Gaseous Detector (MPGD) workshop at CERN. The active area of a single TPC ROC, that is divided in inner and outer chambers (IROC, OROC), is approximately 0.89 m^2. The area is composed of 4 different GEM foils, one for the IROC and 3 for the OROC, simply because a single GEM foil of this size can not be produced so far. The TPC is composed of 18 ROCs on each side. One single ROC consists of 4×4 GEM foils. In total we get 576 single GEM foils that need to be tested or approximately 128 m^2. Due to an expected production yield of 85% and the need for spare foils, an additional production of 25% - 50% is envisaged.

3 Quality assurance scheme

Motivation for the QA scheme follows from the experience of other experiments using GEM detectors for example COMPASS, TOTEM or KLOE with yields between 65-90%. The rejection of unsuitable foils at earliest possible stage and selection of best foils in terms of gain uniformity, stability and number of defects is important. A quality monitoring after each production step is necessary. The overall ROC production work flow, includes many production sites on two continents. Therefore defects have to be taken into account resulting from transport and foil handling at the different production places.
Several methods have been developed within the ALICE TPC upgrade collaboration to test the GEM foils in all relevant aspects. All information and data collected by the different QA methods will be fed into a central database. Each foil will have a full history available for the ROC assembly and for a later forensic autopsy in case of detector failure.
The QA is divided into *basic* QA and *advanced* QA. The basic QA will be done directly at the production site of the GEM foils and consists of

- Cleaning procedure under HV,
- Coarse optical inspection,
- Leakage current measurement.

The advanced QA that will only be done at dedicated QA centers, consists of

- Advanced HV tests,
- High resolution optical scanning,
- Gain uniformity test.

A *traffic light system* is introduced to classify the usability of GEMs. In case the foil did not pass the basic production QA a *red light* is given. A *yellow light* indicates that the foils passed the basic QA, i.e. no fatal defects, but did not show sufficient uniformity in gain, or shows non-optimal leakage currents. In case the foils passes all tests, a *green light* is awarded. A continuous quality monitoring has to guarantee that the quality is maintained after each production step.

HV cleaning procedure

All HV sectors of the GEM foils are immediately ramped up to 600 V with a high current limit in the order of several tenth of μA. Through the sparks dust as well as chemical remnants from the production process are burned and therefore removed. Foils with several sparks occurring at the same position should be discarded as it could be a sign of defects in the GEM foils.

Coarse optical inspection

Macroscopic defects of GEM foils should be spotted by eye and the help of a movable microscope. Such defects could be strong over-etching of holes, chemical remnants, deep scratches, etc. Foils holding such defects will be sent for re-cleaning or discarded.

Leakage current measurement and advanced HV tests

At production sites a leakage current test is foreseen. Each sector of the GEM foil will be ramped up in parallel to 500 V in humid air and leakage current will be measured using multi-channel picoammeters. For a good foil the leakage current should settle fast to under 500 pA. Only a small number of sparks, yet to be defined, will be allowed. At dedicated QA centers more advanced HV tests are planned, e.g. to measure the leakage current between neighboring HV sectors, to determine a quality factor and to perform long term HV tests. The HV test system will be introduced in more detail in Sec. 4.

High resolution optical scanning

Each foil will be optically scanned with an automated microscopic camera system mounted onto an xyz-table. The system is able to measure the size and shape of each individual hole in a GEM foil as well as to detect microscopic defects such as over-etched holes, covered holes, etc. Detailed maps and histograms will be generated of each GEM foil showing geometrical properties and distribution of those. The individual selection criteria still need to be defined. The optical scanning system will be discussed in Sec. 5.

Gain uniformity test

The gain uniformity is planned to be measured for at least one GEM foil of each type and for all GEM foils that exceed the limit of predicted gain uniformity derived from the high resolution scanning. The GEM foil will be mounted on top of a MWPC with a 2 dimensional readout. The local gain of the test foil will be measured by using the primary ionization above and below the foil using an ^{55}Fe *X*-ray source. Foils should show a gain uniformity within 10% RMS. The gain uniformity setup will be described in Sec. 6.

4 HV test system

The HV test system built at HIP consists of a 24 channel high precision HV power supply [7], a *Picologic* 24

Figure 2. HV test setup including HV power supply, picoammeters, HV box with gas flow and control pc.

Figure 4. High resolution optical scanning system working on an IROC GEM foil mounted inside a stretching frame.

channel picoammeter [8] and a Acrylic box with gas flow holding the GEM foil, as shown in Fig. 2. The test system is controlled via custom made automated software based on Labview [9]. Scripts for certain test procedures can be implemented with data written out to a data file. As an example the results of an IROC GEM foil measurement using this HV test system is shown in Fig. 3. Each of the 18 HV sectors was ramped up in parallel in steps of 100 V every 60 s till 500 V was reached. The voltage was kept constant for 30 min before ramping it down. The criteria

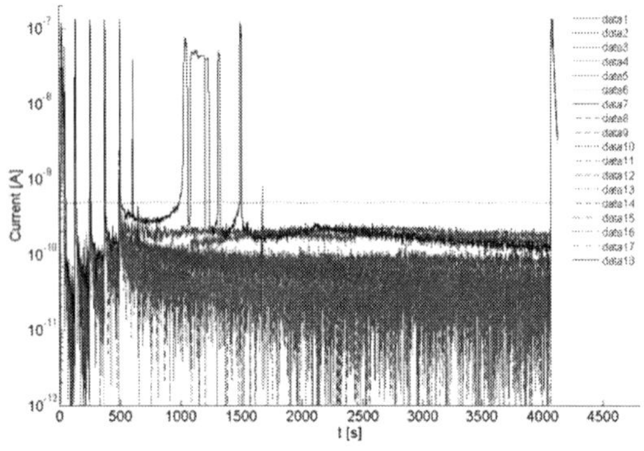

Figure 3. Example leakage current measurement of an IROC test GEM foil.

for the foil to pass this test is a stable leakage current below 500 pA for at least 30 min. A small number of random sparks right after reaching the voltage flattop are allowed. All sectors except except one passed the test. Channel 18 showed some non stable behavior with currents above 500 pA. The GEM foil was kept under constant nitrogen gas flow.

The HV test system makes it possible to test interconnection currents between neighboring sectors of the foil using a checker board type of HV configuration. Also long term stability tests under HV are foreseen, where the foil is kept at 500 V in nitrogen atmosphere for about 12 hours.

5 High resolution optical scanning system

The optical scanner consists of a computer controlled xyz-robot on a glass table with background illumination, a high resolution camera and telecentric optics with coaxial

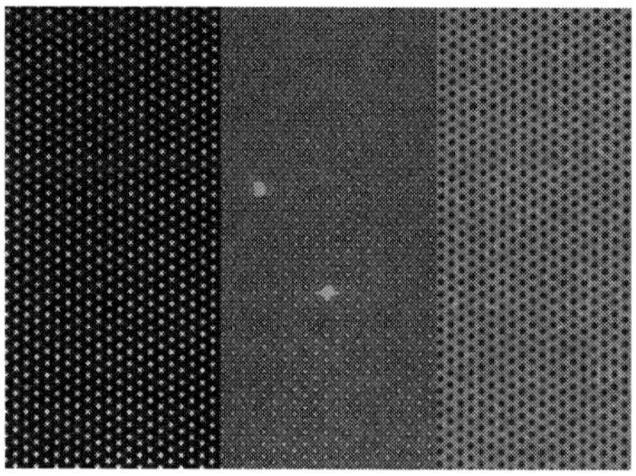

Figure 5. Example image taken by the optical scanner in two exposure mode. Both exposures are shown including an overlay.

lighting, shown in Fig. 4. In addition a LED ring light is mounted. The installed 4.9 megapixel monochrome camera has a $1/2''$ cmos sensor with pixelsizes of 2.2 μm. The resolution depends on the quality and magnification of the available telecentric objectives. The system is controlled via custom made software based on Labview.

The system takes images of the active area of GEM foils in either one or two exposure mode and transfers them to a network data-storage. The GEM foils are stretched before scanning to avoid image quality problems due to the telecentric optics. This also reduces the scanning time due to the reduction in refocusing of the camera. Presently the scanning time for an IROC takes about 14 hours with with a 1$\times$ magnification optics. An image example taken by the presented device is shown in Fig. 5. The left side of the image shows the exposure with background light only, the right side the exposure only with foreground light and the center the overlay of both exposures in different color channels. In addition one can see two clear over-etching defects.

The images are then analyzed offline using a custom made software described in detail in [6]. First the images are pre-processed mainly to enhance the contrast followed by

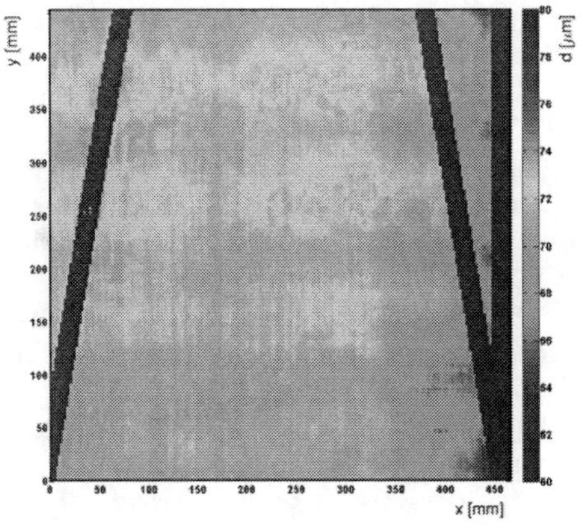

Figure 6. Map of the outer hole diameters of an IROC GEM foil.

segmentation using the edge detection Canny algorithm. The edge contours are then fitted by ellipses from which geometrical parameters are extracted. Classification methods based on neural networks are then used to determine whether the objects are inner or outer holes or defects.

The results are given in histograms and maps such as shown in Fig. 6. Here, the map of the outer hole diameters of an IROC GEM foils is shown. The diameters vary between 64 to 74 μm (not calibrated). The criteria for a foil to pass the optical scanning are based on the uniformity of the hole properties and the number and type of defects. Detailed criteria are yet to be defined as hole size distributions from single mask etching technique have so far not been studied in detail due to the new production method.

6 Gain uniformity test

The gain uniformity test system is based on a MWPC on top of a 2 dimensional readout with custom made electronics. A drawing that shows the construction idea of the system and the principle of the gain measurement is shown in Fig. 7. The test GEM foil is installed on top

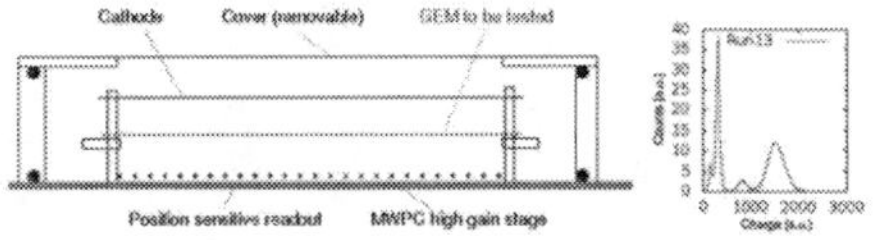

Figure 7. Sketch and principle of the gain uniformity test system.

of the MWPC and covered by a thin cathode. Using an ^{55}Fe X-ray source, that irradiates the full chamber, events with ionization points in the drift fields above and below the test GEM foil are recorded. The local relative gain is extracted as the ratio between both characteristic ^{55}Fe peaks (Fig. 7 right). It was shown that the local relative gain measurement is not affected by varying atmospheric conditions. The prototype test system (Fig. 8) is currently

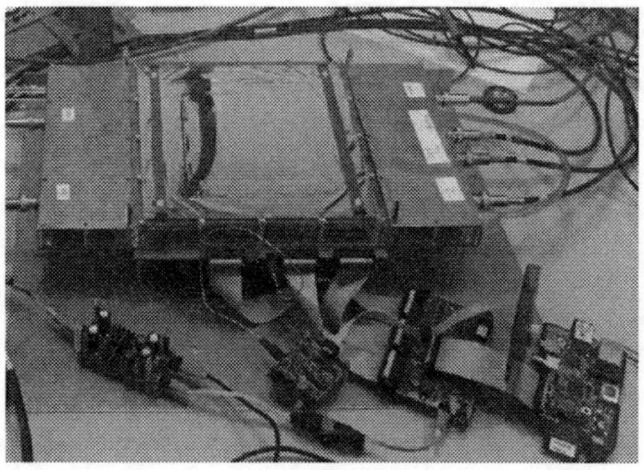

Figure 8. Gain uniformity test system.

replaced by a full size system that can measure the largest ROC GEM foils. The custom electronics with a 12 bit ADC has enough dynamical bandwidth to measure also higher relative gains up to 30.

7 Combination of tools

By comparing the results of the optical scanning, i.e. the individual geometrical properties of the holes, with the local gain measurements of the gain uniformity test system, a distinct correlation was found.

A first successful result was shown in [6]. This study has been successfully continued within the ALICE TPC upgrade collaboration. Preliminary results are shown in [10], where neural networks have been used to predict the gain of a GEM foil based entirely on geometrical foil and electric field properties. The training was done using measurements of several individual GEM foils. The GEM foil used for the prediction was not used in the training. We can conclude that the gain measurement and the prediction from optical scan data are in good agreement. However the method has so far only be used for foils produced with double mask technique and with limited statistic. Therefore it is planned to apply this method on a large data sample of GEM foils produced in single mask technique.

It is foreseen to use this knowledge to predict the gain uniformity to some extent using the geometrical properties only as it might not be possible to measure the absolute gain of each individual foil.

8 Present activity and conclusions

Currently a GEM sample of 15 IROC foils is used to evaluate the QA methodology and to fine tune the QA protocol. The results will be used to fine tune the quality selection criteria such as a suitable leakage current per sector, hole size distribution, etc.

The production phase is anticipated during the first quarter of 2016. The collaboration has built an extensive QA program for GEM foils. The infrastructure and method will be ready for the start of the production. Each individual GEM foil will have a full history record that is put into a database. The goal is to establish a general QA scheme that also solves many open questions on GEM QA R&D.

9 Acknowledgements

We thank Dezso Varga from the Wigner RCP, Budapest, for providing us information about the gain mapping device, the ALICE team of TUM for providing the HV box and the non ALICE personnel of the Helsinki Institute of Physics laboratory that are involved in the infrastructure buildup.

References

[1] G. Charpak, R. Bouclier, T. Bressani, J. Favier, C. Zupancic, *The use of multiwire proportional counters to select and localize charged particles*, Nucl. Instr. Meth. **62** (1968) 262.

[2] F. Sauli, *GEM: A new concept for electron amplification in gas detectors*, Nucl. Instr. Meth. A **386**, (1997) 531.

[3] The ALICE Collaboration, *Technical Design Report for the Upgrade of the ALICE Time Projection Chamber*, CERN-LHCC-2013-020 (2014).

[4] H. Schindler and R. Veenhof, *Garfield++ - simulation of tracking detectors*, http://cern.ch/garfieldpp.

[5] A. Mathis on behalf of the ALICE collaboration, *Study of the dE/dx resolution of a GEM Readout Chamber prototype for the upgrade of the ALICE TPC*, EPJ Web Conf. **174** (2018) 01002.

[6] T. Hildén, E. Brücken, J. Heino, M. Kalliokoski, A. Karadzhinova, R. Lauhakangas, E. Tuominen and R. Turpeinen, *Optical quality assurance of GEM foils*, Nucl. Instrum. Meth. A **770** (2014) 113.

[7] iseg, *EHS | High Precision Versatile HV Module in Multiple Floating Versions*, http://iseg-hv.de/en/products/.

[8] A. Utrobicic, M. Kovacic, F. Erhardt, N. Poljak and M. Planinic, *A floating multi-channel picoammeter for micropattern gaseous detector current monitoring*, Nucl. Instrum. Meth. A **801** (2015) 21.

[9] National intruments, *Laboratory Virtual Instrument Engineering Workbench*, http://www.ni.com/labview/.

[10] T. Hildén and E. Brücken, *New results on hole-size dependence of GEM-foil performance*, MPGD 2015, https://agenda.infn.it/contributionDisplay.py?contribId=33&confId=8839 (April 2016).

Characterization of the water diffusion in GEM foil material

L.Benussi[1], S.Bianco[1], G.Saviano[1,2], S. Muhammad[1,2,3,a], D.Piccolo[1], G.Raffone[1], M.Caponero[1], L. Passamonti[1], D. Pierluigi[1], A. Russo[1], F. Primavera[1], S.Cerbelli[2], A. Lalli[2], M. Valente[2], M. Ferrini[2], B. Teissandier[4], M. Taborelli[4], M. Parvis[5], S. Grassini[5], J. Tirillò[2], F. Sarasini[2], and A.V. Franchi[2]

[1] *Laboratori Nazionali di Frascati - INFN, Frascati, Italy*
[2] *University of Rome "La Sapienza" (IT) - Facoltà di Ingegneria, Ingegneria Chimica Materiali ed Ambiente, Italy*
[3] *National Center for Physics, Quaid-i-Azam University Campus, Islamabad, Pakistan*
[4] *CERN- Vacuum, Surfaces and Coatings (VSC) TE Department, Switzerland*
[5] *Politecnico di Torino-Dipartimento di Fisica (DIFIS) Corso Duca degli Abruzzi, 24, I-10129 Torino, Italy*

Abstract.
Systematic studies on the GEM foil material are performed to measure the moisture diffusion rate and saturation level. These studies are important because the presence of this compound inside the detector's foil can possibly change its mechanical and electrical properties, and in such a way, the detector performance can be affected. To understand this phenomenon, a model is developed with COMSOL Multiphysics v. 4.3 [1], which described the adsorption and diffusion within the geometry of GEM foil, the concentration profiles and the time required to saturate the foil. The COMSOL model is verified by experimental observations on a GEM foil sample. This note will describe the model and its experimental verification results.

1 Introduction

To upgrade the forward muon region of the Compact Muons Solenoid (CMS) at Large Hadron Collider (LHC), the Gaseous Electron Multiplier (GEM) is considered a most suitable detector. Due to its high rate capability 1 MHz/cm^2 [2], excellent position and timing resolution, the GEM detector can easily meet the requirements of the high pseudorapidity region where a high particles flux is expected. The major part of the detector is the GEM foil where charge amplification phenomena occur in the presence of intense electric filed in the conical shape micro holes within a thickness of 60 μm (50 μm kapton cladded with 10 μm copper on both sides) The installation of the triple-GEM detectors into muon stations at the end-cap region, as shown in figure 1, will improve muon momentum resolution, help to reduce the global muon trigger rate, assure a high muon reconstruction efficiency and increase offline muon identification coverage [3]. The GEM foil is exposed to water during production process, and possibly leaks in the detector allowing atmospheric humidity to enter the gas volume. This study aims to measure the diffusion properties of GEM foils in order to obtain useful data to characterize the GEM foils mechanical tensile properties which can be affected by humidity.

[a]Corresponding author; email address: Saleh.Muhammad@cern.ch

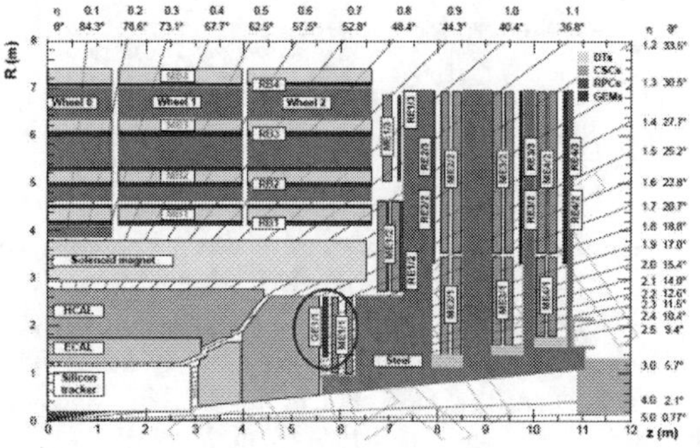

Figure 1. The GEM installation position at the CMS is highlighted in blue circle.

2 Development of a diffusion model using COMSOL Multiphysics

In the analysis of transport phenomena we considered the spread of H_2O only from the holes surface where kapton is exposed to the environment. In fact water cannot penetrate into the polymer from upper and lower surfaces due to copper layers which behave as a waterproof barrier. The presence of any stagnant film around the surface of the hole is considered as negligible because the gas is continuously flushed inside the chamber, allowing to always have a constant and uniform concentration on the surface of the hole: this value represents the maximum concentration of

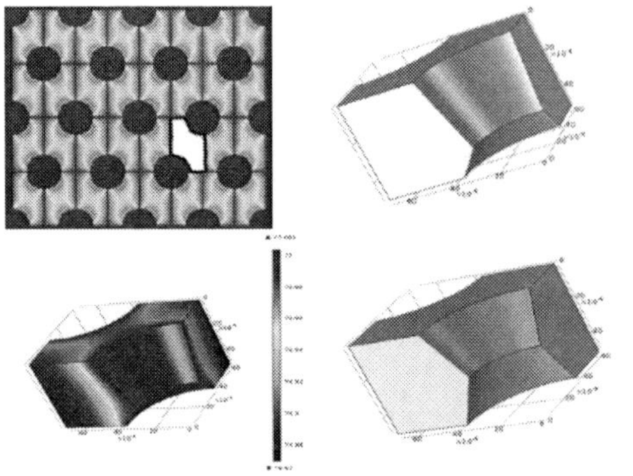

Figure 2. (top left)[4]. The selected area highlighted white (bottom left). The humidity absorption (top right). single mask (bottom right) double mask

water that can get inside the structure, in other words the value in full saturation conditions (after a theoretically infinite time). In general the humidity that comes from the surrounding environment and spreads within the polymer, has condensation on the surface of solid and gas exchange (i.e. on the inner surface of the holes in the sheet). The relation that allows one to obtain the value of this concentration on the surface of a polyimide (kapton) is a function of the external humidity which is derived from the literature [5].

$$c1 = 4.520 * 10^{-4} * RH - 8.319 * 10^{-4} \qquad (1)$$

where c1 is expressed in grams of water to grams of polymer, RH is the relative humidity. The formula used is empirical and it is compatible with the official data released by the manufacturer [6], in case of RH = 100 % the maximum amount of water present in the polymer is equal to a 4.43 %, approximately consistent with the 3 % declared by DuPont for Kapton. The above formula is not applicable for small values of relative humidity; considering, however, that the environment in which the detector will work does not provide RH > 4 %, it is possible to apply the previous formula while limiting to a 10 % error. The value of humidity chosen describes the unrealistic situation of 10 % of leaks inside the gas system of a GEM detector.

The model developed takes into account the actual GEM foil geometry, including size and shape of the holes, for both the single and double mask manufactured process. Based on the problem symmetries, we selected the domain shown in figure 2, as representative periodic unit. This domain was discretized using triangular meshing.

The humidity exchange surfaces are colored in blue (figure 2), these being the only surfaces that allow the water to penetrate inside the foil structure. For all other faces colored in grey (figure 2) it was set up the condition of no flow, this condition is certainly valid for the upper and lower sides (where there is a copper coating).

The diffusion coefficient is taken from the previous laboratory tests, performed at CERN [7], $D = (2.11 \pm 0.18)10^{-10}$ cm^2/s. The results obtained from the two geometries are very similar, the difference of the geometry is basically due to holes formations techniques which are called single and double masks techniques.

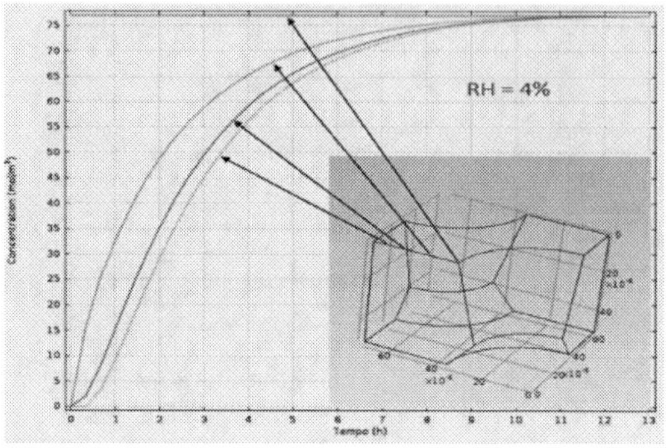

Figure 3. Time vs. concentration for different points of the geometry, with RH = 4%

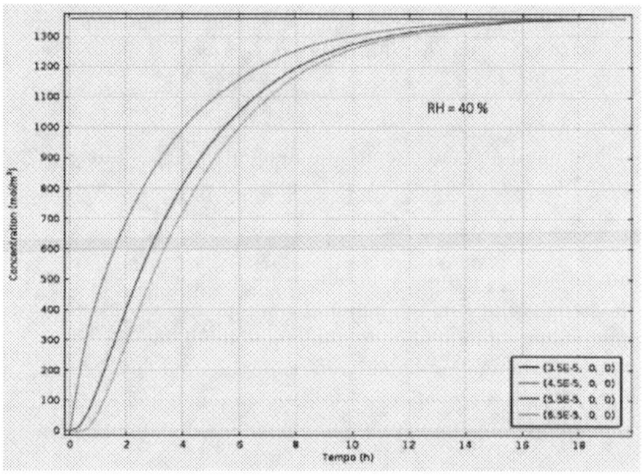

Figure 4. Time vs. concentration for the same points with RH = 40%

The simulation results show that the saturation time is about 10 hours at RH = 4 %, 14 hours at RH = 40 % and 100 % as demonstrated in figures 3, 4, 5, starting from a null value of water inside. The concentrations limits are different 75 mol/m^3, 1380 mol/m^3 and 3450 mol/m^3 at different relative humidity values 4 %, 40 % and 100 % respectively. Along with RH value the saturation time also depends upon the geometry (size and shape) of holes and the value of the diffusion coefficient.

3 Experimental verification of the diffusion model of water inside the GEM foil

The diffusion model was verified to determine if it describes realistically the transport of water within the GEM structure. To achieve this goal, a setup was built at the Frascati Laboratories of INFN to verify and improve the preliminary measurements [7] performed at CERN, partic-

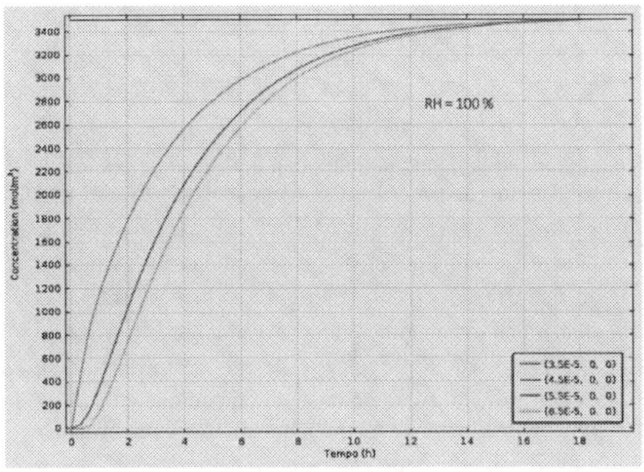

Figure 5. Time vs. concentration for different points of the geometry, with RH = 100%

Figure 6. Data acquisition system (red) and instrumentation for monitoring environmental conditions parameters within the glass container (blue), weighing balance (yellow).

ularly with respect to the weighing mode and the handling of systematic errors.

3.1 Experimental setup

The experimental setup, shown in figure 6, is composed of an analytical balance where the sample is supported by a wire. The analytical balance used for the measurements is a model Gibertini E42 S.The weight measurements are recorded by using a camera in front of the analytical balance display and connected to a computer figure 6 using a custom data acquisition program.

The environment conditions in which the GEM foil is exposed to water vapour are obtained by using the same type of vessel as used in the test held at the CERN laboratories [7]; A mixture of salts and water is used to get almost uniform relative humidity during the test as shown in figure 7. The mixture was composed of 170 ml of water and potassium carbonate.

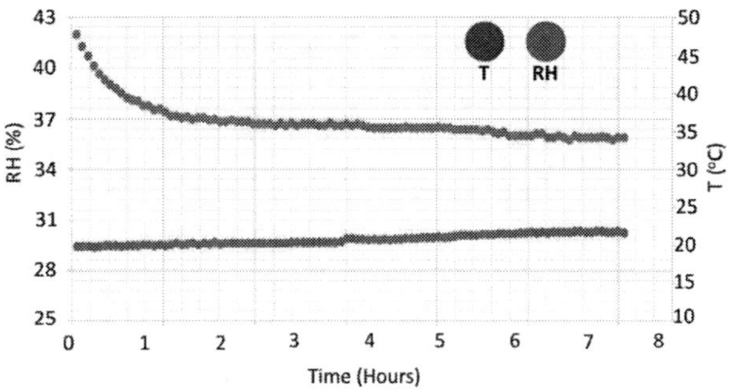

Figure 7. Temperature and relative humidity vs. time within the glass vessel during the test.

Sensors are located inside the glass vessel in order to record temperature and humidity with automatic data acquisition system. The room temperature and humidity where controlled and stabilized. The use of salts provided precise humidity inside the container thus helping reduce the systematic uncertainties during the measurements. Oscillations where reduced by setting an additional weight on the balance plate. The precision reached a tenth of milligram.

3.2 Test procedure

A 105 mm * 50 mm * 60 μm sample of GEM foil is prepared and conditioned in an oven at 105.0 ± 0.5 °C for 36 hours. After the conditioning in oven, the sample is weighed and it is found a value of (0.6575 ± 0.0002) g.

The humidity and temperature range during the test remained RH = 37-38 % and T = 19-20 °C, shown in figure 7, the images are taken with every 5 minutes interval. The trends of the sample saturation are shown in figure 8. In 8-9 hours it is possible to completely saturate the GEM foil. The experimental results (figure 8) are comparable with the results obtained from the COMSOL model as described in figures 3, 4, 5.

Furthermore, the measurements showed the presence of 0.004 g water at saturation level, compatible with what is obtained from the official data sheet by the manufacturer [6] or from the formula (1) for a value of RH = 37.5 %.

We calculated the amount of water absorbed within the foil. For these calculations used the experimental sample dimensions and conditions. The overall volume of the kapton in the sample is $\acute{V}_{Kapton}$=0.2625 cm^3, and V_{Kapton} is the volume of kapton multiplying with factor of holes which is $\alpha = 0.30$, it is the fraction of free space (due to the presence of the holes), ρ_{Kapton}=1.42 g/cm^3 is the density of kapton, P_{Kapton} is the weights of kapton, and PM_{H_2O} is the molecular weight of water.

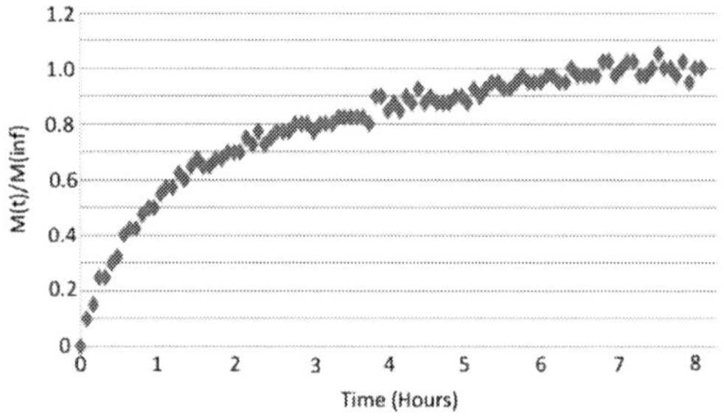

Figure 8. Water saturation trends in the GEM foil sample.

$$V_{Kapton} = \acute{V}_{Kapton} * (1 - \alpha) = 0.1835 \text{ cm}^3 \quad (2)$$

$$P_{Kapton} = V_{Kapton} * \rho_{Kapton} = 0.2609 \text{ g} \quad (3)$$

$$P_{H_2O} = \left(\frac{c1}{PM_{H_2O}}\right) * P_{Kapton} = 0.0042 \text{ g} \quad (4)$$

The calculated number is very close to the experimental results difference is only 4.8 %.

4 Conclusions

A model for describing the diffusion phenomenon within the GEM foil was developed. It is assumed that all the water coming from the humidity condenses on the surface of solid-gas exchange, i.e., the internal surface of the holes where the electronic avalanche will produce. This surface is the only available for the penetration of the water inside the foil, because the copper coating (on the top and bottom) behaves as a water proof barrier. It is also assumed that near the surface of exchange there is not a stagnant film, hypothesis certainly validated by the presence of a gas flow in continuous circulation. Formula (1) is used to estimate the value of the concentration on the surface starting from a value of surrounding humidity; this value is assumed equal to 4% representing a hypothetical situation of 10 % of leaks inside the gas system.

The results obtained from the COMSOL MULTIPHYSICS model showed that the time for having the complete saturation of a single mask GEM foil is about 10 to 14 hours depends upon humidity level.

In order to validate the COMSOL model, an experimental setup is made ready to perform a test at the GEM foil sample. From weigh balance the measurements of the weight are recorded in a well-conditioned system, which is built for the occasion: using a mixture of potassium carbonate and saturated water to get a stable value of RH in the range of 37-38 % inside the container. The sample is hung to a wire, connected to the analytical balance. After approximately 8 to 9 hours the saturation level is achieved and the amount of water absorbed is 4 mg, this value is comparable with the official data sheet released by DuPont.

More studies are going on about the properties changes of the GEM foil due to the water absorption and desorption from the system, in order to establish the standard conditions to obtain the longevity and the best performance of the GEM detector.

References

[1] COMSOL Multiphysics ®; Vers. 4.3, ©1998-2013 by COMSOL, Inc.

[2] A. Colaleo et al., CMS TECHNICAL DESIGN REPORT FOR THE MUON ENDCAP GEM UPGRADE, CERN-LHCC-2015- 012. CMS-TDR-013,<https://cds.cern.ch/record/2021453>.

[3] D. Abbaneo et al 2015 JINST 10 C03039, "Quality control and beam test of GEM Detectors for future upgrades of the CMS muon high rate region at the LHC" <doi:10.1088/1748-0221/10/03/C03039>.

[4] G.Raffone "CMS Trapezoidal GEM Foils Structural Analysis" preprint LNF-10/20(IR).

[5] Sacher E., Susko; "Water Permeation of Polymers Films. I. Polymide"; Material and Engineering Analysis, IBM Corporation System Division Endicott, New York; 1978.

[6] Kapton® HN Technical Info; DuPont, <www.dupont.com/products-and-services/membranes-films/polyimide-films/brands/kapton-polyimide-film.htm>.

[7] A.V.Franchi et al.; "Test for the Measurement od Diffusion Coefficient of Water in Kapton Foils for the GEM Detector of the Upgraded High-Pseudorapidity Muon Detection in CMS" preprint INFN-13-09/LNF.

Investigation of the microstructure of Thick-GEMs with single photo-electrons

Gergő Hamar[1,2,a], Silvia Dalla Torre[2], Shuddha Shankar Dasgupta[2], Stefano Levorato[2], Fulvio Tessarotto[2], and Dezső Varga[1]

[1] *Wigner RCP, Budapest*
[2] *INFN Trieste*

Abstract. Novel Cherenkov detector upgrades favour GEM and ThickGEM based MPGD systems. These detectors have reduced ion backflow, fast signal formation, high gain, and could suppress the MIP signals as well. Their common drawbacks are the inefficiencies of photo-electron collection from the top of the ThickGEM and the local variation of multiplication due to the special geometry.
The developed high resolution scanner using focused UV light gave the possibility to study single photo-electron response of MPGDs in the sub-millimeter scale. Revealing the microstructure of photo-efficiency and local gain provides a new tool to quantitatively compare different ThickGEM geometries and field-configurations, and thus optimize the detector parameters.
The presentation will focus on the key elements of the scanning system; and on the microstructure evolution of different ThickGEM configurations.

1 Introduction

The novel technology of micropattern gaseous detectors (MPGD) opened new promising ways for developing tracking, triggering, and Cherenkov detectors as well [1]. Their complex geometry and the presence of the insulating material makes the optimization of parameters challenging, while the detailed investigation is still an active part of recent R&D.

MPGD-based single photon detection for Cherenkov applications mostly use the first amplification layer as a refractive photocathode, thus using hole-type amplifiers. While in PHENIX HBD [2] GEM [1] foil was used, new RICH upgrades tend to apply ThickGEM [3] technology. The advantages compared to wire chambers have been proven for several aspects [4] by several groups. While the former upgrade plans of ALICE VHMPID [5] considered ThickGEMs, for COMPASS [6] the RICH-1 upgrade [7] will be based on ThickGEM and Micromegas [8] coupled together.

In case of single photon detection, the choice of geometrical parameters as well as electric field configuration are crucial for the high efficiency of the RICH PID. Measuring the behaviour of the detector response in the scale of its micropattern-structure can reveal the inner processes, and could give one a novel way to design the micro-geometry of the detector-elements, and additionally could help in fine-tuning the simulations.

In the Wigner RCP Budapest we have developed a high resolution scanning system [9] to examine the response of MPGDs on single electrons, with the spatial resolution of 30-300um. The system was designed to investigate single ThickGEM layers, and get information on photo-electron detection efficiency and avalanche size, and thus deduce hole-by-hole variations. We have shown that with increased resolution the method can be applied for thin GEM foils as well [10].

Based on the results of the first prototype the idea got supported by the RD51 community and became part of AIDA-2020. The first practical use of the technology was initiated by a common measurement of the Budapest and Trieste groups, with the main aim to compare several different ThickGEM types, that could be used for the mentioned COMPASS RICH Upgrade project.

2 The Leopard System

The high resolution scanner consists of a pulsed UV source, which is focused to a tiny spot (30-300um) onto the surface of the ThickGEM in study. The source is attached to a 3D moving table, to be able to scan the surface and set the proper focal distance. The intensity of the UV light is set to have <<1 photo-electrons (PE) from the surface in average, this way real single electron signals can be measured. To keep the ThickGEM in standard operating condition a postamplification stage is used, this part shall not have its geometrical micro-scale on the same level as the one in study; simple solution can be a wire chamber, or a micromegas. The heart of the system is the combined data acquisition, that takes care of the signal measurement, the moving of the actuator table, setting the

[a] e-mail: hamar.gergo@wigner.mta.hu

high voltage, saving the data, and responsible for the communication with the user. Recent version of the DAQ consists of a custom designed board and a RaspberryPi [11] micro-computer (for details see [10]).

High resolution investigation needs the location of the measurement points about an order of magnitude closer to each other than the pitch of the structure, while reasonable statistics is needed for the single electron spectra. To fulfill these requirements billions of events have to be recorded, making the speed of the DAQ a crucial issue. This upgraded version of the system is working with 130kHz of data taking, while being still a tabletop setup (see photo of the setup on figure 1).

Using single PE signals the photo-electron yield and the gain can be decoupled, and can be calculated for all points. The yield image looks like the fur of the leopard, so it brought the nickname "Leopard" for the whole system.

3 Budapest-Trieste Setup

Combining the expertise on ThickGEMs from the INFN Trieste group, and the capabilities of the Leopard setup from the Wigner RCP Budapest group, a common measurement has been initiated.

The aim was to make comparative studies on ThickGEMs from different production technologies, and via this improve the COMPASS RICH-1 Upgrade. A photo of the whole setup, and a short description of the main parts can be seen on figure 1.

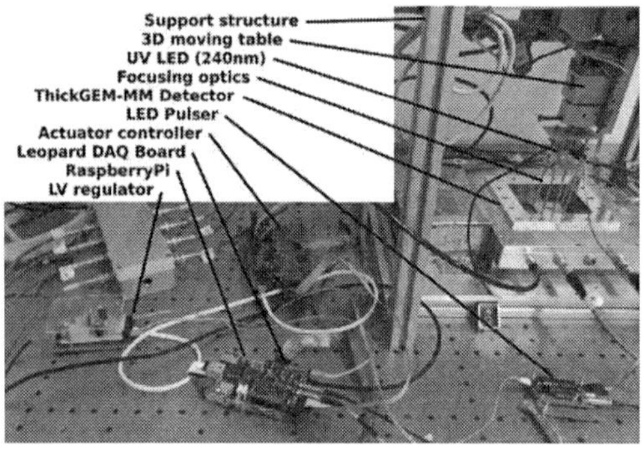

Figure 1. Photo of the Leopard setup in INFN Trieste in 2014. The main components are named and marked on the picture itself.

In case of these measurements the postamplification stage after the ThickGEM was a bulk micromegas (CERN, mesh 45/18, 128um gap), similar as will be used for the upgrade. The list of the ThickGEMs under investigation and their geometrical parameters can be seen in table-1. The chamber was closed with a cathode wireplane and a quartz window.

4 First results

Adjusting the focal distance is necessary to achieve the high resolution information. While rough settings can be done via calculation, the finetuning of the focal distance shall be based on the recorded data itself. In case of ThickGEMs the blur of the image with a 50μm spot size (in focus) is small but not negligible over a 1 mm range, see figure 2.

Table 1. List of the tested ThickGEMs with their main production dimensions.

ThGEM Name	Hole [μm]	Pitch [μm]	Thickness [μm]	Rim [μm]
M1-III	400	800	400	0
DESTRO-I	400	800	400	5
C3HR-II	400	800	400	50
M2.4-G	400	800	600	0
M2.1-II	300	800	400	0

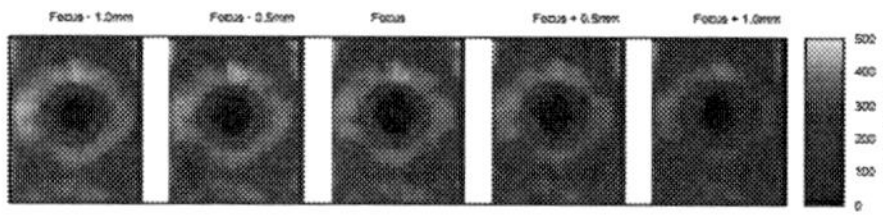

Figure 2. Map of the photo-electron yield with various focal distances. The "sharpest" image corresponds to the best achievable resolution.

One can calculate the photo-electron yield for all points, thus get a map of photo-efficiency; and likewise get map of gain. Within the recorded data the holes can be identified by searching for hole-sized low-yield clusters. Each hole has a well-defined collection area determined by the pattern structure, in our case a hexagon. Gain variation with respect to the entering position of the electron into the hole, can be translated to the emission point of the photo-electron, thus is known in the used Leopard system. Averaging over a large sample of holes we found that the distribution is constant within errorbars for 0.4mm holes, see figure 3 (measured on the ThickGEM : DESTRO-I, see Tab. 1). This is in agreement with the low statistics measurements of the former studies on 0.3mm holes (with large rim) in [9].

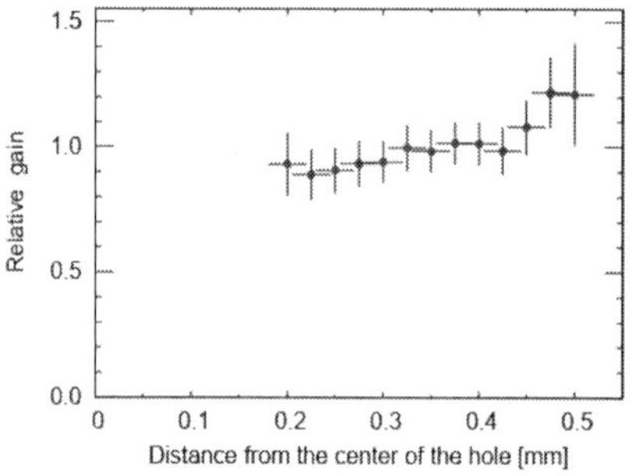

Figure 3. The size of the electron-avalanche as function of the emission point of the photo-electron is consistent with a constant function within the achieved uncertainties.

Based on this information the concept of the hole-gain can be defined. The large number of measured holes in each tested ThickGEMs will make us capable to evaluate the distribution of the hole-gains, and thus compare the different production technologies based on their uniformity in the micro-level.

The displeasing phenomenon of charge-up has non-negligible effect on the measurements: as the scan goes along the line of the tiny points the corresponding holes are charging up slowly, and this can result a slight decrease of gain in each hole towards the direction of the movement. The effect can be overcome via charging up the whole surface beforehand, or during the measurement, or with offline corrections aided by prestudies on the effect.

5 Behaviour of the critical point

An advantage of MPGD-based RICH detectors that with applying a reversed drift field the MIP signals can be suppressed, thus higher gains can be achieved within the range of stability. On the other hand, photo-electrons emitted from the critical/symmetry points and symmery lines are highly sensitive for this parameter [9]. Besides the fact that neither too high normal or reversed fields are good, we have shown that with a small normal field the above effect can be eliminated.

To perform fine scans in voltage within a large range, the time consuming 2D-mapping-measurements has been replaced with single point measurements in the critical/symmetry point; and single-line-measurements connecting hole centers and symmetry points, let us call it the critical-line. On the critical-line-scans one can observe the change of photo-electron yield evolution with various drift fields in the critical points, hole-perimeters, and intermediate regions as well. Sample of the critical-line scan can be seen on figure 4.

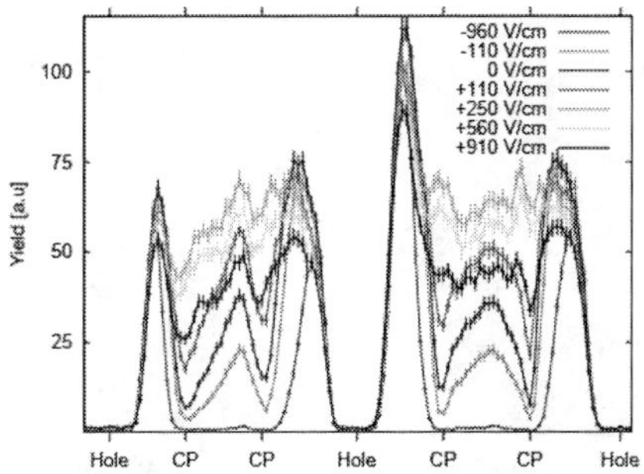

Figure 4. Photo-electron yield along the critical-line scans show the change of the function for several of the applied drift bias. The influence is most crucial at the critical point.

We have shown that PE yield of the critical/symmetry point is extremely sensitive to the applied drift field. It has a well defined maximum at the several hundreds of V/cm normal field, below which the function starts to fall rapidly in a close-to exponential way. For higher fields than the optimum of a small plateau is followed by a slight decrease. This behaviour has been checked for several critical/symmetry points, while the describing parameters seemed to be similar for a given piece of ThickGEM, as shown in figure 5. The plateau for the ThickGEM with 0.3mm holes was significantly smaller.

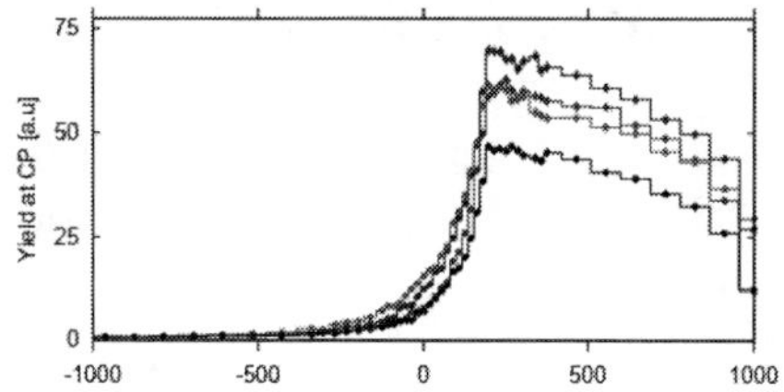

Figure 5. Photo-electron yield of the critical point highly depends on the applied drift field above the ThickGEM (which is responsible for the photoconversion itself).

6 Summary

The Leopard system is designed to investigate the microstructure of MPGD detectors, using single photo-electron response. The combined fast DAQ was developed as an RD51 common project, and was used to check thin GEM foils, and for detailed study of different ThickGEMs. Usability for precise measurements, quality assurance, optimization, and production technology evaluation has been proven. The system could provide even detailed answer on how to set the drift field in case photon-detection is an issue. The evaluation of the performed comparative measurements of the Trieste ThickGEMs are still ongoing, while preliminary results are even now proving the usability of the methods and the system.

This work has been supported by the Lendület (Momentum) grant of the HAS gn. LP2013-60, by INFN, RD51, and the H2020 project AIDA-2020, GA no. 654138.

References

[1] F.Sauli, Nucl.Inst.Meth. A **477** 1 (2002)
[2] Z.Fraenkel et al., Nucl.Inst.Meth. A **546** 466 (2005)
[3] A.Breskin et al., Nucl.Inst.Meth. A **598**, 107 (2009)
[4] V.Peskov, M.Cortesi, R.Chechik, A.Breskin, JINST **5**, P11004 (2010)
[5] A.Di Mauro et al., Nucl.Inst.Meth. A **639**, 274 (2011)
[6] P. Abbon et al., Nucl.Inst.Meth. A **577**, 455 (2007)
[7] M.Alexeev et al., Nucl.Inst.Meth. A **617**, 396 (2010)
[8] I.Giomataris, P.Rebourgard, J.P.Robert, G.Charpak, Nucl.Inst.Meth. A **376**, 29 (1996)
[9] G.Hamar, D.Varga, Nucl.Inst.Meth. A **694**, 16 (2012)
[10] G.Hamar, D.Varga, Proc.o.Science **TIPP2014**, 056 (2014)
[11] RaspberryPi, *http://www.raspberrypi.org/*

Development of μ-PIC with resistive electrodes using sputtered carbon

Fumiya Yamane[1,a], Atsuhiko Ochi[1], Yasuhiro Homma[1], Satoru Yamauchi[1], Noriko Nagasaka[1], Hiroaki Hasegawa[1], Tatsuo Kawamoto[2], Yosuke Kataoka[2], and Tatsuya Masubuchi[2]

[1] *Kobe University*
[2] *Tokyo ICEPP*

Abstract. The Micro Pixel Chamber (μ-PIC) has been developed for a hadron-collider experiment. The main purpose is detecting Minimum Ionizing Particles (MIP) under high-rate Highly Ionizing Particles (HIP) environment. In such an environment, sufficient gain to detect MIP is needed, but continuous sparks will be caused by high-rate HIP. To reduce sparks, cathodes are made of resistive material. In this report, sputtered carbon was used as a new resistive cathode. Gas gain $>10^4$ was achieved using an ^{55}Fe source. This value is sufficient to detect MIP without GEM or other floating structures. Also, thanks to production improvement, pixels are well aligned in the entire detection area.

1 Introduction

The Micro Pixel Chamber (μ-PIC) [1] [2] is a 2-D gaseous imaging detector produced with PCB/FPC technology. μ-PIC has been developed for many applications, such as the Electron-Tracking Compton Camera (ETCC) [3], directional dark matter search (NEWAGE experiment) [4], neutron imaging [5], and space dosimeter (PS-TEPC) [6]. Our goal includes a particle-tracking detector in a hadron-collider experiment.

One of our development targets is the new ATLAS forward muon detector called the Muon Tagger, which is being considered for installation between the end-cap calorimeter and the JD (The Disk Shielding) near the beamline ($2.7 < |\eta| < 4.0$) in the long shutdown 3 of the LHC (year 2023) [7]. There is a very small area in the ATLAS high-η region to measure the muon track, and there is large background radiation scattered near the beamline such as fast neutron and gamma ray. Requirements for the Muon Tagger are as follows.

- Granularity of a few hundred μm for multiple track separation.
- Stable operation in high-rate radiation background (>100 kHz/cm^2).
- Multiple layers installed within a thickness of 5 cm.

Taking into account the thickness requirement, MPGDs, or silicon detectors, are considered. In case MPGDs are used, high-rate hadronic particles, mainly gas nuclei recoiled by fast neutrons, will cause continuous sparks [8]. Those particles are called Highly Ionizing Particles (HIP). Using high-resistive electrodes is the best way to reduce the sparks. In the case of μ-PIC, strong spark reduction was shown by using resistive cathodes [9]. Sparks are suppressed by voltage drop on the resistive electrode. If the resistivity is high, sparks are strongly reduced, but the voltage drop becomes large. If HIP are incident at too high a rate, the voltage supply cannot supply enough current, a voltage drop is generated and the electric field is reduced. Thus, resistivity should be adjusted to the radiation environment precisely. Also, with respect to the granularity, resistive electrodes should be patterned finely on the scale of a few μm. For those requirements, sputtered carbon [10] is a very promising material. In this paper, we report the detector design and gas gain measurement of μ-PIC with resistive cathodes using sputtered carbon.

[a] e-mail: fumiyamane@stu.kobe-u.ac.jp

2 Detector concept

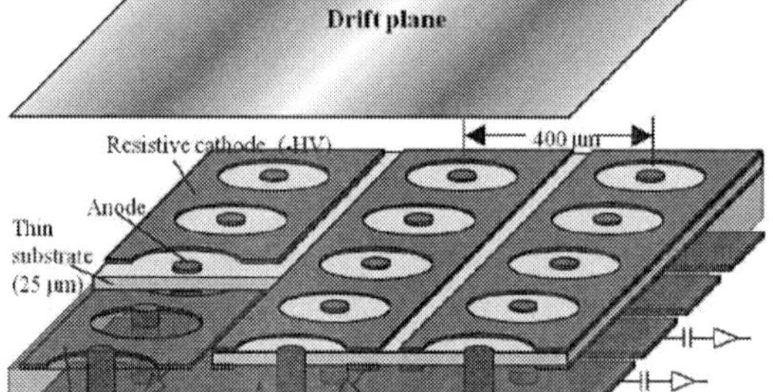

Figure 1. Schematic view of μ-PIC with resistive cathodes.

Figure 1 shows a schematic view of the μ-PIC with resistive cathodes. Isolated pixels are arranged by 400 μm pitch with holes 250 μm in diameter on the substrate. Each pixel has an anode pin, which is connected to the readout strip through the thick substrate, and a surrounding cathode. Cathode electrodes are made of resistive material and pickup electrodes are placed under cathode electrodes. Charges are induced to pickup electrodes from resistive cathodes. Because this capacitive readout does not require any capacitors for AC coupling, detector construction is simplified. In our previous research [9], the μ-PIC, in which cathodes were made by carbon-loaded polyimide as a resistive material, showed strong spark reduction in the fast neutron irradiation test. The spark counting rate was $10^{3\sim5}$ times lower than normal μ-PIC with gas multiplication around 10^4. This means that resistive μ-PIC has high potential for both MIP detection and HIP tolerance.

More granularity and precise control of resistivity are needed for usage as a Muon Tagger; however, carbon-loaded polyimide is not so uniform, a feature that makes difficult the fine patterning and the precise control of the resistivity. As mentioned, sputtered carbon is a very promising material. It has been developed for the resistive strips of ATLAS Micromegas, and its properties have been studied well [10]. Fine patterning is available (<50μm) using a liftoff process. By varying thickness (a few hundred to a few thousand Å) and nitrogen doping, resistivity can be controlled between 50 kΩ/sq. to 4 GΩ/sq. Figure 2 shows 3-D images of a pixel of the μ-PIC with resistive cathodes made of carbon-loaded polyimide and sputtered carbon, respectively.

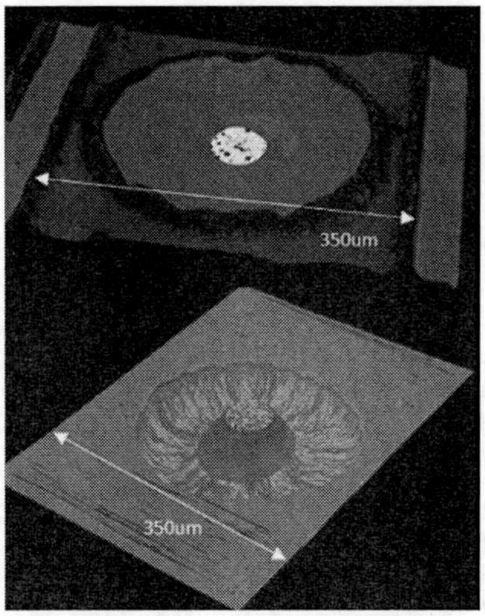

Figure 2. 3-D images of pixels of resistive μ-PIC. Top: carbon-loaded polyimide. Bottom: Sputtered carbon. Compared to carbon-loaded polyimide, great uniformity and flatness of sputtered carbon can be seen.

3 Production improvement

Manufacturing processes were done by Raytech Inc., and carbon sputtering was done by Be-Sputter Co., Ltd. Figure 3 shows our first trial. Cathode electrodes and pickups were formed by double-sided mask; then anode pins were formed by polyimide etching and nickel plating. This layer is called the top substrate. The width of cathode pickups is smaller than the resistive cathodes in order to reduce the dependence from the electrical field. Next, the thick substrate was laminated under the top substrate. This layer is called the bottom substrate. The bottom surface was covered by a copper layer for anode electrodes, which were connected to top anode pins by laser drilling and nickel plating. This process caused a misalignment of pixels because the top substrate could not be seen from the bottom side. Thus, the anode electrodes were not connected to top anode pins. To solve this problem, we used transparent dry resist for the bottom substrate. We could see the top substrate and pixels resulted aligned well by the photolithographic process. Connection to top anode pins and patterning of anode electrodes were done by copper sputtering and nickel plating. These improved processes are shown in Figure 4.

New prototypes have a $10 \times 10\text{cm}^2$ detection area with 256 anode/cathode strips. Figure 5 shows a microscopic image of the detector surface. Pixels are well aligned, and carbon sputtered cathodes are properly formed in the entire detection area. A 3-D image of a pixel is also shown in Fig 2, bottom.

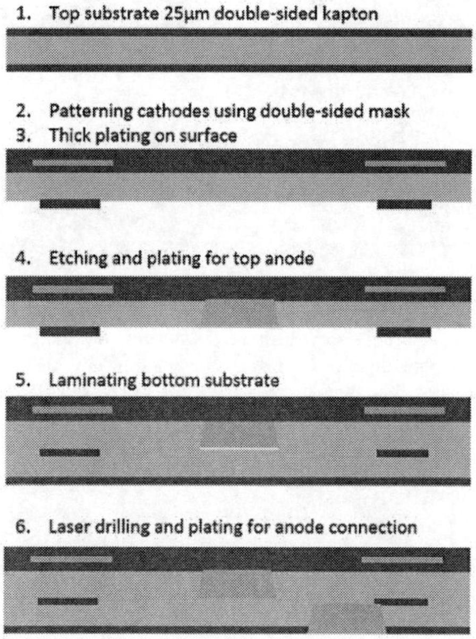

Figure 3. Previous manufacturing processes. In the laser drilling process, the top substrate cannot be seen; thus some pixels are misaligned.

https://doi.org/10.1051/epjconf/201817404001

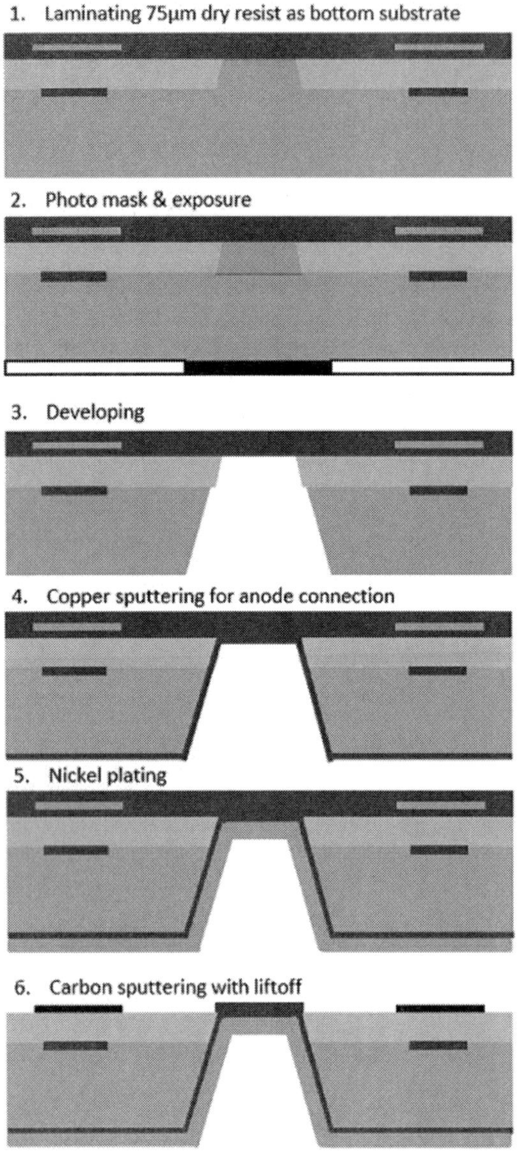

Figure 4. Improved processes. Manufacturing processes of the top substrate are the same as before. By using transparent dry resist and photolithography, pixels are well aligned in the entire detection area.

Figure 5. Microscopic image of a new prototype. The readout pitch is 400 μm. Pixels are well aligned, and carbon sputtered cathodes are properly formed in the entire detection area.

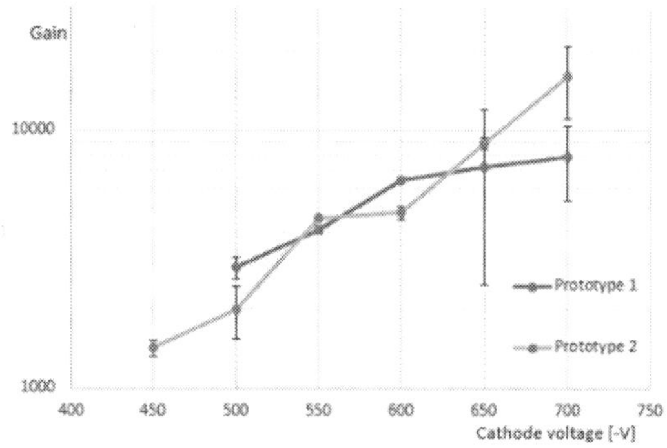

Figure 6. Gas gain of new prototypes versus cathode voltage. Anode voltage was set to 0. The gas mixture was Ar:C_2H_6=90:10. Drift field was set to 2 kV/cm. Signals were read by cathode pickup electrodes. Gain >10^4 was obtained.

4 Operation test

The gas gain measurement of new prototypes was performed using an ^{55}Fe 5.9keV X-ray source. The gas mixture was Ar:C_2H_6=90:10. Drift field was set to 2 kV/cm with 5mm drift gap. Negative HV was supplied to the resistive cathodes and the potential of the anodes was set to 0. Signals were read by cathode pickup electrodes, amplified by ATLAS ASD preamplifier [11], and then measured by Amptek's MCA-8000D Digital Multichannel Analyzer. Figure 6 shows the result of the measurement. The maximum gain is around 10^4. Although plots were finished at (-700 V, ~ 10^4), this is not due to discharge effects but to the ASD saturation.

Although signals were obtained from the cathode pickups, they could not be obtained from the anodes. Additionally, supplying positive HV to the anodes and setting the cathodes to 0 (this operation is rather typical for μ-PIC), signals could not be obtained by both anodes and cathode pickups. Gain reduction by continuous irradiation was also observed. After an irradiation for a few hours, the gas gain decreased by a factor two (Figure 7).

The observed problems are caused by anode strips not connected to the top anode because dry resist remained around it. Figure 8 shows an electron microscopic image of the cross section of the new prototypes. Plating of the anode was properly formed but did not reach the top anode. Thus, gain variation was caused by the charging-up effect on the top anodes. Recently, processes of anode connection have been improved, and new detectors have been produced.

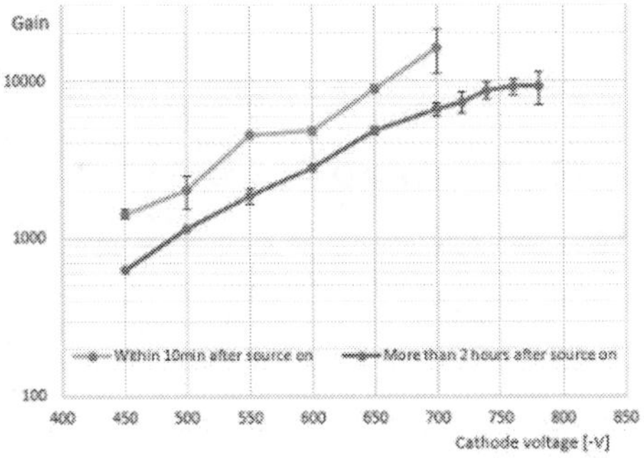

Figure 7. Gain variation due to charging-up effect on top anodes. Gain reduction occurred during several hours from source on.

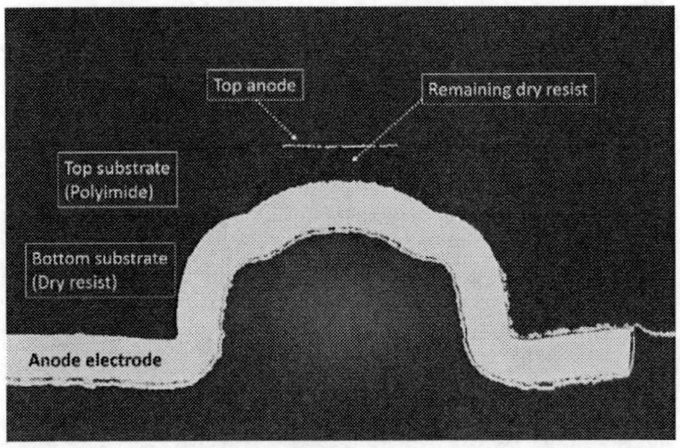

Figure 8. Electron microscopic image of cross section of a new prototype. Anode strips were not connected to top anode due to dry resist remaining around it.

5 Conclusion

μ-PIC with resistive cathodes using sputtered carbon was developed and tested. By using dry resist as the bottom substrate and photolithography, pixels were well aligned in the entire detection area. High gas gain ($>10^4$) was achieved using an ^{55}Fe 5.9keV X-ray source and discharge did not occur. This means that this new detector has the potential to detect MIP without preamplification by GEM or other floating structures such as foils or meshes. However, these prototypes are not perfect because the anode strips were not connected to top anodes. Processes were improved, and new prototypes were produced. We are planning to perform a fast-neutron irradiation test to check HIP tolerance in 2016.

We would like to thank RAYTECH Inc. and Be-Sputter Co., Ltd. This work was supported by JSPS KAKENHI Grant Numbers 26104707, 26610069.

References

[1] A. Ochi et al., NIM **A 471**, 264 (2001)
[2] A. Ochi et al., NIM **A 478**, 196 (2002)
[3] T. Tanimori, The Astrophysical Journal **810**, 28 (2015)
[4] K. Nakamura, PTEP **043F01**, (2015)
[5] J. D. Parker, NIM **A 497**, 23 (2013)
[6] Y. Kishimoto, NIM **A 732**, 591, (2013)
[7] The ATLAS Collaboration, *ATLAS Phase-II Upgrade Scoping Document* (CERN-LHCC, 2015)
[8] H. Raether, *Electron avalanches and breakdown in gases* (Butterworth, London U.K., 1964)
[9] A. Ochi et al., JINST **9 C01039**, 1 (2014)
[10] A. Ochi et al., PoS **TIPP2014**, 351 (2014)
[11] ATLAS TGC collaboration, *Amplifier-Shaper-Discriminator ICs and ASD Boards* (Production Readiness Review Report, 1999)

Ageing Studies on the First Resistive-MicroMeGaS Quadruplet at GIF++

Preliminary Results

B. Alvarez Gonzalez[1,a], M. Bianco[1], E. Farina[1,2], P. Iengo[1], F. Kuger[1,3], T. Lin[4], L. Longo[5,6], G. Sekhniaidze[6], O. Sidiropoulou[1,3], M. Schott[4], C.Valderanis[4], and J. Wotschack[1] on behalf of the ATLAS Muon Collaboration

[1]*CERN*
[2]*University of Pavia (IT)*
[3]*Bayerische Julius Max. Universitaet Würzburg (DE)*
[4]*Johannes-Gutenberg-Universitaet Mainz (DE)*
[5]*University of Salento (IT)*
[6]*INFN (IT)*

Abstract. A resistive-MicroMeGaS quadruplet built at CERN has been installed at the new CERN Gamma Irradiation Facility (GIF++) with the aim of carrying out a long-term ageing study. Two smaller resistive bulk-MicroMeGaS produced at the CERN PCB workshop have also been installed at GIF++ in order to provide a comparison of the ageing behavior with the MicroMeGaS quadruplet. We give an overview of the ongoing tests at GIF++ in terms of particle rate, integrated charge and spatial resolution of the MicroMeGaS detectors.

1 Introduction

A resistive-MicroMeGaS quadruplet built at CERN, serving as prototype of the ATLAS [1] MicroMeGaS (MM) for the New Small Wheel upgrade of the Muon Spectrometer [2], has been installed at the new CERN Gamma Irradiation Facility (GIF++) with the aim of carrying out a long-term ageing study. This detector has four active layers about 0.3 m^2 each equipped with 1024 read-out strips and sputtered resistive layer for spark protection [3]. Two smaller resistive bulk-MicroMeGaS, 10x10 cm^2, produced at the CERN PCB workshop have also been installed at GIF++ in order to provide a comparison of the ageing behavior with the MicroMeGaS quadruplet.

The detectors are exposed to an intense gamma (γ) irradiation of instantaneous flux of about 50 MHz/cm^2 (depending on the attenuation filters) provided by the 13.9 TBq ^{137}Cs source of GIF++, corresponding to 10 times more than the highest expected counting rate at the High-Luminosity Large Hadron Collider (LHC). The resistive-MicroMeGaS chambers will be exposed during several months at GIF++ in order to accumulate an integrated charge up to ~1 C/cm^2 which corresponds with the highest expected flux in the NSW for about 25 years of HL-LHC operation.

An overview of the ongoing tests at GIF++ in terms of particle rate, integrated charge and spatial resolution of the MicroMeGaS detectors is given in this document.

[a]e-mail: balvarez@cern.ch

2 Gamma Irradiation Facility at CERN

The new Gamma Irradiation Facility [4, 5] is located in the North Area of the CERN Super Proton Synchrotron (SPS) and it has been fully operational since March 2015. GIF++ is motivated by strong needs from the LHC detectors and accelerator communities in order to perform long-term ageing studies. It is a unique place where high energy charged particle beams (mainly muons) are combined with a flux of photons (662 keV) from a 13.9 TBq ^{137}Cs source. A filter system permits attenuating the photon rate in several steps to reach attenuation factors of several orders of magnitude (~10^4 - 10^5).

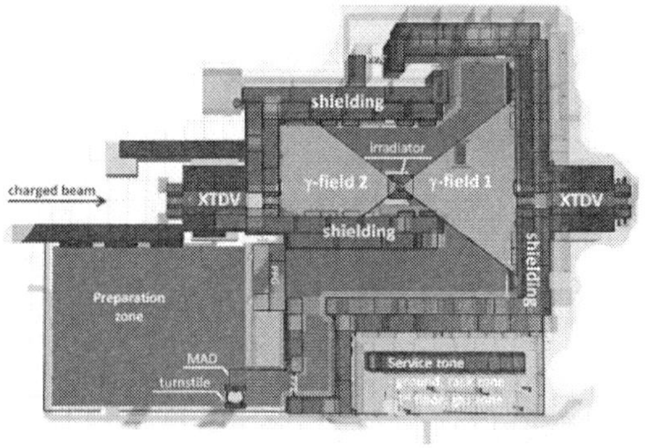

Figure 1. GIF++ bunker layout. The MicroMeGaS detectors are situated in the upstream area (γ-field 2) at 1 m from the source.

The high source activity produces a very intense background gamma field allowing to accumulate doses equivalent to High Luminosity LHC (HL-LHC) experimental conditions in a reasonable time.

Figure 1 corresponds to the GIF++ area with the service and preparation zones and the inside of the bunker where the irradiator is placed. The MicroMeGaS detectors are situated in the upstream area (γ-field 2) at 1 m from the source (position U1) exposed to a γ flux up to 64 MHz/cm^2, see in Fig. 2.

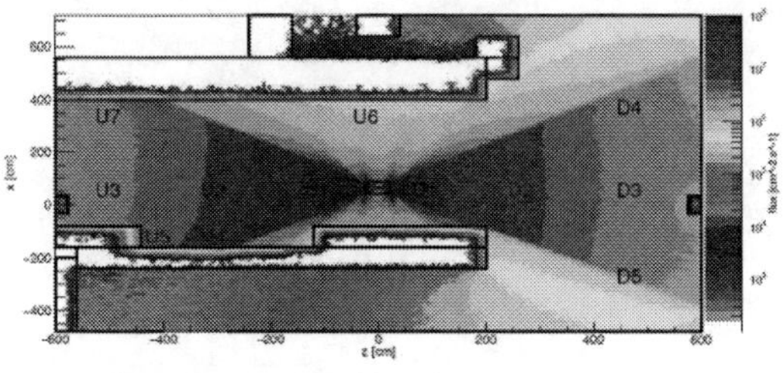

Figure 2. Total dose calculated from simulated flux. The values of the flux range from 10^3 to 10^8 Hz/cm^2. 'D' stands for positions downstream, and 'U' for positions upstream. The MicroMeGaS detectors are situated in the upstream area at 1 m from the source, position U1 (0.65 m, 2.0 m, -1,1 m).

3 Experimental Set-up

The detectors used for these studies are gaseous particle detectors detecting the traversing particles by amplifying the electrons that have been created by ionization in the gas volume. This gas volume is divided in two, the conversion gap and the amplification gap, by a metallic micro-mesh. Figure 3 shows the outline of the MM detectors with a 5 mm conversion gap and 128 μm amplification gap.

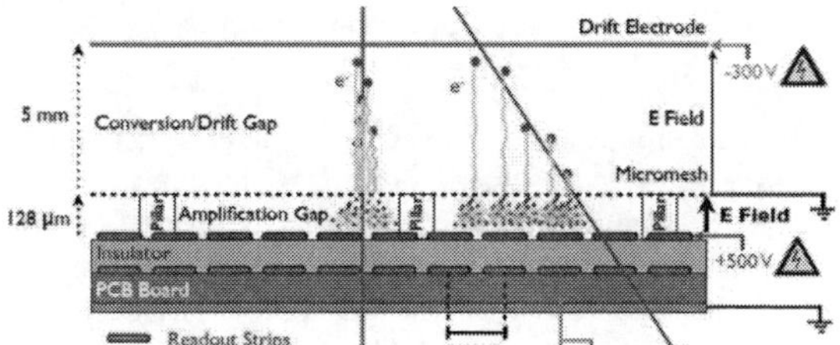

Figure 3. MicroMeGaS detector schema. HV numbers and dimensions correspond to the typical working conditions of ATLAS NSW MMs.

Ageing studies are mandatory to assess the capability of the detectors to the level of radiation at the HL-LHC. For the NSW MicroMeGaS, a very intense gamma source has been selected to perform the radiation studies.

Gamma radiation ionizes via three processes: photoelectric effect which is the dominant process for γ with energies below 50 keV, Compton scattering, which is the dominant contribution in the photon energy range between 0.1 and 10 MeV and is the main contribution at GIF++, and finally pair production which becomes important at energies over 5 MeV.

There are two types of MM detectors used in GIF++ for the studies described in this document. One corresponds to the prototype for the ATLAS New Small wheel (NSW) [2], MicroMeGaS Small Wheel (MMSW2), shown in Fig. 4. It is made of four active layers ~0.3 m^2, 4096 read-out strips and sputtered resistive layers, with mechanical floating mesh and strip pitch of 415μm. It was exposed to the radiation source for one week during the muon beam period in August 2015.

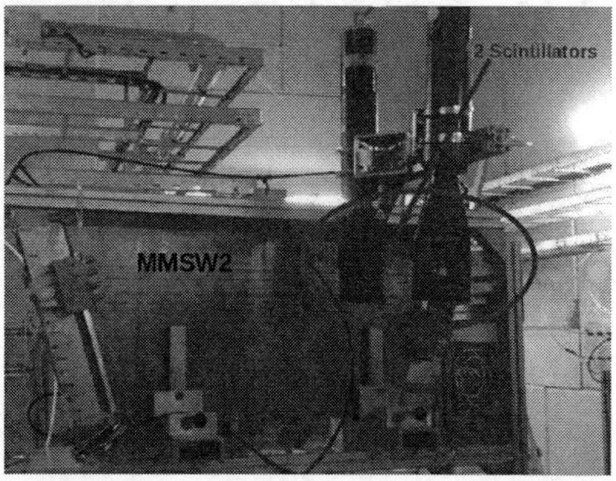

Figure 4. MMWS2 detector plus the two scintillators used for the trigger signal.

The other detectors are two resistive bulk-MicroMeGaS chambers, with an active area of 10x10 cm^2 and strip pitch of 400μm, called T5 and T8, shown in Fig. 5. Those chambers have been exposed at GIF++ for more than 3 months and more detailed studies described in Sec. 6 have been performed.

Figure 5. T8 in the frame during the GIF++ data-taking.

For the trigger signal during muon beam, two 10x10 cm^2 standard scintillators (highlighted in Fig. 4) were used in coincidence with a third scintillator located about 2 m away upstream.

4 Measurements

The measurements are performed varying the amplification voltage and the attenuation filters. The attenuation factors used for our measurements range from 1 to 10^4 in different steps; for instance we used the following values [1, 2.2, 4.6, 10, ..., 10^4]. The amplification voltages used for the MMSW2 chamber go from 480 to 570 V in 10 V steps and for the bulk chambers from 420 to 540 V in 10 V steps as well. The voltage difference between the chambers is due to MMSW2 having a mechanical floating mesh.

As mentioned already above, there is the possibility of combining the source with a high energy muon beam which was used for some resolution studies presented in the following section.

The working conditions during the measurements were as follows:

- Gas type: Ar/CO_2 93%, 7%
- Gas Flow: 5 l/h
- Amplification Gap: 128 μm
- Amplification Field: 30-45 kV/cm
- Drift Gap: 5 mm
- Drift Field: 600 V/cm
- Operating Gain: ~ 5x10^3

In total, more than 25 runs with 25k events each were collected and consequently analyzed.

5 Results for MMSW2

In Fig. 6 the current for the four layers of the MMSW2 chamber (L1, L2, L3 and L4) as a function of the amplification voltage is shown, in this case an attenuation factor of the order of 22 was used corresponding to an incoming photon flux of 3 MHz/cm^2 on the MMSW2 detector. The curves show an exponential behaviour in all cases as expected.

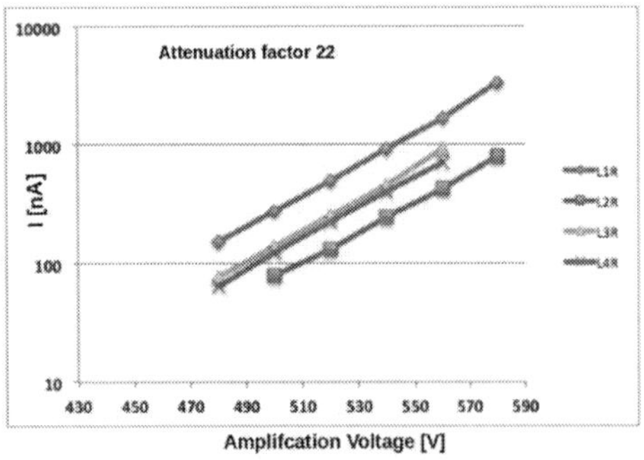

Figure 6. Current [nA] as a function of the amplification voltage [V] for the four layers of the MMSW2 chamber using an attenuation factor of the order of 22 corresponding to an incoming photon flux of 3 MHz/cm^2.

To estimate the spatial resolution (for these preliminary results), the first step is to reconstruct the clusters for each event and for each layer of the MMSW2 chamber. Later a tracking algorithm is applied in order to define the track of a muon candidate using at least one cluster in all the layers. Then the distance between the cluster positions of a possible muon on L1 and L2 is calculated.

The distribution of the distance between the clusters on L1 and L2 is displayed in Fig. 7 for the maximum flux, no attenuation factor. The resolution is evaluated performing a gaussian fit to the distance distribution. Assuming that both layers have the same resolution, the final resolution is taken as the sigma of the gaussian fit divided by $\sqrt{2}$ leading to ~100 μm resolution. This resolution is evaluated by taking charge centroids in the clusters using perpendicular tracks. The results are not corrected by the effect of multiple scattering and beam divergence.

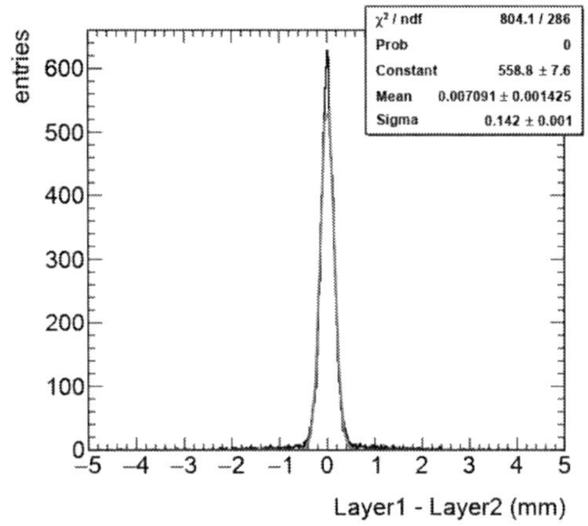

Figure 7. The distribution of the distance between the clusters on L1 and L2 for MMSW2 for the maximum flux, no attenuation factor. The fit is performed using a gaussian distribution function and the result of the fit is provided in the statistics box.

One can repeat the analysis for all attenuation points and study the resolution as a function of the flux as shown in Fig. 8. The resolution is flat up to the incoming flux of ~10 MHz/cm^2 which means that the algorithm used to select the muon candidates works successfully and provides the same resolution under high radiation conditions. The highest expected rate for the HL-LHC is about 15 kHz/cm^2 [2]; taking into account the detector sensitivity to photons which is about 2x10^{-3}, a source rate of ~7.5 MHz/cm^2 would give the equivalent highest expected HL-HLC flux. The flux has been defined as the expected flux from simulation in the position of the chamber inside the GIF++ bunker. In addition, it has been proven that simulation agrees well with the measurements.

6 Results for Bulk MicroMeGaS

During the data-taking at GIF++, several runs were collected with several attenuation filters to estimate the particle rate. The particle rate in this case is the average number

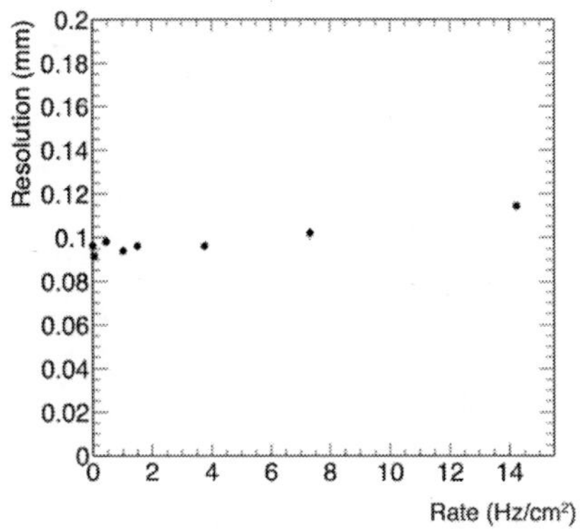

Figure 8. Resolution as a function of the incoming flux for L1 of MMSW2.

of photons (hits per event) counted in a time windows of 625 ns divided by the time window length.

Figure 9 shows the measured particle rate (Hz/cm^2) as a function of the amplification voltage per attenuation factor. This measured rate includes the detector sensitivity to

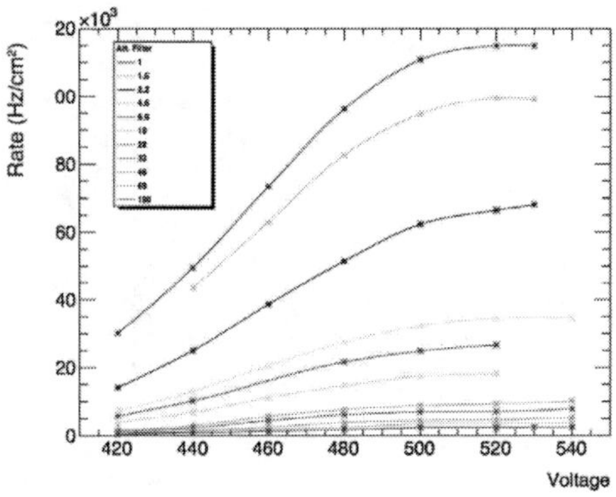

Figure 9. The measured particle rate as a function of the amplification voltage. The legend shows the different attenuation fa

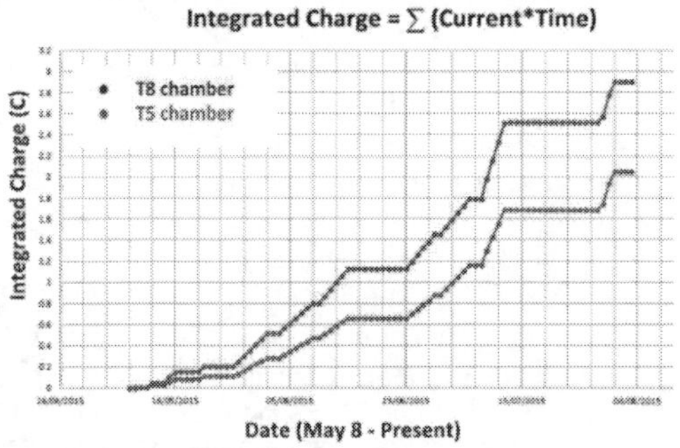

Figure 10. Integrated Charge as a function of time for T5 and T8.

photons that is about 2x10^{-3}, the sensitivity accounts for the difference to the simulated flux values (per attenuation factor) and the measured particle rate.

Unlike the MMSW2 chamber, the T5 and T8 chambers have been exposed to the radiation in GIF++ for several months. Fig. 10 shows the integrated charge (C) from May to the beginning of August 2015; with a total of ~3 C (0.03 C/cm^2) no degradation of the detector performance was observed.

7 Prospects and Plans

The plan is to expose of the resistive-MicroMeGaS chambers up to ~1 C/cm^2 in order to accumulate the highest expected flux in the NSW for about 25 years of HL-LHC operation and study the performance in this very high particle rate environment. The detector ageing evolution will be checked in terms of dark and amplification currents, efficiency and noise stability as a function of the integrated charge.

References

[1] ATLAS Collaboration, JINST 3 S08003 (2008).
[2] ATLAS collaboration, ATLAS-TDR-20-2013.
[3] P. Iengo et al., Construction of a large-size four plane MicroMeGaS detector. PoS (TIPP2014) 058.
[4] M.R. Jakel et al., CERN GIF++: A new irradiation facility to test large-area particle detectors for the high-luminosity LHC program, PoS(TIPP2014)102.
[5] http://ph-dep-dt.web.cern.ch/irradiation-facilities/gif

Multi-GEM Detectors in High Particle Fluxes

P. Thuiner[1,2,a], F. Resnati[1,b], S. Franchino[1], D. Gonzalez Diaz[1,3], R. Hall-Wilton[4], H. Müller[1], E. Oliveri[1], D. Pfeiffer[1,4], L. Ropelewski[1], M. Van Stenis[1], C. Streli[2], and R. Veenhof[3]

[1] CERN, European Organization for Nuclear Research, 1211 Geneva, Switzerland
[2] Technische Universität Wien, Atominstitut, Stadionallee 2, 1020 Vienna, Austria
[3] Uludağ University, Department of Physics, 16059 Bursa, Turkey
[4] ESS, European Spallation Source, 22100 Lund, Sweden

Abstract. Gaseous Electron Multipliers (GEM) are well known for stable operation at high particle fluxes. We present a study of the intrinsic limits of GEM detectors when exposed to very high particle fluxes of the order of MHz/mm^2. We give an interpretation to the variations of the effective gain, which, as a function of the particle flux, first increases and then decreases. We also discuss the reduction of the ion back-flow with increasing flux. We present measurements and simulations of a triple GEM detector, describing its behaviour in terms of accumulation of positive ions that results in changes of the transfer fields and the amplification fields. The behaviour is expected to be common to all multi-stage amplification devices where the efficiency of transferring the electrons from one stage to the next one is not 100%.

1 Introduction

Gaseous Electron Multipliers (GEM) [1] are known to maintain stable gains when exposed to 6 keV X-ray fluxes of at least 10^6 Hz/mm^2 for a single stage device operated at a gain of about 10^2 [2], and at least 10^5 Hz/mm^2 for double stage devices operated at a gain of 2×10^4 [3]. For triple stage devices it was observed that at high fluxes of soft X-rays the effective gain as a function of the flux first increases and then decreases [4]. This effect was assumed to be due to space-charge effects resulting in field distortions in the detector. Considerable efforts have been made to reduce the back-flow of positive ions into the conversion volume of multi GEM detectors to reduce space-charge induced field distortions [5][6]. Further studies [7] show a decrease of the ion back-flow of a triple GEM detector due to ion space-charge when exposed to high X-ray fluxes.
In sections 2 and 3 we will explain the experimental setup and methods used. Section 4 contains measurements and models describing the observed effects. Conclusions and outlooks are found in section 5.

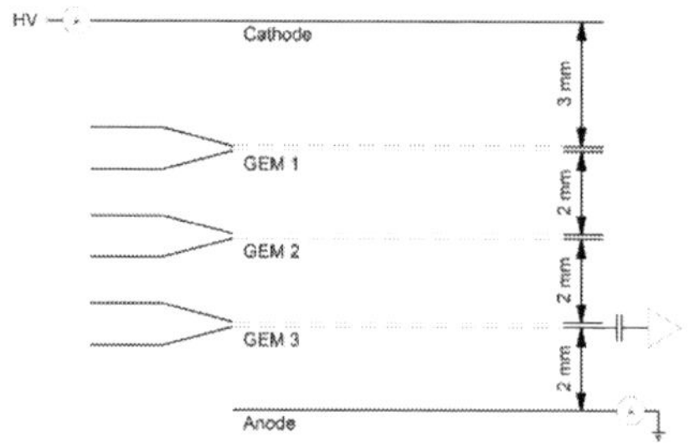

Figure 1: Schematic representation of the setup.

2 Experimental setup

A *standard* triple GEM detector with double-conical holes of 70 μm (50 μm) outer (inner) hole diameter and pitch of 140 μm in an active area of 10×10 cm^2 is used. The drift gap is 3 mm long, all other gaps are 2 mm in length. A schematic representation of the detector is shown in figure 1. All electrodes of the detector are powered either by resistor divider (results in figure 2) or individually (results in figure 3) using high current resolution power supplies with a resolution of 50 pA for all stages. The drift cathode is powered individually for all results discussed in the following sections. The read-out anode is read with a pico-amperemeter (Keithley 6487) with a resolution of approximately 5×10^{-5} times the monitored current. To facilitate event by event measurements of the signals, a decoupling capacitor of 2.2 nF connects the bottom electrode of the third GEM to a charge pre-amplifier (ORTEC 142) and a shaper (ORTEC 572) chain connected to a Multichannel Analyzer (AMPTEK MCA-8000D). The detector is continuously flushed with an Ar/CO_2 70/30 gas mixture at 5L/h. Collimated 8 keV X-rays of about 1 mm^2 beam diameter generated by a copper X-ray tube are used to produce the primary ionisation charges of approximately 290 primary electrons in the conversion volume. The X-ray

[a] e-mail: patrik.thuiner@cern.ch
[b] e-mail: filippo.resnati@cern.ch

flux is adjustable by changing the X-ray tube intensity or using calibrated attenuators.

3 Experimental methods

The observed effects are checked to be unrelated to global effective gain variations, e.g. due to changing of ambient parameters. A reference position P_{ref} is continuously irradiated with an Fe-55 source. The peak position of the 5.9 keV peak in the pulse height spectra is continuously monitored when the shutter of the X-ray tube is closed. All electrodes are powered for at least one hour before starting the measurements to ensure stability of the currents monitored. Charging-up effects [8] are excluded as the origin of the observed behaviour by waiting until the currents $I_{i,on}$ and the peak position of the 8 keV peak in the pulse height spectra are stable after starting the irradiation at low flux, and by repeated measurements going from low to high X-ray flux and vice versa. The currents collected at the drift cathode, GEM electrodes and read-out anode are calculated as $I_i = I_{i,on} - I_{beamoff}$ by measuring the currents $I_{beamoff}$, taking into account the charge collected when the detector is continuously exposed to low flux X-ray of the Fe-55 source as well as leakage currents from electrodes to ground, and $I_{i,on}$ during irradiation with high X-ray fluxes of the Cu X-ray tube.
The effective gain is calculated as $G_{eff} = I_A/(Rn_pe)$ with the current I_A collected at the read-out anode, the estimated interaction rate R, the number of primary electrons $n_p \approx 290$ for the gas mixture used and the electron charge e. The ions per primary electron are calculated in a similar way as $G_{ion} = I_C/(Rn_pe)$ with the current I_C collected at the drift cathode. The ion back-flow is defined as the ratio $IBF = I_C/I_A$ of the currents collected on cathode and anode.

4 Measurements

As shown in figure 2 the effective gain, as a function of the X-ray flux is stable up to a few kHz/mm^2 for the nominal effective gain studied. Increasing the flux results in a slight rise of effective gain, which becomes prominent when increasing the flux above a few hundred kHz/mm^2. When reaching a flux of about 1 MHz/mm^2 the effective gain reaches a peak before decreasing significantly at a flux of the order of 10 MHz/mm^2.
The number of ions per primary electron show an inverse behaviour up to 1 MHz/mm^2, with the value decreasing. When the effective gain reaches the peak and start decreasing again, the ions further decrease, with the slope of the curve getting flatter with increasing X-ray flux.
The ion back-flow starts decreasing significantly when the effective gain increases. At a flux of the order of 4×10^6 Hz/mm^2, the ion back-flow reaches a plateau.
The observed change in effective gain is strongly dependent on the nominal effective gain of the detector (Fig. 3).

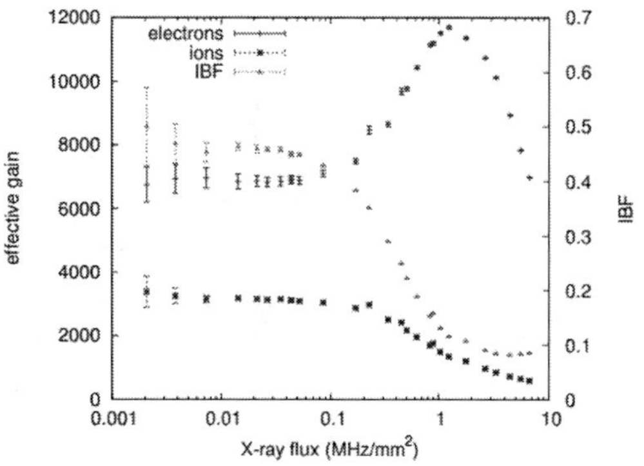

Figure 2: Dependence of the effective gain, the number of ions per primary electron, and ion back-flow on flux of 8 keV X-rays for a nominal effective gain of approx. $7{\times}10^3$ in Ar/CO_2 70/30 gas mixture.

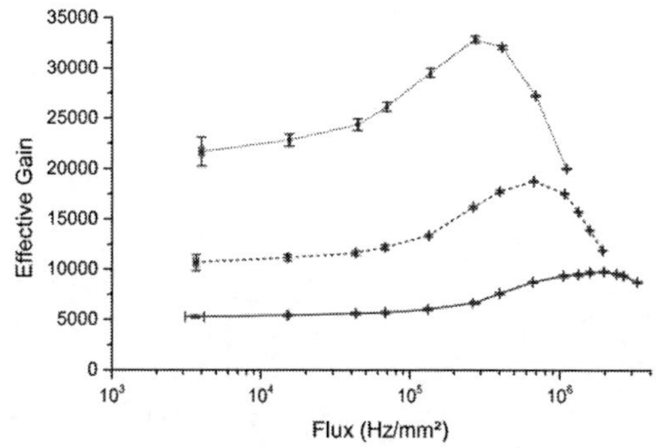

Figure 3: Dependence of effective gain on 8 keV X-ray flux for nominal effective gains of $5{\times}10^3$, 10^4 and $2{\times}10^4$.

Since the effect is related to the charge density in the detector, increasing the nominal effective gain will lead to the appearance of the observed effects at lower particle fluxes while a decreased gain will shift the effect towards higher fluxes.

The behaviour of the triple GEM detector is simulated using the Finite Element Analysis software package COMSOL Multiphysics® [9]. The electron and ion distributions in the detector are described macroscopically by numerically approximating the solutions of a set of partial differential equations of second order: the drift and diffusion equations for electrons and ions, and the Poisson equation for the potential distribution. Electron transport coefficients are computed using Magboltz 8.97 [10], and the amplification is calculated with the Townsend coefficient corrected for Penning transfer [11]. The ion mobilities are taken from [12]. A two-dimensional axial-symmetric geometry is used to simulate the effect in a triple GEM detector. The modelled GEMs therefore only contain a single GEM hole, with the holes of all three GEMs perfectly aligned.

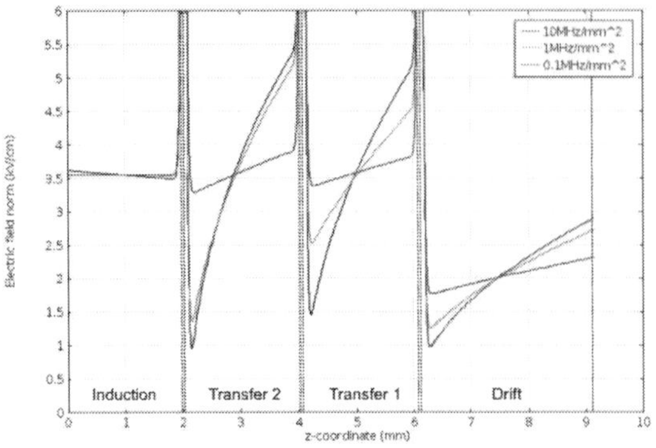

Figure 4: Computation of the field distortions in the transfer gaps for X-ray fluxes of 0.1 MHz/mm^2 (red), 1 MHz/mm^2 (green) and 10 MHz/mm^2 (blue) along the hole axis. The nominal fields are 2 kV/cm in the *Drift* gap and 3.5 kV/cm in the other gaps.

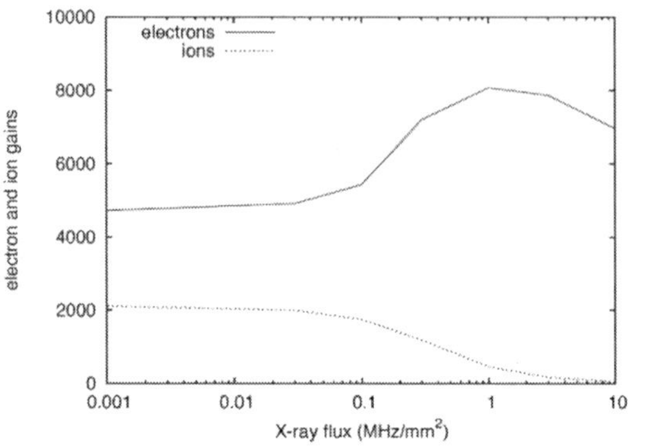

Figure 7: Computation of effective gain and number of ions per primary electron for different X-ray fluxes at a nominal effective gain of approx. 5×10^3.

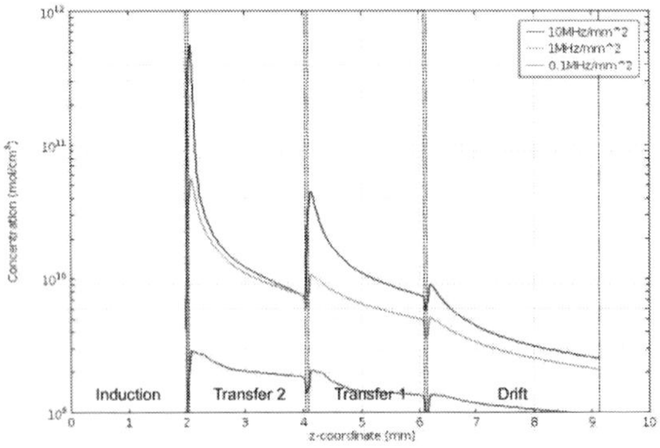

Figure 5: Computation of the ion concentrations in the detector for X-ray fluxes of 0.1 MHz/mm^2 (red), 1 MHz/mm^2 (green) and 10 MHz/mm^2 (blue) along the hole axis.

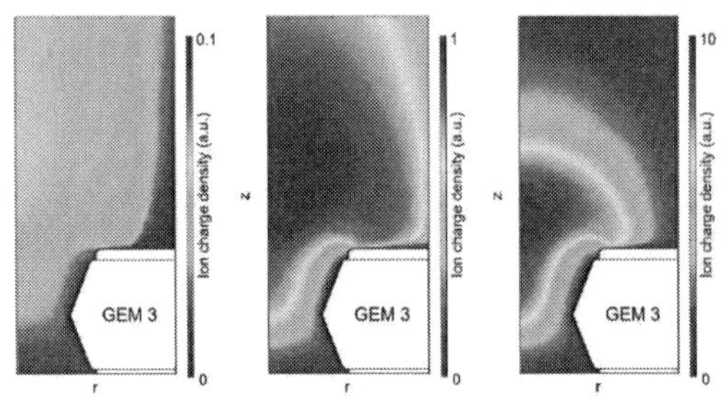

Figure 6: Computation of the ion charge density close to the bottommost GEM for X-ray fluxes of 0.1 MHz/mm^2 (left), 1 MHz/mm^2 (center) and 10 MHz/mm^2 (right) along the hole axis.

The ions created in the GEM holes during the electron avalanches drift upward into the transfer and conversion volumes. The dynamic equilibrium of production and evacuation results in an accumulation of positive ions in these volumes. This leads to a steady space-charge distribution that modifies the electric field. For increasing X-ray flux the fields get distorted in a way that the fields close to the top of the GEM electrodes are reduced and the fields close to the bottom electrodes are increased (figure 4), with the nominal field as the average over the length of the gap. The effect becomes stronger for each successive stage of amplification, with the smallest distortion in the *Drift* gap and the largest in the *Transfer 2* gap. The distortions in the *Induction* gap are minimal due to an absence of ions in this region.

The electron collection and extraction efficiencies of a GEM are strongly dependent on the fields applied above and below the GEM, respectively [13]. This results in two effects: first, both the collection and extraction efficiencies for electrons are increased. With the efficiencies for electrons not being 100% for the nominal fields applied, this results in an observable increase in effective gain up to the point where the collection and extraction efficiencies reach 100%. Secondly, at the same time the collection and extraction efficiencies for ions are decreased. This leads to a significant reduction of the ion back-flow. Effective gain and ions per primary electron for the simulations are calculated as already described in section 3. As can be seen in figure 7, the simulations are semi-quantitatively comparable to the measurements shown in figure 2.

After reaching the effective gain peak at approximately 1 MHz/mm^2, space-charge effects in the amplification regions become dominant. With the ion extraction decreasing significantly, the ions start accumulating close to the top of the GEM holes. Due to the presence of the ions the fields are further decreased, additionally reducing the ion extraction. Ions generated during an avalanche are now accumulating in great number in the GEM holes, thus reducing the fields and decreasing the effective gain of the detector. This effect is simulated and can be seen in figure 5. At the same time the transfer fields are modified in a way, such that the evacuated ions get collected by the top electrode of the GEM in increasing number, leading to a reduced ion back-flow. This effect can be seen in figure 6. While the effects described are observed in all stages of the triple-GEM detector, the last GEM foil is the one dominating the behaviour of the detector.

5 Conclusions

Space-charge effects for high fluxes of 8 keV X-rays in a Ar/CO_2 70/30 gas mixture were studied. At high particle flux ion space-charge modifies the transfer fields and the amplification fields such that, as a function of the flux, the effective gain first increases and then decreases. The ion back-flow is decreasing with increasing particle flux. The field distortions are dependent on the flux and the nominal effective gain of the detector. Increasing the nominal effective gain leads to the appearance of the observed effects at lower X-ray flux while a decreased gain shifts the effect towards higher flux. The effects are modelled for standard triple-GEM detectors and quantitatively understood. Further systematic studies of the energy resolution will be soon published elsewhere.

References

[1] F. Sauli, Nucl. Instr. Meth. A **386**, 531 (1997)
[2] J. Benlloch et al., IEEE Trans. Nucl. Sci. **45**, 234 (1998)
[3] A. Bressan et al., Nucl. Instr. Meth. A **425**, 262 (1999)
[4] P. Everaerts, Master Thesis, Gent University (2006)
[5] F. Sauli, L. Ropelewski, P. Everaerts, Nucl. Instr. Meth. A **560**, 269 (2006)
[6] A. Breskin et al., Nucl. Instr. Meth. A **553**, 46 (2005)
[7] M. Ball, K. Eckstein, T. Gunji, Journal of Instrumentation **9**, C04025 (2014)
[8] M. Alfonsi et al., Nucl. Instr. Meth. A **671**, 6 (2012)
[9] *Comsol multiphysics®*, `http://www.comsol.com`
[10] S. Biagi, Nucl. Instr. Meth. A **421**, 234 (1999)
[11] Özkan Şahin, T.Z. Kowalski, R. Veenhof, Nucl. Instr. Meth. A **768**, 104 (2014)
[12] H.W. Ellis et al., Atom. Data and Nucl. Data Tables **17**, 177 (1976)
[13] S. Bachmann et al., Nucl. Instr. Meth. A **438**, 376 (1999)

Sealed operation, and circulation and purification of gas in the HARPO TPC

M. Frotin[1,a], P. Gros[1,a], D. Attié[2], D. Bernard[1], V. Dauvois[3], A. Delbart[2], D. Durand[3], Y. Geerebaert[1], S. Legand[3], P. Magnier[2], P. Poilleux[1], and I. Semeniouk[1]

[1] LLR, Ecole Polytechnique, CNRS/IN2P3, 91128 Palaiseau, France
[2] IRFU, CEA Saclay, 91191 Gif-sur-Yvette, France
[3] DEN/DANS/DPC/SECR/LRMO, CEA Saclay, 91191 Gif-sur-Yvette, France

Abstract. HARPO is a time projection chamber (TPC) demonstrator of a gamma-ray telescope and polarimeter in the MeV-GeV range, for a future space mission. We present the evolution of the TPC performance over a five month sealed-mode operation, by the analysis of cosmic-ray data, followed by the fast and complete recovery of the initial gas properties using a lightweight gas circulation and purification system.

1 Introduction

High-energy astronomy is hindered by the sensitivity gap between the energy ranges over which the Compton and the pair telescopes are efficient, making the γ-ray sky in the range 0.1 – 100 MeV almost unexplored. Furthermore no polarimeter with a decent sensitivity above 1 MeV has ever been flown in space.

On the low-energy side of the gap, a number of projects are aiming at an improved sensitivity with respect to the COMPTEL telescope [7].Of these, those using stacks of silicon wafers without any tungsten converter can also detect pair conversions [1–3] with an angular resolution of $\approx 1.5°$ @ 100 MeV, a significant improvement over that of *Fermi*-LAT ($\approx 6°$ @ 100 MeV [4]). These stacks can have some sensitivity to polarisation if the wafers are thin enough and provided that matching can be performed in the first wafer after the conversion wafer, i.e. that each of the hits in the x direction can be assigned to one of the two hits in the orthogonal (y) direction. In subsequent wafers, the azimuthal information carried by the pair is washed out by multiple scattering [5].

In the quest for even thinner wafers, it is tempting to use an even less dense detector such as a gas detector. With a spatial resolution of 100 µm, and with an optimal tracking in the presence of multiple scattering provided by a Kalman filter, an angular resolution of $\approx 0.4°$ @ 100 MeV can be achieved [6]. This provides a point-like source sensitivity as low as 10^{-6}MeV/(cm^{2}s) @ 100 MeV with a 10 kg gas time projection chamber (TPC), while that of COMPTEL plateaued at $\approx 10^{-4}$MeV/(cm^{2}s) [7]. In such gas detectors the dilution of the azimutal information carried by the pair is close to unity, so that the sensitivity to polarisation is excellent [8]. Finally track matching can easily be performed by comparing the distribution of the energy deposited along the tracks in the x and y directions thanks to the large fluctuation in the local deposition of energy along the track [9].

These unique features come at the cost of the need to maintain a good-quality gas in which the collection of the ionizing electrons is not affected by impurities with a large electron attachment. For example the EGRET gas (99.5% neon, 0.25% argon, and 0.25% ethane) became contaminated by the breakdown of ethane by spark chamber firings and to a small degree by residual outgassing. The five planned gas exchanges were carried out once per year, after which the lack of fresh gas became a serious issue [10]. These concerns have triggered the development of detectors built out of low outgassing rate materials. For example, the design of the GEMS X-ray polarimeter TPC involves gas electron multipliers (GEM) made of copper clad liquid crystal polymer (LCP), and a dielectric structure of polyetheretherkeytone (PEEK) on a gold-plated titanium frame. This gives it an expected lifetime of 23 year [11].

We report here on a gas purity study performed with the HARPO detector, a TPC prototype of a $\gamma \rightarrow e^+e^-$ telescope built with techniques much better suited to our actual budget and that we have studied in the laboratory with radioactive sources [12] and with cosmic rays [13], and that we have tested recently [14] in a γ-ray beam.

2 Description of the HARPO TPC

HARPO is a 30 cm cubic TPC surrounded by 6 scintillator plates, each equipped with two wavelength shifter bars and two photomultipliers (PMTs) for trigger and background rejection. The gas container is an aluminum cylinder designed to be operated from few mbar up to 5 bar in sealed mode (Fig. 1). With a mixture of 95% of argon 5.0 and 5% of iC_4H_{10} 3.5 and at a 220 V/cm drift electric field, the electron drift velocity is about 3 cm/µs. The TPC endplate is a "hybrid" MicroPattern Gas Detector (MPGD)

[a] corresp. authors: frotin@llr.in2p3.fr, philippe.gros@llr.in2p3.fr

composed of two GEM foils[15] with 2 mm spacing above a 128 μm amplification gap bulk-micromegas[16]. The anode collection plane is segmented in 2×288 strips in the two orthogonal x, y directions and the signal is read by an electronic chain composed of two T2K/TPC Front-End cards[17], two FEMINOS back-end cards and a Trigger Clock Module [18]. A simple coincidence between scintillators is used here for triggering on cosmic rays.

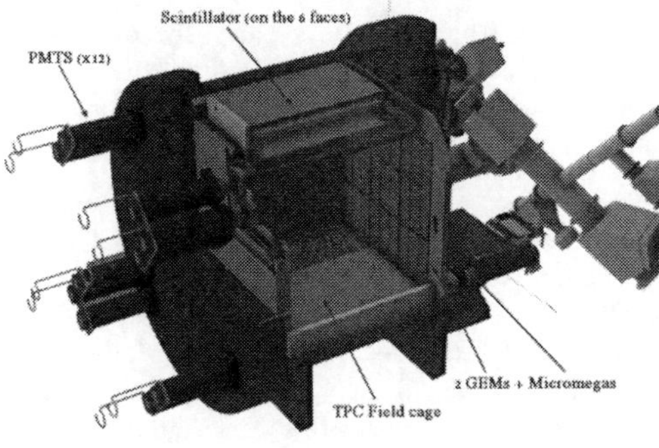

Figure 1. A 3D CAD drawing of the HARPO demonstrator.

3 Description of the circulation and purification system (CAPS)

We used a circulator composed of a simple turbine (Fig. 2, left). Full schematics of the circulation system is shown in Fig. 3 The gas was lead through an Oxisorb purification cartridge that can absorb up to 0.5 L H_2O and 0.1 L O_2, with output purity as low as < 5 ppb for O_2 and 30 ppb for H_2O [19]. We measured the gas flow with a F-111B Bronkhorst mass flowmeter (Fig. 3, I10) that was calibrated for our gas mix at 2 bar and 20 °C. We monitored the gas pressure and temperature with a Keller EV-120/PA(A)-33X sensor (Fig. 3, I4 and I5) with ranges of 0 − 50 °C and 0 − 10 bar and resolutions of 0.01 °C and 1 mbar. The outgassing areas inside the vessel amount to 0.82 m^2 Aluminum, 3.50 m^2 PVC, 0.92 m^2 PCB, 1.63 m^2 PVT, for a gas volume $V_{det} = 84$ L.

We tested the gas circulation system with a helium leak detector (ASM 310) down to $3.3 \cdot 10^{-8}$ mbar · L/s, after which we opened the connection between the CAPS and the detector. This caused a pressure drop from 2015 mbar to 1960 mbar at 27 °C, with a flow of 1.5 L/h. On July 20 2015, we opened the Oxisorb purificator in vacuum which lead to a new pressure drop to 1932 mbar when the connection to the detector was opened. The CAPS then ran continuously, with a flow through the filter of $f_{circ} = 1.4$ L/h. In these conditions we expect the water and oxygen content in the gas, C, to evolve following a law of the form:

$$C = C_0 e^{-t/\tau_{purif}} + K_{cont}\tau_{purif}, \qquad (1)$$

with $\tau_{purif} = V_{det}/\epsilon_{purif} f_{circ}$, where C_0 is the original content and K_{cont} is the contamination rate from leaks and outgassing. In case the purification efficiency of the filter is $\epsilon_{purif} = 1$, the purification rate is equal to $\tau_{purif} = 60$ h.

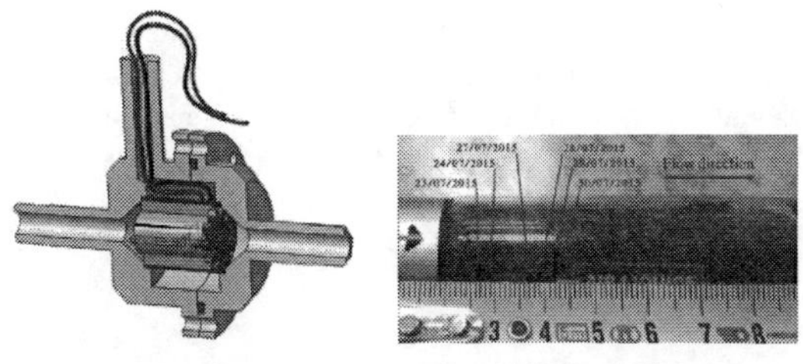

Figure 2. (Left) Sectional view of the circulator (CEA patent BD15730SG). (Right) Picture of the Oxisorb cartridge. The progression of the purple-coloured region over 10 days of continuous use is indicated.

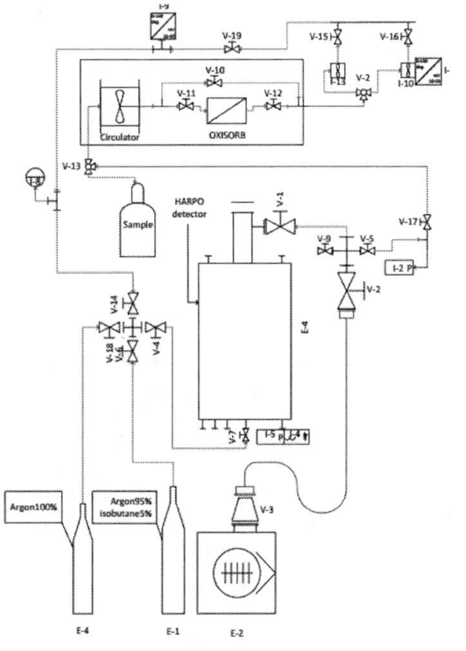

Figure 3. Layout of the gas system.

The Oxisorb purificator color ranges from blue to dark brown with the quantity of H_2O and O_2 absorbed. At the beginning of the purification, we observed a purple ring of 3 mm thickness growing 3 mm per day during the first 4 days. It stabilised at 12 mm after a few days (Fig. 2). This indicates that the cartridge was never saturated.

Several gas samples were taken for analysis, which lead to density drops (lines 1, 2, and 3 on Fig. 7). Before and after every operation that affected the gas in the vessel, data was taken with the TPC for consistency checks.

4 Data taking and analysis

We took cosmic-ray data once per week from February to July, with the detector set so that the drift field was vertical. Most runs were one hour long, recording a few 10,000 events. After starting the purification on July 20 we took data more frequently. The high voltage was only turned on when taking data. Due to the vertical detector orientation, most of the recorded tracks are approximately aligned with the drift direction. Figure 4 shows the X, t and Y, t projections for a typical event. We select events with

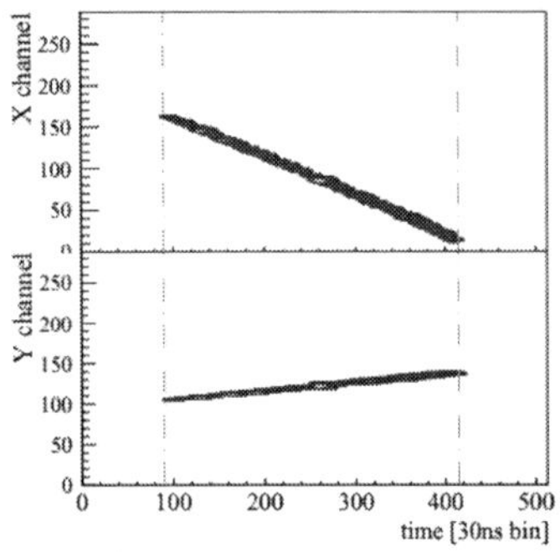

Figure 4. A cosmic-ray event in HARPO. The "maps" show the charge distributions in the projection X-time and Y-time.

a single track in each projection, with a high momentum muon selection based on the χ^2 cut of a linear fit. For each time bin t_i (30 ns $\approx$ 1 mm), we obtain the charge deposited by a track, $Q(t_i)$, summing over all the channels associated with the track. After an angle-effect correction using the direction $\vec{u}$ of the track,

$$\frac{dE}{dx}(t_i) = Q(t_i) \times u_z, \qquad (2)$$

we obtain the $dE/dx(t_i)$ in ADC units, that is up to an overall factor that includes the gain of the amplification of both the MPGDs and that of the readout electronics. The result-

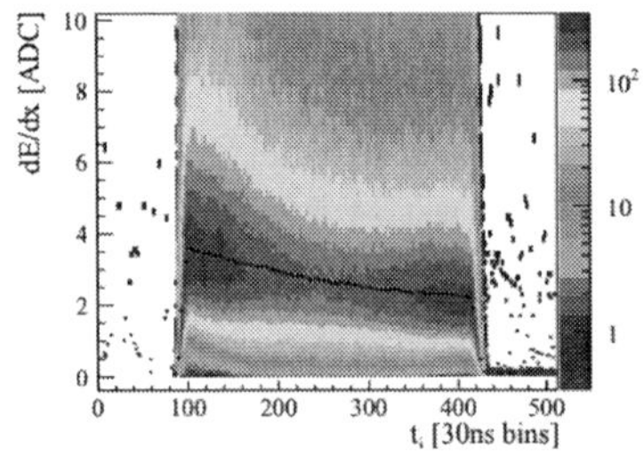

Figure 5. Normalised charge as a function of the drift time for a 6000 s cosmic-ray run.

ing distribution (Fig. 5) is also affected by threshold and saturation effects. To mitigate them, we estimate the most probable value (MPV, solid line in Fig. 5) of the charge for each time bin with a Landau-function fit. From this distribution we extract :

The drift velocity. As the tracks are selected to traverse the full TPC length, the drift-time duration provides a measurement of the drift velocity.

The gain. The height of the distribution at short drift times is proportional to the amplification gain.

Table 1. Gas content from mass spectrometry (volume/volume). Note that the set-up was not sensitive to H_2O.

Sample # / Compound		1 July 08	2 Aug. 27	3 Sept. 17
iC_4H_{10}	%	5.10	4.42	4.49
O_2	ppm	180	<20	<20
CO	ppm	190	250	130
CO_2	ppm	120	160	130
N_2	ppm	620	890	850

The attenuation. The signal loss along the drift from electron capture is visible in Fig. 5. We fit the central part of the distribution, which is less affected by the threshold and saturation effects, with an exponential function.

These measurements are affected by systematic effects. Since the detector configuration was kept unchanged throughout the data taking, they cancel in the relative value compared to a reference run, chosen to be run 2008 (Feb. 18th), i.e. right after the fresh gas fill. Figure 6 shows the relative charge distribution as a function of the drift distance for several runs taken at a few weeks interval.

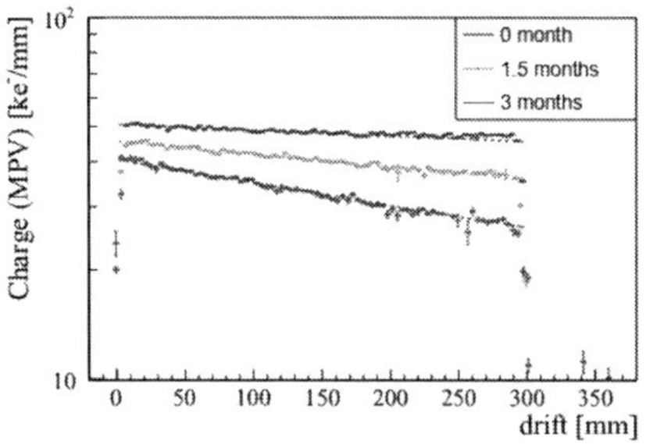

Figure 6. Charge as a function of the drift distance, for three runs, 6 weeks apart, normalised to the reference run 2008.

5 Results

For the first five months, while the TPC was sealed, we observe a steady degradation of the gain, of the drift velocity and of the electron attenuation (Fig. 7). After we turned the CAPS on, the drift velocity and the attenuation got back to their initial values. We corrected for the gain changes due to the density decrease induced by our extraction of gas samples and extension of the volume with the CAPS circuit, by adjusting the MPGD voltages. The pressure change also affected the drift velocity but to a negligible extent as shown by Garfield simulations[20], compared to the observed effect.

Gas samples were collected before (#1, line 3 on Fig.7), and respectively five weeks (#2) and eight weeks (#3) after activating the purification system. They were analysed in a high resolution (R = 2800) direct-injection THERMO MAT 271 mass spectrometer (Fig. 8, Table 1).

The relative amount of O_2 and N_2 before purification is consistent with that of air, due to either leaks or outgassing. Contaminations by CO and CO_2 probably originate from outgassing from plastic elements.

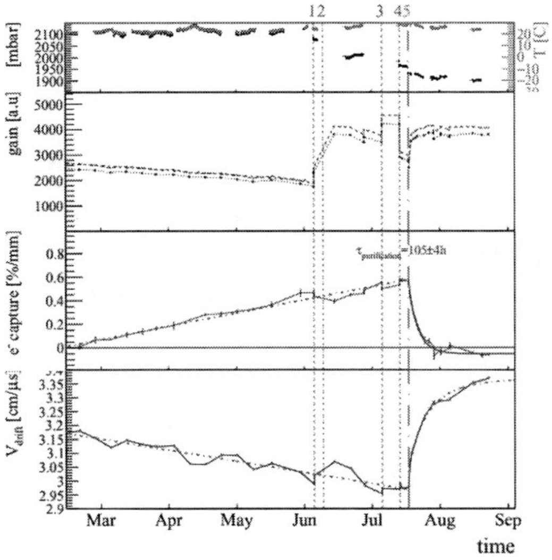

Figure 7. Time evolution of the gain, of the attenuation and of the drift velocity. We can see several variations of the gain due to the pressure loss when gas samples were extracted. For the first five months during which the TPC was sealed we see a regular degradation of all three parameters. On July 20 (line 5), the CAPS was turned on.

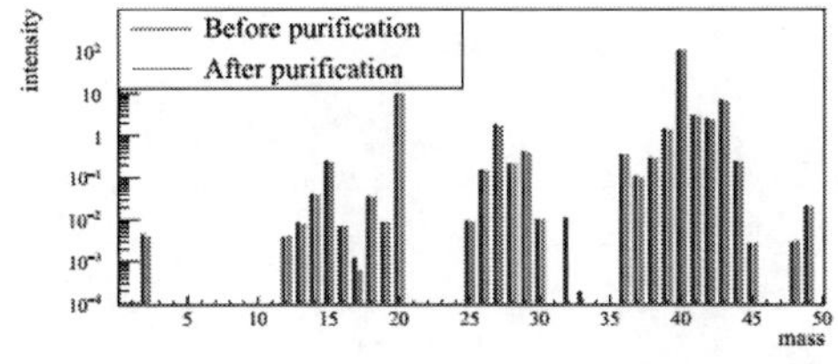

Figure 8. Mass spectroscopy results for samples taken before and after purification. The overall composition is similar, except for N=32 (O_2), that disappeared in the second sample.

After purification, the residual oxygen content is found to be lower than the detection threshold of the spectrometer. We also note a small but significant decrease of the amount of isobutane content upon purification after which it stabilized, an effect that is not understood.

6 Discussion

Garfield++ simulations [20] show that the degradation of the attenuation during sealed operation, attributed to electron attachment, is proportional to the oxygen content. After purification was started the effect disappears exponentially with a characteristic time $\tau_{\text{purif}} = 105 \pm 4\,\text{h}$ (Fig. 7). The observed time constant is larger than the computed value, most likely due to the limitation of gas circulation inside the almost closed field cage. We also observe a decrease of the drift velocity, which cannot be associated with the oxygen contamination and which is recovered after purification, and therefore that can be attributed to water contamination.

After purification, the drift velocity was measured to be $6.0 \pm 0.5\%$ higher than its initial value. From the Garfield simulation, this can be explained by the cumulative effects of the reduced gas pressure from 2.1 to 1.9 bar (+4%) and of the measured decrease of the isobutane content from 5.1% to 4.5% (+3%).

7 Conclusion

We have successfully operated a sealed TPC over five months, with a clear but manageable gas degradation. We used the data from cosmic rays in the TPC to monitor the relevant parameters of the gas (drift velocity, gain, attenuation). We then purified the gas with a simple, low-power circulation system and observed a fast and complete recovery of the initial parameters. If operated routinely, for example on a space mission, the system should therefore enable a stable performance for several years.

This work is funded by the French National Research Agency (ANR-13-BS05-0002)

References

[1] "ASTROGAM", proposal submitted for the ESA M4 Mission Programme, Jan. 2015, http://astrogam.iaps.inaf.it/
[2] A. A. Moiseev *et al.*, arXiv:1508.07349 [astro-ph.IM].
[3] X. Wu *et al.*, Proc. SPIE **9144**, 91440F (2014) [arXiv:1407.0710 [astro-ph.IM]].
[4] M. Ackermann *et al.* [Fermi-LAT Collaboration], Astrophys. J. Suppl. **203** (2012) 4 [arXiv:1206.1896 [astro-ph.IM]].
[5] D. Bernard, The gamma-ray sky with astrogam, Second ASTROGAM Workshop, March 2015, Paris.
[6] D. Bernard, Nucl. Instrum. Meth. A **701**, 225 (2013) [Erratum-ibid. A **713**, 76 (2013)] [arXiv:1211.1534 [astro-ph.IM]].
[7] V. Schönfelder, New Astr. Rev. 48 (2004) 193.
[8] D. Bernard, Nucl. Instrum. Meth. A **729** (2013) 765, [arXiv:1307.3892 [astro-ph.IM]].
[9] D. Bernard, 12th Pisa Meeting on Advanced Detectors, Isola d'Elba, Italy, May 2012, Nucl. Instrum. Meth. A **718**, 395 (2013), [arXiv:1210.4399 [astro-ph.IM]],
[10] J. A. Esposito *et al.*, Astrophys. J. Suppl. **123** (1999) 203.
[11] J. E. Hill *et al.*, Proc. SPIE **8859**, San Diego, United States, Aug. 2013, 885900.
[12] P. Gros *et al.*, TIPP14, 2-6 June 2014 Amsterdam, proceedings PoS(TIPP2014)133
[13] D. Bernard *et al.*, SPIE **9144**, June 2014, Montréal, Canada, 91441M; arXiv:1406.4830 [astro-ph.IM].
[14] S. Wang *et al.*, TPC2014, Dec. 2014 Paris, J.Phys.Conf.Ser. 650 (2015) 012016, arXiv:1503.03772 [astro-ph.IM].
[15] F. Sauli, Nucl. Instrum. Meth. A **386**, (1997) 531.
[16] I. Giomataris *et al.*, Nucl. Instrum. Meth. A **560**, (2006) 405.
[17] P. Baron *et al.*, IEEE Trans. Nucl. Sci. **55** (2008) 1744.
[18] D. Calvet, IEEE Trans. Nucl. Sci. **61** (2014) 675.
[19] Oxisorb, air liquide, ref 123342.
[20] R. Veenhof *et al.*, http://garfieldpp.web.cern.ch/garfieldpp/

https://doi.org/10.1051/epjconf/201817405003

Study of the performance of Micromegas detectors in magnetic field

Sampsonidis Dimitrios[1,a], On behalf of the ATLAS Muon Collaboration

[1] *Aristotle University of Thessaloniki, Thessaloniki 54124, Greece*

Abstract. Resistive Micromegas (MICRO MEsh GAseous Structure) detectors have been chosen by the ATLAS collaboration at LHC for the high luminosity upgrade, due to their capability to maintain full efficiency and high spatial resolution at high occupancy, for tracking muons in the forward region of the detector. The Inner Muon Station, in the high-rapidity region, the so called New Small Wheel (NSW), will be composed of micromegas detectors that will have to maintain good performance in the presence of magnetic field of up to about 0.3 T. The response of micromegas detectors is affected by the magnetic field, where the deflection of the drift electrons is described by the Lorentz angle, resulting in a bias in the reconstructed track position. Several test-beam campaigns have been performed to test the behaviour of small size resistive micromegas prototypes (10x10 cm^2) in magnetic fields up to 1 T, using high momentum muon and hadron beams at CERN. These studies are performed in order to validate the capability of the chambers to provide unbiased tracks in the NSW conditions. Measurements of the Lorentz angle and drift velocity as a function of the magnetic field are presented and both are compared to expectations based on Garfield-Magboltz simulations. Several methods to correct the position bias are applied, based on the chamber configuration or on the knowledge of the local value of the magnetic field. The results of these studies are presented together with an overall discussion of the Micromegas tracking capability in magnetic field.

1 Introduction

The LHC will be upgraded again during the second long shutdown (LS2) in 2018 [1]. After the LS2 the accelerator luminosity will increased to $2x10^{34}$ $cm^{-2}s^{-1}$, allowing ATLAS to collect approximately 100 fb^{-1}/year of data. The ATLAS detector [2], in order to benefit from the expected high luminosity performance, needs to be upgraded following the same schedule as the LHC. The Phase-I upgrade of the ATLAS muon spectrometer [1] focuses on the end-cap region, namely on the replacement of the inner station of the ATLAS muon end-cap system (Small Wheel, SW). The New Small Wheel (NSW) will have to operate in a high background radiation region while, providing muon tracks with high precision, as well as information for the Level-1 trigger [3].
The performance criteria demand for the precision track reconstruction a spatial resolution of about 100 μm and for the Level-1 trigger, on-line track segments with an angular resolution of approximately 1 mrad.
Two detector technologies have been chosen by the ATLAS collaboration for the NSW: small-strip Thin Gap Chambers (sTGC) as the primary trigger and Micromegas (MM) as the primary precision tracker [4].
With the proposed NSW the ATLAS muon system will maintain its excellent muon tracking in its entire acceptance region, even at the highest luminosities expected at the LHC . At the same time the Level-1 low p_T (typically $p_T > 20$ GeV) single muon trigger rate, will be kept at an acceptable level.

[a] e-mail: Dimitrios.Sampsonidis@cern.ch

2 The New Small Wheel Layout

The NSW, following the present structure of the Muon Spectrometer, is divided in sixteen sectors, eight large and eight small sectors. Each sector comprises sixteen layers of trigger and precision tracking detectors, namely sTGCs and MM covering a total area of 1200 m^2. Fig. 1 shows the MM segmentation in the large and the small sectors of the NSW [5].

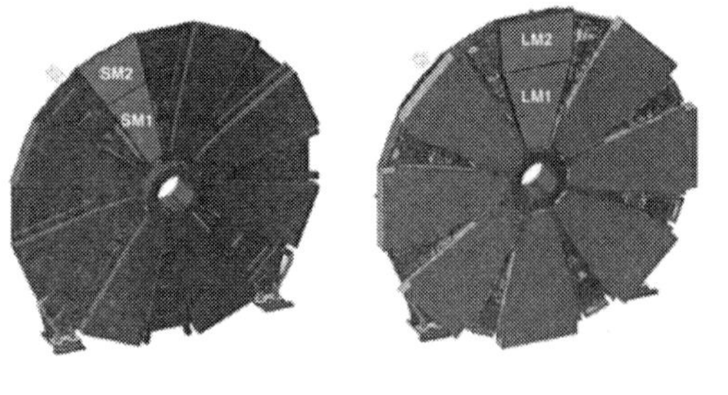

Figure 1. The two sides of the NSW showing the micromegas chambers segmentation. The Small sector (left) has two types of chambers, the SM1 and SM2 while the Large sector (right) has the LM1 and LM2 chambers

For both technologies, each module is a multiplet of four active layers each (hence quadruplets) and is arranged in the following order: sTGC-MM-MM-sTGC. The arrangement of the sTGC and MM modules will maximize the distance between the two sTGCs multilayers for improved track segment angular resolution at the trigger level. In total sixteen points will be measured for each track crossing the NSW. The choice of eight planes per detector will provide a robust, fully functional detector system over its whole lifetime.
In the radial direction the MM chambers are segmented in two. For the large sectors there are LM1 and LM2 while for the small sectors SM1 and SM2 respectively, where 1 and 2 refer to the small and the large radii.

3 The Micromegas chambers

The MM chambers have trapezoidal shape, sizes between 2 and 3 m^2 and they are arranged in quadruplets in order to provide four independent points.
Within a quadruplet the four MM layers are grouped in two pairs. In each pair the detectors are mounted back-to-back onto stiffening panels with internal structure of aluminium honeycomb as shown in Fig. 2. The readout strips in the first pair of detector layers are parallel to the base of the trapezoid in order to measure the precision coordinate (η-strips). The readout strips, in the other two layers, are inclined by ± 1.5 (stereo strips) with respect to the η- strips, to provide the second coordinate (azimuthal) measurement. The strip pitch for the small sectors will be 425 μm while for the large ones 450 μm resulting in about 2x106 readout channels in total for the MM system of the NSW.

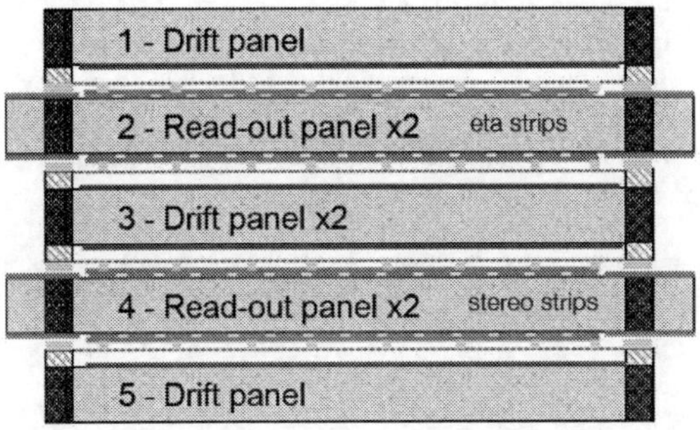

Figure 2. The arrangement of the MM layers within a quadruplet. The drift and readout electrodes are mounted back-to-back onto stiffening panels.

4 The Micromegas detector

The MM detector has been developed in the 1990's [4]. The principles of operation of the MM chambers chosen for the NSW are shown in Fig. 3, as well as the baseline parameters of the ATLAS-NSW chambers.
It consists of planar (Drift) electrode, a gas gap of a five millimetres thickness acting as conversion and drift region, and a fine metallic mesh at 128 μm distance from the readout electrode, creating the amplification region. The readout strips are covered by a layer of resistive strips on top of a thin insulator. In such a way the MM is protected from the sparks and able to operate in high rate environment [6]. The signals are capacitively coupled to the readout electrode that is not anymore directly exposed to the charge created in the amplification region. For the ATLAS NSW, the mesh has been grounded and positive HV was applied to the resistive strips.

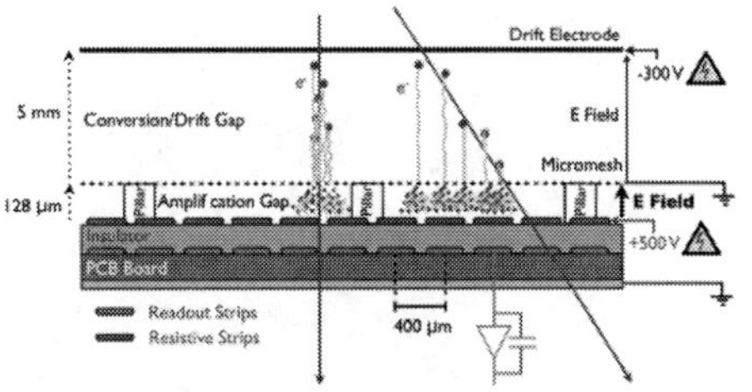

Figure 3. Schematic view of the Micromegas detector (not to scale) and the principles of operation.

When the ionising particle traverses the active volume of the micromegas it causes ionization of the gas. The electrons produced by the ionization drift towards the mesh due to the electric field of a few hundred V/cm in the drift region. Having an electric field of 40-50 kV/cm in the amplification region the mesh is almost transparent to the electrons. The electron avalanche takes place in the amplification region above the strips. The drift of the electrons in the conversion gap is a slow process and depends on the gas Ar:CO_2 (93:7), the drift distance and the drift field takes several tens of nanoseconds while the amplification process happens in a fraction of a nanosecond, resulting in a fast pulse of electrons on the readout strip.
The MM will provide spatial resolution better than 100 μm independent of track incident angle. The performance of MM detectors was studied with prototypes of various sizes with particles in beams. These studies are based on the properties of the reconstructed clusters. The cluster is defined as a group of fired strips with charge above a threshold. The cluster position is evaluated utilizing a combination of two algorithms. For small incident angles, a simple cluster centroid method is employed where the position is calculated as the average of strip position weighted by the strip charge. For larger incident angles, the micro-Time-Projection-Chamber (μTPC) method is used. This method exploits the single strip time information. A track is reconstructed inside the drift gap by converting the measured drift time of charges arriving on individual strips to the position from which the drift electrons originate based on the drift velocity. The cluster position is calculated as the interpolation of the track in the middle of the drift gap. The μTPC method has been successfully applied in several test-beam experiments at CERN.

 https://doi.org/10.1051/epjconf/201817405003

5 Micromegas in a Magnetic Field

The MM chambers of the NSW will be exposed to stray magnetic fields from the barrel and end-cap toroids with field strengths of up to about 0.3 T with different orientations with respect to the chamber planes but with a sizable component orthogonal to the MM electric field. The effect of the magnetic field on the detector operation has been studied with test beam data and simulations.
Fig. 4 shows the ϕ component of the magnetic filed in the ATLAS detector at a distance z=7800 mm from the interaction point [[7].

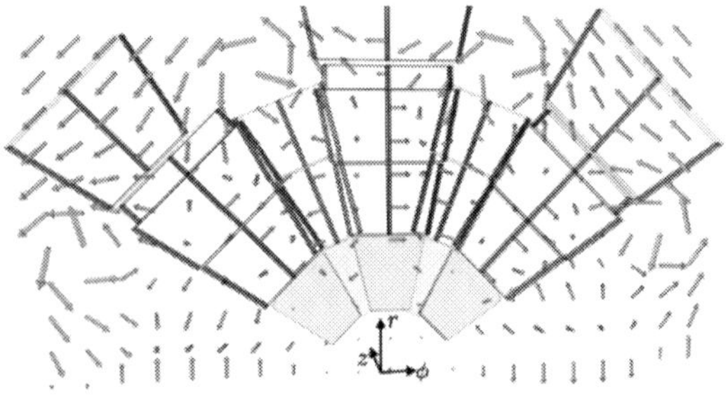

Figure 4. The ϕ component of the magnetic field at a distance z =7800 mm from the interaction point. The currently installed detectors of the Small Wheel (MDT, CSC) are displayed for reference. The plot has been created with Persint ATLAS software [8].

The electron drift velocity and the Lorentz angle as a function of the electric field for the gas mixture, Ar:CO_2 (93:7), were simulated under different magnetic fields using the Garfield [8] and the results are shown in Fig. 5(a) and 5(b) respectively.

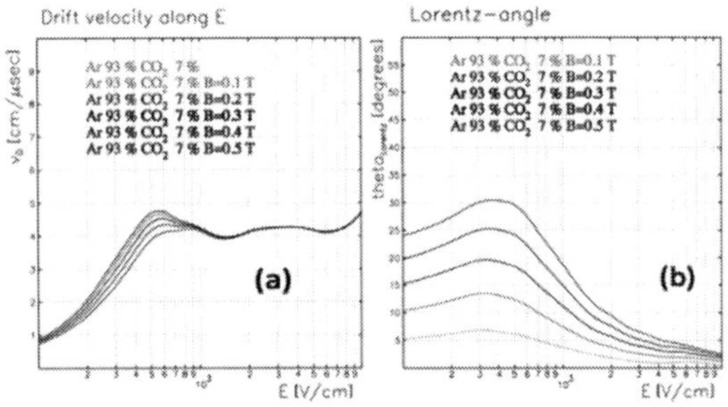

Figure 5. Drift velocity along the electric field (a) and Lorentz angle (b) as a function of the drift field for different values of the magnetic field.

The effect of the magnetic field on the detector operation has been studied with test beam data and simulations. Figure 6 illustrates the effect of the magnetic field on a MM chamber. The drift direction of the ionization electrons is tilted with respect to the electric field direction by the Lorentz angle θ_L; tanθ_L is proportional to the magnetic field intensity B. The tilt of the drift direction gives a sizable shift (x) of the reconstructed hit position (Fig. 6a). This shift affects both centroid and μTPC reconstruction and reaches values up to 400 μm. Singular configurations are expected when $\theta = \theta_L$, θ being the inclination angle of the track (Fig. 6b).
In this configuration the ionization cluster is confined to a very small number of strips and the centroid method can be applied. On the other hand, when θ and θ_L have different signs, the ionization cluster is spread over a larger number of strips. A combination algorithm can be applied to have a constant resolution through all configurations.

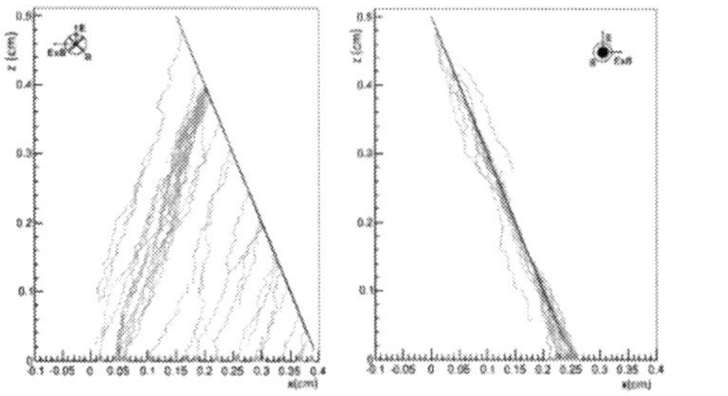

Figure 6. In the magnetic field the drift direction of the ionization electrons is tilted with respect to the electric field direction by the Lorentz angle θ_L. Two cases with the magnetic field in opposite directions from Garfield simulation [8].

6 Test Beam

Several test-beam campaigns have been performed to test the behaviour of small size resistive micromegas prototypes (10x10 cm^2) as well as large prototypes in magnetic fields up to 1 T using high momentum muon and hadron beams at CERN. Figure 7 shows the MM prototypes inside the "Goliath" dipole magnet at H2 area at CERN during the December 2014 test beam run. The magnet was able to provide a magnetic field up to 1 T in two polarities. These studies were performed in order to validate the capability of micromegas chambers to operate and get unbiased tracks in the NSW conditions.

Figure 7. The micromegas chambers installed in the experimental area H2 at CERN in the Goliath magnet during December 2014 test beam run.

The systematic shifts of the reconstructed position due to the magnetic field can be corrected in two ways: i) by measuring accurately the magnetic field on the NSW with an adequate number of magnetic field sensors, ii) by using the average point measured in a doublet in back-to-back configuration, which is free of systematics, as

demonstrated with simulations.
Figure 8 shows the reconstructed incident angle, the so call effective angle λ, which does not represent the real track direction because of the Lorentz angle. If the Lorentz angle is known the true incident ϕ can be computed. The back-to-back configuration of the doublets is exploited by defining "SuperPoints". Since the shifts of the reconstructed hits in a doublet are equal and in opposite directions, a SuperPoint is defined as the "average" of the two hits. The SuperPoint coordinate x_{SP} is defined by:

$$x_{SP} = \frac{x_1 + x_2}{2} \tag{1}$$

where x1 and x2 are either centroids or μTPC precision coordinates. x_{SP} is provided if the magnetic field (B), the electric field (E) and the drift gap are the same for both chambers in the back-to-back configuration

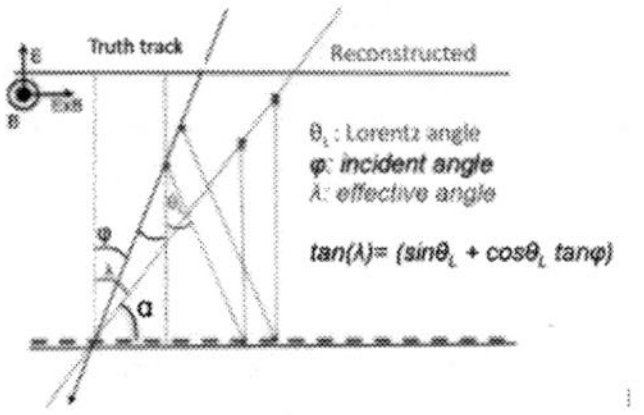

Figure 8. The reconstructed angle of incident λ, is biased due to the Lorentz angle θ_L. Application of the correction formula can give the true incident angle ϕ.

The values of the Lorentz angle θ_L can be extracted from the data by measuring the shift of the cluster position distributions with respect to no magnetic field case, at the different magnetic field intensities. In Fig. 9 the measured values of θ_L are compared to Garfield values [9].

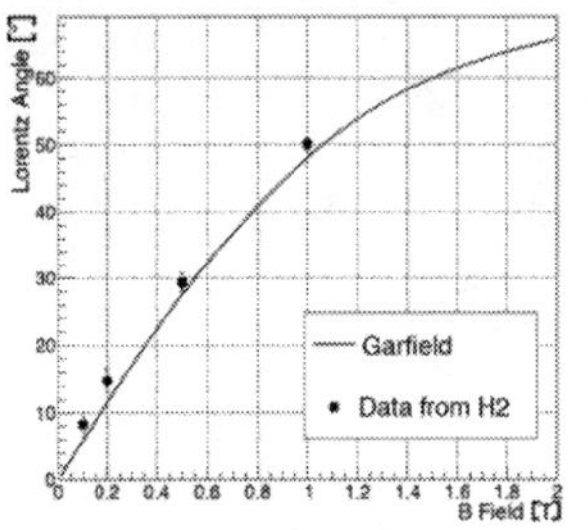

Figure 9. The Lorentz angle as function of the magnetic field (B) calculated with Garfield simulation and experimental points from test beam results.

The component of the drift velocity along E field can be measured, from the distribution of the arrival times of the single clusters knowing the gap size and the applied B field. Knowing the Lorentz angle the Drift Velocity can also be calculated [7].

Figure 10 shows the comparison of the drift velocity with data from the test beam (left) and the component of drift velocity along the electric field as a function of the magnetic field compared to expectations based on Garfield-Magboltz simulations. Results are consistent with Garfield prediction.

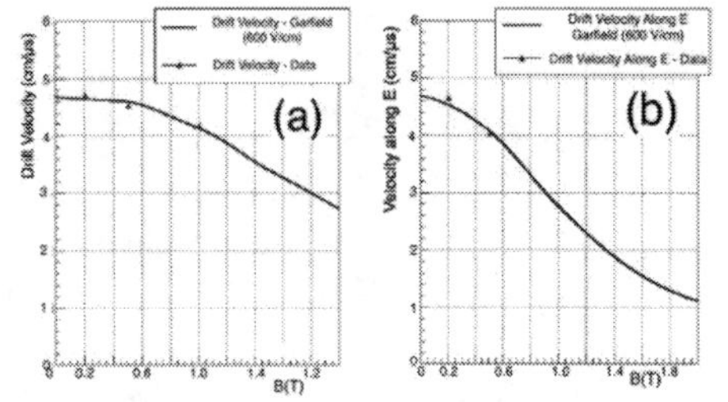

Figure 10. Comparison of the data from test beam with Garfield simulation predictions for the drift velocity (a) and drift velocity along the electric field (b).

7 Conclusions

The Micromegas technology chosen for the ATLAS NSW upgrade will enable the Muon Spectrometer to retain its excellent performance. These will be the largest Micromegas detectors ever built for large area chambers in a particle physics experiment and in a high rate environment. Optimised reconstruction techniques have been developed in the context of the test beam studies in order to improve the performance of the Micromegas chambers in magnetic field. Studies inside magnetic field do not show any sign of degraded performance.

References

[1] LHC Collaboration, The CERN Large Hadron Collider : Accelerator and Experiments, JINST 3 S08001 (2008).
[2] ATLAS Collaboration, The ATLAS Experiment at the CERN Large Hadron Collider, JINST 3 S08003 (2008).
[3] ATLAS Collaboration, ATLAS New Small Wheel TDR, CERN-LHCC-2013-006 ATLAS-TDR-020-2013.
[4] Y. Giomataris, P. Rebourgeard, J.P. Robert and G.Charpak, Micro-Pattern Gaseaus Detectors, Nucl. Instrum. and Meth. A376 29-35 (1996) .
[5] M. Bianco, Development of large size Micromegas detectors for the upgrade of the ATLAS experiments, PoS (EPS-HEP 2013) 488 (2013).
[6] T. Alexopoulos, et al., A spark-resistant bulk-micromegas chamber for high-rate applications, Nucl.Instrum.Meth. A640 110-118 (2011).
[7] G. A. Iakovidis, Research and Development in Micromegas Detector for the ATLAS Upgrade, CERN-THESIS-2014-148, (2014).
[8] R. Veenhof, Garfield simulation of the gaseous detectors, urlhttp://garfield.web.cern.ch/garfield/.
[9] K. Ntekas, Design and performance studies of micromegas chambers for the ATLAS muon spectrometer upgrade, ATL-MUON-PROC-2014-011.

https://doi.org/10.1051/epjconf/201817405004

Candidate eco-friendly gas mixtures for MPGDs

L. Benussi[1], S. Bianco[1], G. Saviano[1,2,a], S. Muhammad[1,2,3], D. Piccolo[1], M. Ferrini[2], M. Parvis[4], S. Grassini[4], S. Colafranceschi[5], J. Kjølbro[6], A. Sharma[5], D. Yang[6], G. Chen[6] and Y. Ban[6], and Q. Li[6]

[1] *Laboratori Nazionali di Frascati - INFN, Frascati, Italy*
[2] *University of Rome "La Sapienza" (IT) - Facoltà di Ingegneria, Ingegneria Chimica Materiali ed Ambiente, Italy*
[3] *National Center for Physics, Quaid-i-Azam University Campus, Islamabad, Pakistan*
[4] *Politecnico di Torino-Dipartimento di Fisica (DIFIS) Corso Duca degli Abruzzi, 24, I-10129 Torino, Italy*
[5] *CERN, Geneva, Switzerland*
[6] *Peking University, Beijing, China*

Abstract. Modern gas detectors for detection of particles require F-based gases for optimal performance. Recent regulations demand the use of environmentally unfriendly F-based gases to be limited or banned. This review studies properties of potential eco-friendly gas candidate replacements.

1 Introduction

Many currently used refrigerant gases have a great impact on the environment since they either contribute largely to the greenhouse gas effect, or because they tear the ozone layer, or both. In an attempt to protect the environment, regulations preventing the production and use of certain refrigerant gases have been implemented [1]. Gas detectors are wide spread for detection, tracking and triggering of charged particles such as muons in Nuclear and High Energy Physics (HEP). A large part of gas muon detectors used in HEP operates with mixtures containing the regulated refrigerants as quenching medium in applications where excellent time resolution and avalanche operation are necessary. Therefore, actions towards finding new mixtures must be undertaken. Gas Electron Multiplier (GEM) [2] detectors operate in experiments such as CMS (Compact Muon Solenoid) at the LHC (Large Hadron Collider) with an argon/CO_2 mixture [3]. However, for high time resolution applications an argon/CO_2/CF_4 mixture is used [4], where CF_4 has a Global-Warming Potential (GWP) of 7390 [5]. Resistive Plate Counters (RPC) [6] currently operate with a F-based R134a/Isobutane/SF_6 gas mixture, with typical GWP of 1430. Investigations into new gas mixtures have to be performed in order to keep the mixture properties while complying with the regulations. A few industrial refrigerant industrial replacements were proposed [7] as alternatives to R134a. A study of transport properties of currently used gas mixtures in HEP, and evaluation of transport properties of freon-less gas mixtures, was recently published.

The aim of this paper is to discuss some of the important properties of gases for particle gas detectors, to list and summarize basic properties of eco-friendly refrigerants from the literature, to discuss their properties for materials compatibility and safe use, and make a prediction on selected parameters crucial for the performance of gas detectors considered. While this study is aimed to GEM and RPC detectors, its findings can be considered for selection of ecogas replacement for other gas detectors.

2 Gas properties

For a gas mixture to be appropriate in a gas detector, first of all it has to comply with the regulations. Furthermore, its properties must also be appropriate for the specific type of detectors. For example, a gas that is suitable for the RPC detectors may not be fully optimized for the GEM detectors. To better find the appropriate gas for a detector, an understanding of the influence of different parameters is required. This section aims to clarify the most essential parameters for gases.

In order to estimate the impact of a refrigerant on the environment, the effects have to be quantified. Two important effects are the contribution to the greenhouse effect and the depletion of the ozone layer. The first mentioned effect is measured in Global- Warming Potential (GWP), and is normalized to the effect of CO_2 (GWP = 1), while the effect on the ozone layer is measured in Ozone Depletion Potential (ODP), normalized to the effect of CCl_3F (ODP = 1). The effects of selected refrigerant candidates are listed in table 1. When a particle passes through a medium, energy is transferred from the particle to the surroundings. The energy lost is typically defined as the stopping power expressed as $\frac{1}{\rho}\left(\frac{dE}{dx}\right)$ where ρ denotes the density of the medium, E denotes energy, and x is length.The radiation length X_0 is a characteristic length of a medium. It describes both the mean distance

[a] Correspondingauthor; e-mail address:
Giovanna.Saviano@uniroma1.it -and- *Giovanna.Saviano@cern.ch*

required for a high energy electron to lose all but e^{-1} of its energy due to bremsstrahlung, and $\frac{7}{9}$ of the mean free path of a $e^{+}e^{-}$ produced by a high-energy photon [8]. When an incoming particle passes through a medium, it will eventually interact with the medium and transfer some of its energy to ionize atoms. In this process, a pair consisting of an ionized atom and a free electron is produced. The number of ionizations produced by an incoming particle per unit length is denoted byN_P , in units of cm^{-1} . Each produced ion pair will have an initial kinetic energy and can itself produce an ion pair, called secondary ion pair production. In the absence of electric field, electrons move randomly in all direction having an average thermal energy$\frac{3}{2KT}$. In presence of an electric field, the electrons start to drift along the field direction with mean drift velocity v_d (the average distance covered by the drift electron per unit time). The average distance an electron travels between ionizing collisions is called mean free path and its inverse is the number of ionizing collision per centimetre (the first Townsend coefficient). This parameter determines the gas gain of the gas. Many refrigerants may constitute danger for the user and its environment. The greatest dangers involved are the flammability and toxicity. The Health Material Hazardous Material Information System (HMIS), rates Health/ Flammability/ and Physical hazards from 0 (low) to 4 (high). Some refrigerants are incompatible with certain materials, and can either react violently, or have long term effect. Some refrigerants may even produce toxic decomposition and/or polymerization.

The aging phenomenon is very complex and depends on several parameters. The commonly used variables include the cross-sections, electron/photon energies, electrostatic forces, dipole moments, chemical reactivity of atoms and molecules, etc.

3 Estimation of Gas Parameters

Quantities such as the minimum ionization energy can be found if the stopping power is known. An approximate expression for moderately relativistic particles in the momentum region $0.1 \leq \gamma\beta \leq 1000$ can be found using the Bethe-Bloch equation, given by

$$\frac{1}{\rho}\left(\frac{-dE}{dx}\right) = Kz^2\frac{Z}{A}\frac{1}{\beta^2}\left(\frac{1}{2}ln\frac{2m_ec^2\beta^2\gamma^2T_{max}}{I^2} - \beta^2 - \frac{\delta(\beta\gamma)}{2}\right) \quad (1)$$

Where $\left(\frac{-dE}{dx}\right)$ is the mean energy loss per length, ρ denotes the density of the medium, I is the mean excitation energy and $\delta(\beta\gamma)$ is the density effect correction function to ionization energy loss. K is a constant given by $4\pi N_A r_e^2 m_e c^2 r^2$ and T_{max} is the maximum energy transfer in a single collision, given by

$$T_{max} = \frac{2m_ec^2\beta^2\gamma^2}{1 + \frac{2\gamma m_e}{M} + (m_e/M)^2} \quad (2)$$

where M is the mass of the incoming particle. The radiation length X_0 of an atom can be found by [8],

$$X_0 = \frac{716.405(cm^{-2}mol)A}{Z^2(L_{rad} - f(z)) + Z\acute{L}_{rad}} \quad (3)$$

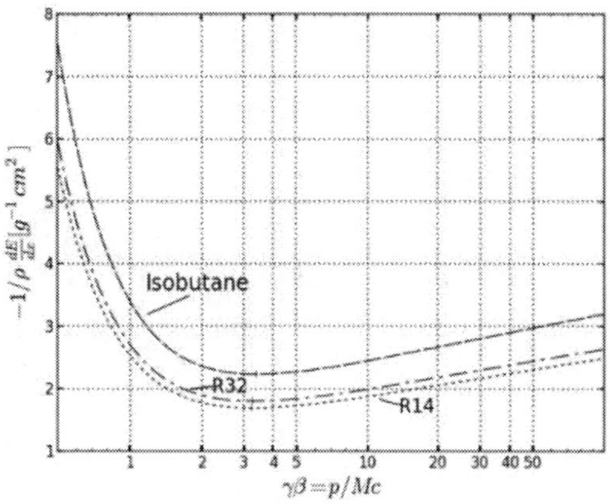

Figure 1: Energy loss as a function of the relativistic time dilation factor $\gamma\beta$ for various refrigerants.

Table 1 describes the important parameters of some refrigerants. An approximate correlation between primary ionization and atom number has been found based on experimental data by [9]

$$N_p = 3.996\left(\frac{Z_m}{Z^{0.4}}\right) - 0.025\left(\frac{Z_m}{Z^{0.4}}\right)cm^{-1} \quad (4)$$

The above expression holds at normal temperature and pressure (NTP) (1atm, 20^{0}C). For different pressure and temperature, the number scales with the density. This value should only be taken as a rough estimation though. This formula has proven to work best for hydrocarbons and worst for molecules consisting mainly of fluorine, differing as much as 30% from the experimental value for CF_4.

4 Molecules and their optimized geometries

The freon gases we have been interested in are R134a (CH_2FCF_3), R152a ($C_2H_4F_2$), HFO1234ze ($CFHCHCF_3$), HFO1234yf (CH_2CFCF_3), CF_3I and HFO1233zd ($CHClCFCF_3$). Meanwhile we choose R12(CCl_2F_2) to be the standard freon gas model. CH_4 and CF_4 are the molecules we were used to make a comparison between the NWC_{HEM} calculated results and experimental results, in order to check the reliability of NWC_{HEM}. Figure 2 shows the optimized ground state geometries of gas molecules, where the green balls stand for carbon atoms, the grey balls stand for hydrogen atoms, the indigo balls stand for fluorine atoms, the brown balls stand for chlorine atoms, and purple ball for iodine. Figure 3 shows the highest occupied molecular orbitals (HOMO) of gas molecules.

5 Calculation of absorption spectrum

The excitation energy of a molecule is one of the fundamental properties of molecular interactions one can get

Table 1: Minimum ionization, radiation length and number of primary ion pair creation for the considered refrigerants, as well as the approximated mean ionization energy used.

Name	I [eV]	$\left(\frac{-dE}{dx}\right)$ $[MeV\frac{g}{cm^2}]$	X_0 $[\frac{g}{cm^2}]$	N_p $[cm^{-1}]$
R32	89.3602	1.80973	35.4581	49.2
R7146	127.401	1.67833	28.6027	92.0
R600a	47.848	2.24057	45.2260	81.0
R1234yf	91.9674	1.7734	35.8204	89.5
R152a	78.1889	1.88706	37.0969	67.1
R1234ze	91.9674	1.7734	35.8204	89.5
R115	116.695	1.69178	29.2197	98.4
R1233zd	106.689	1.73915	29.7636	105
R290	47.0151	2.26184	45.3725	65.2
R1311	271.737	1.42486	11.5399	272
R134a	95.0294	1.76439	35.1542	81.6
R14	107.127	1.69909	33.9905	63.6
R123	125.275	1.69722	25.5416	98.4
R143a	87.8152	1.8126	35.8928	74.8
R744	88.7429	1.81124	36.1954	37.2
R23	99.9508	1.7402	34.5214	56.9
R116	105.075	1.70566	34.2947	93.3
RC318	101.578	1.71721	34.8435	123
R218	104.13	1.70873	34.439	117

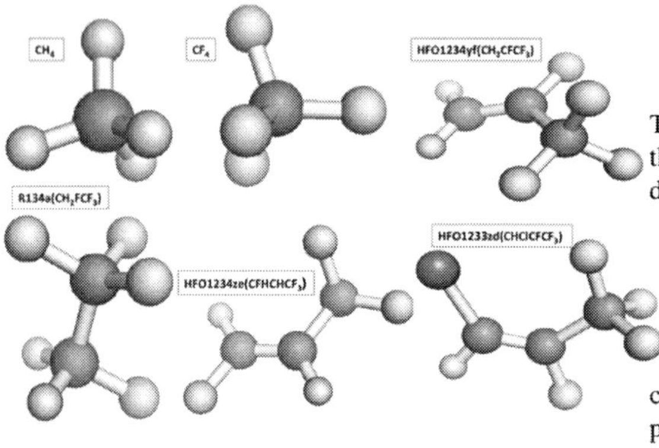

Figure 2: Optimized ground state geometries of gas molecules.

from experiment. To study these properties, we used the framework introduced in Ref.[10] to simulate the time-dependent response of molecules under external fields using quantum chemical calculations. The framework implemented is the Real-Time Time-Dependent Density Functional Theory (RT-TDDFT) [11] method within the NWCHEM , making it capable of doing the simulation beyond small perturbation from the ground state. In our case, we are mostly interested in low excitations of small molecules. We therefore adopt the procedure as described in section 3 of the reference [10]. The choice of external field is δ-function-like electric field "kick"

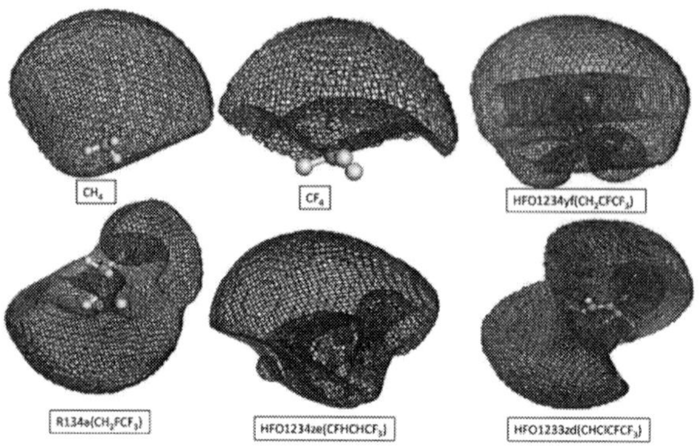

Figure 3: Highest Occupied Molecular Orbitals (HOMO) of gas molecules.

$$E(t) = k.exp\left[\frac{-(t-t_0)^2}{2\omega^2}\right]\hat{d} \tag{5}$$

where t_0 is the centre of pulse, ω is the pulse width, which has dimesnions of time, $\hat{d} = \hat{x}, \hat{y}, \hat{z}$ is the polarization of the pulse, and k is the maximum field strength. The system is then evolved in time, and the dipole moment can be calculated with respect to the added dipole coupling term

$$V^{app}_{\mu\nu}(t) = -D_{\mu\nu} \cdot E(t) \tag{6}$$

where D is the transition dipole tensor of the system. Then we Fourier transform the dipole signals to construct the complex polarizability tensor $\alpha_{i,j}(\omega)$, and finally the dipole absorption spectrum is

$$S(\omega) = \frac{1}{3}Tr[\sigma(\omega)] = \frac{4\pi\omega}{3c}Tr[Im[\alpha_{i,j}(\omega)]] \tag{7}$$

We first validate our calculation by repeating the calculation of CH_4 lowest excitation energy as given in the paper [10]. With the same basis set (6-311G) and functional (B3LYP), our calculation gives 11.16 eV excitation energy, which is consistent with the 11.13 eV in the paper. The small difference can come from other minor uncertainty sources like the choice of time separation in the simulation.

6 Estimation of the first Townsend parameter

Following Ref. [12], we express the dependance of the first Townsend parameter α as a function of the reduced electric field $\frac{E}{p}$, and $\frac{\alpha}{p} = Aexp\left[\frac{-Bp}{E}\right]$, The reference [12] shows that the first Townsend coefficient at high reduced electric field depends almost entirely on the mean free path of the electrons. The mean free path, which is defined as $\lambda_m = \frac{1}{n\sigma}$. where n is the number of atoms per unit volume and σ is the total cross section for electron collision

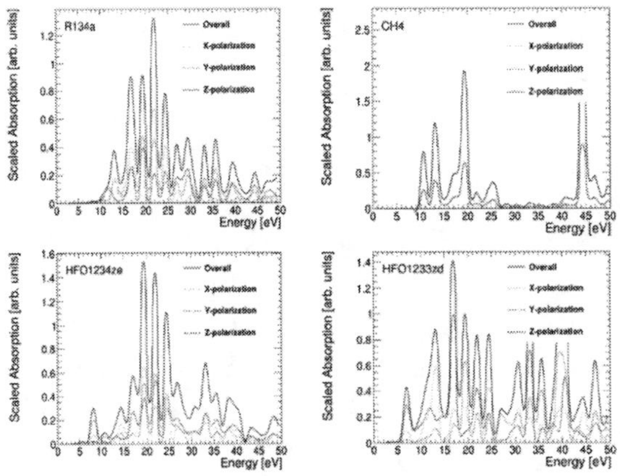

Figure 4: Absorption spectrum of interested gas molecules. The calculation is performed in the Atomic Orbital (AO) basis, where the solid line corresponds to the overall dipole strength and the dashed lines correspond to polarizations in the coordinate basis directions.

with atoms, can be calculated if the environment of gas molecule is provided. For real gases, we cannot take the free path lengths as a constant and the ionization cross section is only a fraction of the total cross section. In that case the estimation needs to be modified and use the following equation:

$$\alpha(r) = Ap{\cdot}exp\left(\frac{-Bp}{E(r)}\right)\left(1 - exp\left(\frac{-I_0 n\sigma}{eE(r)}\right)\right) + n\sigma_i{\cdot}exp\left(\frac{-I_0 n\sigma}{eE(r)}\right) \tag{8}$$

Numerous measurements of the Townsend coefficients are available for standard gas mixtures, such as those reviewed in [13].

7 Conclusions

Currently used F-based gases today used in HEP gas detectors are being phased out by industry and replaced by eco-friendly substitute gases. This study has reported on a general survey of industrially available replacements for HEP gases, discussed their physical properties, materials compatibility and safety issues. Parameters of interest for their use in HEP detectors have been computed following different approaches ranging from parametrizations to quantum chemical calculations: ionization energy, electronegativity, number of primary pairs. Statistical methods to compute amplification parameters of the ionization shower production such as the Townsend coefficients were investigated and preliminary results reported. Promising candidates with lower GWP are identified for further studies.

References

[1] L. Benussi et al., "Properties of potential eco-friendly gas replacements for particle detectors", LNF note INFN-14-13/LNF, CERN-OPEN-2015-004 [arXiv:1505.00701 [physics.ins-det]].

[2] F. Sauli and A. Sharma, "Micropattern gaseous detectors," Ann. Rev. Nucl. Part. Sci. 49 (1999) 341.

[3] A. Sharma, "Muon tracking and triggering with gaseous detectors and some applications," Nucl. Instrum. Meth. A 666 (2012) 98.

[4] M. Alfonsi, G. Bencivenni et,al., Aging measurements on triple- GEM detectors operated with CF-4 based gas mixtures," Nucl. Phys. Proc. Suppl. 150 (2006) 159.

[5] International Panel on climate changes. http://www.ipcc.ch/publications and data/ar4/wg1/en/ch2s2-10-2.html

[6] R. Santonico and R. Cardarelli, "Development of Resistive Plate Counters," Nucl. Instrum. Meth. 187 (1981) 377.

[7] M. Abbrescia, Updates from the DESY upgrade meeting, presented at ECFA meeting, June 19th, 2013, DOI http://10.5072/oar.it/1430579044.36.

[8] Y. S. Tsai, "Pair Production and Bremsstrahlung of Charged Leptons," Rev. Mod. Phys. 46 (1974) 815 [Erratum-ibid. 49 (1977) 521].

[9] I. B. Smirnov, "Modeling of ionization produced by fast charged particles in gases," Nucl. Instrum. Meth. A 554 (2005) 474.

[10] K. Lopata and N. Govind, Journal of Chemical Theory and Computation 7(5), 1344 (2011), URL http://dx.doi.org/10. 1021/ct200137z, http://dx.doi.org/10.1021/ct200137z.

[11] Y. Takimoto, F. D. Vila, and J. J. Rehr, The Journal of Chemical Physics 127(15), 154114 (2007), URL http://scitation. aip.org/content/aip/journal/jcp/127/15/10.1063/1.2790014.

[12] Y. I. Davydov, IEEE Transactions on Nuclear Science 53, 2931 (2006), physics/0409156.

[13] A. Sharma and F. Sauli, "A Measurement of the first Townsend coefficient in argon based mixtures at high fields," Nucl. Instrum. Meth. A 323 (1992) 280.

Numerical Investigation on Electron and Ion Transmission of GEM-based Detectors

Purba Bhattacharya[1,a], Sumanya Sekhar Sahoo[1], Saikat Biswas[1], Bedangadas Mohanty[1], Nayana Majumdar[2], and Supratik Mukhopadhyay[2]

[1] School of Physical Science, National Institute of Science Education and Research, Jatni, Khurda, India 752050
[2] Applied Nuclear Physics Division, Saha Institute of Nuclear Physics, 1/AF, Bidhannagar, Kolkata, India 700064

Abstract. ALICE at the LHC is planning a major upgrade of its detector systems, including the TPC, to cope with an increase of the LHC luminosity after 2018. Different R&D activities are currently concentrated on the adoption of the Gas Electron Multiplier (GEM) as the gas amplification stage of the ALICE-TPC upgrade version. The major challenge is to have low ion feedback in the drift volume as well as to ensure a collection of good percentage of primary electrons in the signal generation process. In the present work, Garfield simulation framework has been adopted to numerically estimate the electron transparency and ion backflow fraction of GEM-based detectors. In this process, extensive simulations have been carried out to enrich our understanding of the complex physical processes occurring within single, triple and quadruple GEM detectors. A detailed study has been performed to observe the effect of detector geometry, field configuration and magnetic field on the above mentioned characteristics.

1 Introduction

A Time Projection Chamber (TPC) [1] is an ideal device for three-dimensional tracking, momentum measurement and identification of charged particles. They are used in many on-going experiments, including ALICE [2]. Owing to the enormous particle multiplicity per event, very specific requirements are made on the performance of the detectors in harsh radiation environments. Different R&D activities are currently concentrated on the adoption of the Gas Electron Multiplier (GEM) [3] as the gas amplification stage of the ALICE-TPC upgrade version. Despite the promise, several issues related to the operation of the GEM have to be resolved before it can be finally considered as an option. For example, to keep distortions due to space-charge at a manageable level, a lower ion feedback in the drift volume is required [4]. Again, for a substantial detector gain, it is important that a large fraction of primary electrons partcipate in the avalanche process and contribute to the signal generation. Thus, a proper optimization of the detector geometry, field configuration and gas mixtures are required to have a high electron transparency and low ion backflow.

In the present work, extensive numerical simulations have been carried out to estimate the effects of detector geometry, electric field configurations and magnetic field on electron transparency and ion backflow fraction. To begin with, single and triple GEM configurations have been studied in detail. A good understanding of the complex physics process in these two devices has allowed us to deal with relative ease the quadruple GEM configurations, which is being considered as a possible option for the ALICE TPC. In this manner, it has been possible for us to achieve a understanding of the likely optimum configurations of the quadruple GEMs.

2 Simulation Tools

The Garfield [5] simulation framework has been used in the present work. In this framework neBEM [6] is used to carry out 3D electrostatic field simulation. Besides neBEM, the Garfield framework provides interfaces to Magboltz [7] for computing drift, diffusion, Townsend and attachment coefficients.

Table 1: Design parameters of GEM-based detectors.

Polymer substrate	50 μm
Copper coating thickness	5 μm
Hole diameter (copper layer)	70 μm
Hole diameter (Polymer substrate)	50 μm
Hole to hole pitch	140 / 280 μm
Drift Gap	3 mm
1st Transfer gap	2 mm
2nd Transfer gap	2 mm
3rd Transfer gap	2 mm
Induction gap	2 mm

[a] e-mail: purba.bhattacharya85@gmail.com

3 Simulation Models

The design parameters of GEM-based detectors, considered in the numerical work, are listed in table 1. The model of a basic GEM cell built using Garfield, is shown in figure 1(a). It represents a GEM foil, having two bi-conical shaped holes placed in a staggered manner along with a readout anode and a drift plane on either sides of the foil. The distance between top surface of the GEM and the drift plane is called the drift gap whereas that between the lower surface and the readout plate is named induction gap.

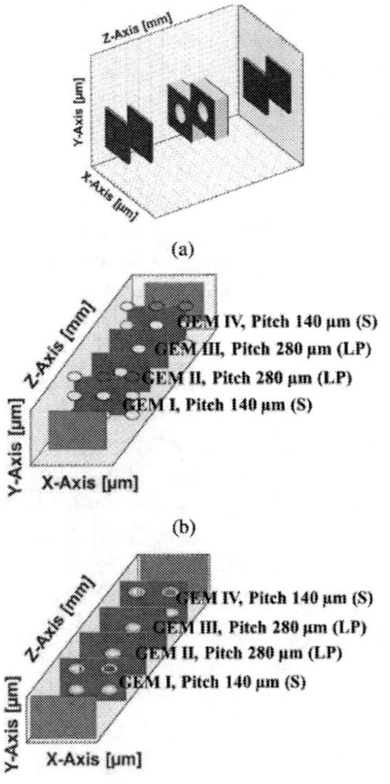

(a) (b) (c)

Figure 1: Simulation model for (a) single GEM and quadruple GEM with (b) aligned holes (denoted as QGemI) (c) misalgned holes (denoted as QGemII)

In comparison to single GEM, in case of multi GEM detector, several GEM foils are placed in between the drift and the read-out plane. The naming scheme used in this work, numbers the foils in the order of the passage of electrons coming from the drift region. The top most GEM is called GEM I and the others are GEM II, GEM III and so on. The gap in between GEM I and II is called transfer gap I and that between GEM II and III is called transfer gap II etc. For example, the simulation models of two different quadruple GEM devices are shown in figure 1. Among the four foils, GEM I and GEM IV, have the pitch of 140 μm (denoted as S), whereas the middle two foils have a larger pitch of 280 μm (denoted as LP). This arrangement, denoted S-LP-LP-S, allows to block ions efficiently by employing asymmetric transfer fields and foils with

Table 2: Voltage configuration of triple, quadruple GEM.

	Triple GEM	Quadruple GEM
Drift Field	0.4 kV/cm	0.4 kV/cm
ΔV_{GEMI}	250 V	275 V
Transfer Field I	5 kV/cm	2 kV/cm
ΔV_{GEMII}	280 V	240 V
Transfer Field II	0.2 kV/cm	3 kV/cm
ΔV_{GEMIII}	325 V	254 V
Transfer Field III		1 kV/cm
ΔV_{GEMIV}		317 V
Induction Field	4.5 kV/cm	4.5 kV/cm

low optical transparency. An increasing sequence of gas gains down the GEM stack helps reducing the ion backflow since ions created in the inner two layers are blocked more efficiently. Two different geometrical variation in case of four GEM foils have been considered. In the first case (QGEMI), the central hole of the basic unit from all the four GEM foils are perfectly aligned (figure 1(b)). In the other case (QGemII), as shown in figure 1(c), the first and the last foils (S) are aligned with each other whereas the second and third foils (LP) are misaligned with them. In comparison to the quadruple GEM devices, the foils of triple GEM detectors have the same pitch of 140 μm and the holes are aligned perfectly. The basic cell structure then has been repeated along both positive and negative X and Y-Axes to represent a real detector. With the help of these models, the field configuration of the detectors have been simulated using appropriate voltage settings. These are followed by the simulation of electron transmission and ion backflow fraction in $Ne/CO_2/N_2$ (90/10/5) gas mixture.

For the calculation of electron transmission, 10,000 electrons have been injected in the drift gap in random positions. These electrons are made to drift towards the GEM foil. The electron transmission has been estimated as the ratio between the number of electrons that reach the anode plate to the number of electrons created in the drift volume.

The electrons during their drift produce avalanche inside the GEM foil. The primary ions in the drift region and the ions created in the avalanche have been considered for the estimation of the backflow fraction, calculated as the ratio between the ions drift back to drift volume to the total number of ions.

4 Results

4.1 Electron Transmission

Electron transmission can be presented as a function of two mechanisms: electron focusing and transverse diffusion. Electron focusing depends not only on the field ratio, but also on different geometrical parameters. On the other hand, the transverse diffusion is mainly affected by the electric field and the gas composition. Besides that, the

Table 3: ϵ_{coll}, ϵ_{ext} and ϵ_{tot} of triple, quadruple GEM detectors.

Geometry	B (T)	ϵ_{collI}	ϵ_{extI}	ϵ_{collII}	ϵ_{extII}	$\epsilon_{collIII}$	ϵ_{extIII}	ϵ_{collIV}	ϵ_{extIV}	ϵ_{tot}
Triple	0	0.996	0.329	0.324	0.042	0.989	0.402			0.002
QGemI	0	0.996	0.348	0.184	0.343	0.144	0.226	1.0	0.429	0.0003
QGemI	0.5	0.996	0.349	0.175	0.361	0.2	0.326	1.0	0.357	0.0005

electron attachment coefficient, can also influence transmission.

For a single GEM detector, the total electron transmission (ϵ_{tot}) can be identified as the multiplication of two efficiencies, the collection efficiency (ϵ_{coll}) and the extraction efficiency (ϵ_{ext}).
The collection efficiency has been defined as:

$$\epsilon_{coll} = \frac{\text{Electrons reached inside the GEM foil}}{\text{Electrons created in drift volume}} \quad (1)$$

The extraction efficiency has been defined as:

$$\epsilon_{ext} = \frac{\text{Electrons reached the readout plane}}{\text{Electrons present inside the GEM foil}} \quad (2)$$

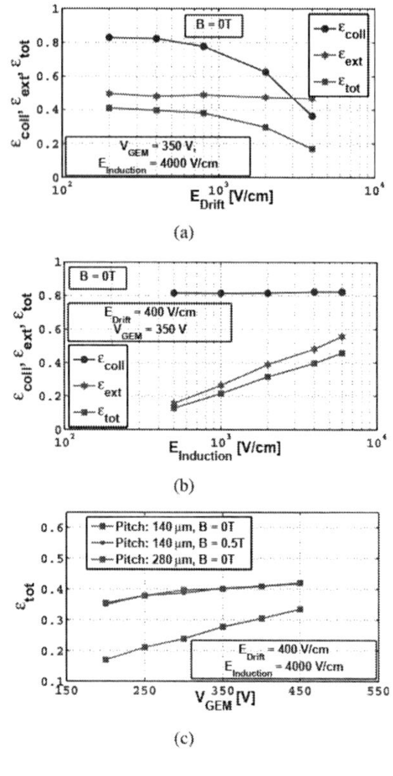

Figure 2: Variation of ϵ_{coll}, ϵ_{ext}, ϵ_{tot} with (a) E_{Drift} and (b) $E_{Induction}$. (c) Variation of ϵ_{tot} with V_{GEM}. Effect of pitch and magnetic field are shown in (c).

The variations of ϵ_{coll}, ϵ_{ext} and ϵ_{tot} under different field configurations have been plotted in figure 2 . For a fixed V_{GEM} and $E_{Induction}$, ϵ_{coll} and thus ϵ_{tot}, decrease with the increase of the drift field, whereas no significant effects of drift field on ϵ_{ext} has been observed (figure 2(a)). Similarly, at a fixed V_{GEM} and E_{Drift}, the increase of induction field, increases ϵ_{ext} as shown in figure 2(b). The change of V_{GEM} only has effect on ϵ_{coll} and thus ϵ_{tot} (figure 2(c)). It is also seen from figure 2(c), for the same voltage configuration, the smaller pitch GEM foils is better in terms of higher electron transmission, whereas no significant effect of 0.5 T magnetic field has been observed.

For the present work, the voltage configuration for triple and quadruple GEM detectors are taken from [4] and [2], respectively and are listed in table 2. For the multi-GEM detectors, the electron transmission can be also identified as the multiplication of collection and extraction efficiencies of the individual GEM foils. ϵ_{coll} and ϵ_{ext} of individual GEM foils for a triple GEM and a quadruple GEM detectors (QGemI) have been listed in table 3. As is evident for triple GEM detector, the maximum loss happens at GEM II. This is because, the transfer field II is low, which acts as the induction field for the GEM II. Due to this low induction field ϵ_{ext} for GEM II is significantly low and thus affect ϵ_{tot}. For the quadruple GEM detector, for the present voltage configuration, no such low transfer field has been considered. But increase of number of GEM foil worsens the total transmission. ϵ_{tot} for the multi-GEM devices is also affected significantly by the variation in geometry. No significant effect of the magnetic field on ϵ_{tot} has been observed till now.

4.2 Ion Backflow Fraction

As mentioned earlier, the ions created in the avalanche process, drifting back to the drift volume, can disturb the homogeneity of the drift field and, thus, distort the behaviour of the detector. The backflow fraction mainly depends on the field ratio and the transverse spread of the electron avalanche. Thus, a proper optimization of the field in the drift volume, GEM hole and induction regions is necessary to prevent those ions from entering the drift volume.

Table 4: ion collection of triple GEM detectors.

GEM I	GEM II	GEM III	Drift
0.0553	0.0248	0.8368	0.0814

The ion backflow of a single GEM can be reduced by decreasing E_{Drift} because less number of field lines will get out of the hole into the drift volume. At higher E_{Drift}, the

https://doi.org/10.1051/epjconf/201817406001

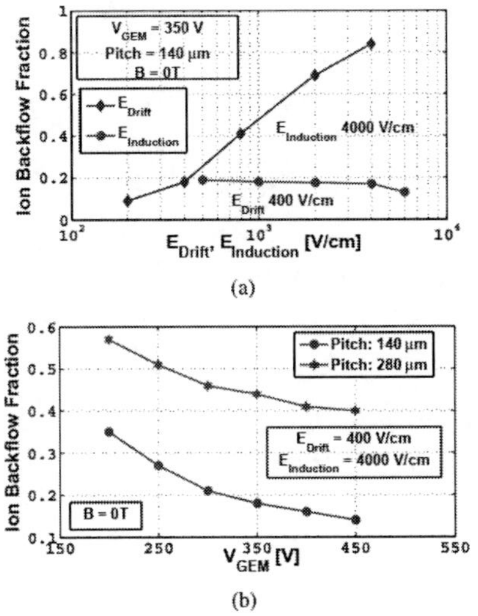

Figure 3: Variation of ion backflow fraction with (a) E_{Drift} and $E_{Induction}$, (b) V_{GEM}. Effect of pitch is shown in (b).

ratio between E_{Drift} and E_{GEM} is large resulting in the drift of more number of ions into the drift volumm. At higher E_{GEM}, the ratio between E_{Drift} and E_{GEM} is small and thus a large fraction of ions is collected at the top surface of the GEM foil. No significant effect of $E_{Induction}$ has been observed except at the higher $E_{Induction}$. From figure 3, it is also seen that the GEM foil with smaller pitch is better in terms of lower backflow fraction, whereas no significant effect of 0.5T magnetic field has been observed.

A better suppression of the ion backflow is known to be achieved by using multiple GEM structures. For the triple GEM detector, the ions collection efficiency of three individual GEM foils is listed in table 4. Most of the ions are collected on GEMIII and only 5.4% ions are able to travel back towards the drift volume. The collection efficiencies of four GEM foils for the two different geometry have been listed in table 5. The backflow for the first case is ~ 1.7%, whereas the misaligned holes decrease its value to ~ 0.2%. The effect of magnetic field has been studied in conjuction with the case when there is no magnetic field. No effect of 0.5T field on overall backflow fraction has been observed.

5 Conclusion

In the upgraded version of the ALICE TPC, the amplification device has been planned to be based on the GEM. Despite its excellent resolution and rate handling capabilities, ion backflow leading to possible space charge accumulation continues to be a source of likely problem for this TPC. In this work, an attempt has been made to numerically model and optimize the geometrical and electrical configuration of related amplification devices in terms of electron and ion transmission. Study of single GEM detectors shows that higher electron transmission and lower backflow fraction can be obtained with higher GEM voltage, lower drift field and higher induction field. GEM foils with larger pitch gives better electron transmission, as well as less backflow fraction. Multi-GEM devices are found to be better in terms of lower backflow fraction though the electron transmission is affected adversely. Several studies on triple and quadruple GEM detectors with various geometry and field configuration which are likely candidate for the ALICE TPC upgrade, have been performed. No significant effect of 0.5T magnetic field has been observed. Further work is necessary to achieve a comprehensive understanding as well as to find out an optimal geometry and field configuration in the scenario of ALICE upgrade. Important details such as space charge and charging up effects, estimates of manufacturing tolerances and defects have been left out of the present computations. In future, we hope to make progress in these areas in order to achieve an even better understanding of these devices.

Acknowledgements

This work has partly been performed in the framework of the RD51 Collaboration. We wish to acknowledge the members of the RD51 Collaboration for their help and suggestions. We would also like to thank the members of the ALICE-TPC collob-oration for their valuable suggestions. We thank our respective Institutions for providing us with the necessary facilities.

Table 5: ion collection of quadruple GEM detectors.

	GEMI	GEMII	GEMIII	GEMIV	Drift
QGemI 0T	0.056	0.006	0.15	0.78	0.023
QGemII 0.0T	0.13	0.008	0.207	0.65	0.0021
QGemII 0.5T	0.128	0.008	0.201	0.66	0.0023

References

[1] V. Lepeltier, Jour. Phy. Conf. Ser. **65** 012001 (2007).
[2] ALICE Collaboration; TDR **020** (2014).
[3] F. Sauli, Nucl. Instrum. Meth. A **386** 531 (1997).
[4] M. Ball et al., Jour. Instrum. **9** C04025 (2014).
[5] http://cern.ch/garfield.
[6] http://cern.ch/neBEM.
[7] http://cern.ch/magboltz.

Influence of water on the surface of graphene

Yunus Kaya[1,a], Yalçın Kalkan[2], and Rob Veenhof[2,3]

[1]Department of Chemistry, Faculty of Arts and Sciences, Uludağ University, 16059 Bursa, Turkey
[2]Department of Physics, Faculty of Arts and Sciences, Uludağ University, 16059 Bursa, Turkey
[3]RD51, CERN

Abstract. We have studied how water modifies the surface of graphene and in particular how the surface conductivity of graphene is affected. According to the literature, two types of interactions should be distinguished: physical, where a water molecule remains intact and is located at some distance from the mesh, and chemical, where a water molecule is imbricated in the graphene bond structure. We have developed theoretical models for both types of interactions using the density functional theory (DFT) with the B3LYP hybrid functional combined with the 6-31G(d) basis set. Our calculations show that the surface conductivity of graphene is reduced in the presence of water.

1 Introduction

Graphene, a two-dimensional sheet of sp^2-hybridised carbon atoms, is considered to be a next-generation electronic material due to its unique characteristics, such as high conductivity [1], optical transparency [2, 3], and ultra-high mechanical strength [4]. Graphene has been investigated for applications in small and fast electronic devices, more efficient batteries, and stronger materials, amongst others. In addition, the exceptional surface properties of graphene hold promise for applications in biomaterials, medical instruments, microfluidic devices, sensors and energy devices [5, 6]. For example, a large hydrophilic electrode surface allows reactants to maximise the reaction surface area in aqueous systems for better electrocatalysis. The surface properties are governed by both chemical composition and geometrical structure, which can also significantly affect the performance of nano-materials [7].

Both experimental observations [8] and theoretical studies [9] have revealed that graphene is a strongly hydrophobic material, which limits its applications in contact deposition in solution and may contaminate the nano-electro-mechanical systems [10]. In addition, materials with controllable wetting properties are widely used in advanced multifunctional systems, such as biomaterials and microfluidic devices [10, 11]. In particular, the wettability of graphene is essential for its utilisation in hybrid organic systems [12]. Although these hydrophobic properties, the graphene interacts with water molecules, and then the water may condense on the graphene changing its properties. In this study, we will examine how water modifies the surface conductivity of graphene.

2 Computational methods

All calculations were conducted using density functional theory (DFT) as implemented in the

[a]e-mail: ykaya@uludag.edu.tr

GAUSSIAN 09 program package [13]. The molecular geometries and vibrational frequencies of all molecules in the ground state have been calculated using the B3LYP [14] hybrid functional with the 6-31G(d) basis set. Vibrational frequencies were used for calculation of the Gibbs free energies.

3 Result and discussion

Graphene may also find application in MPGDs, GEMs in particular. The hope is that a graphene-coated GEM lets electrons through but not ions. As a result, there would be no space charge due to ions drifting from their production point into the drift volume.

Graphene makes two types of interaction with water: physical and chemical. These interactions are shown in figure 1.

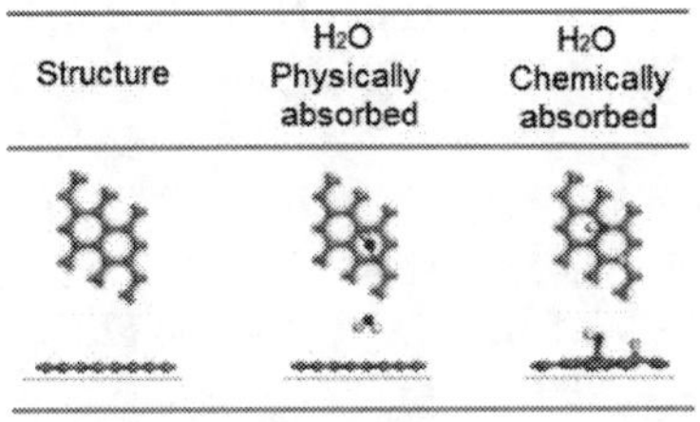

Figure 1. Atomic structures of a H_2O molecule adsorbed on graphene with different types of defects.

In addition, there are four interaction orientations of the physical type. These orientation are shown in figure 2.

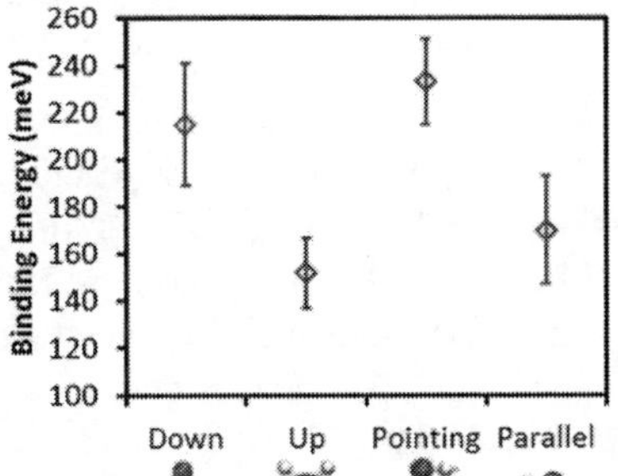

Figure 2. Various interaction orientations for water.

The binding energy is ca 180 meV and has a weak water-orientation dependence.

Attempts to reproduce the gas phase data of graphene with an increasing number of rings showed that theoretical band gap tends to decrease. When the number of rings is 81, the band gap decreases to 0.139 eV. This result is in agreement with experimental values, which are in the range 0-0.25 eV [15]. All calculation results are shown in figure 3. We have used a molecule with 81 rings for our calculations of graphene-water interactions.

The structure of the graphene sheet used in the present study are illustrated in figure 4. In the present study, we examined the physical and chemical interactions of a water molecule with graphene mesh as seen figure 4 b and figure 4 c. To determine the effect of water molecules on the surface conductivity of graphene, all modeled geometries were fully optimised at the DFT/6-31G(d) level.

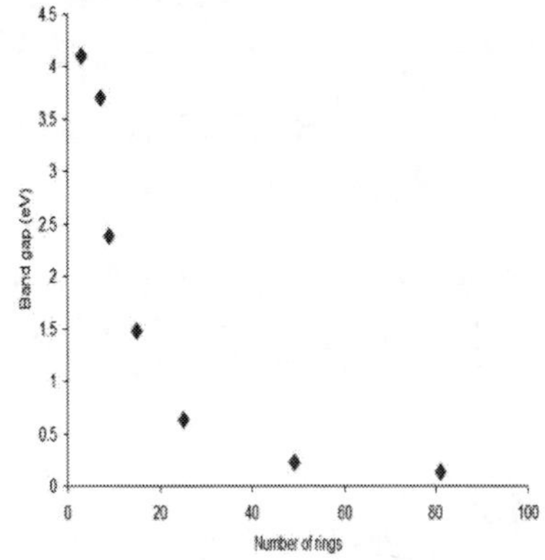

Figure 3. Theoretical band gap calculations of graphene through phenyl rings using 6-31G(d) basis set.

In addition, the band gaps for graphene, physically adsorbed graphene and chemically adsorbed graphene were calculated with the TD-DFT method. The band gaps were used for explaining the conductivity. It was found that the band gap of graphene 0.139 eV. On the other hand, the band gaps were calculated at 0.143 eV and 0.221 eV for physically and chemically adsorbed water molecules, respectively. These results indicate that the band gap increases and the surface conductivity of graphene decreases in the presence of water molecules.

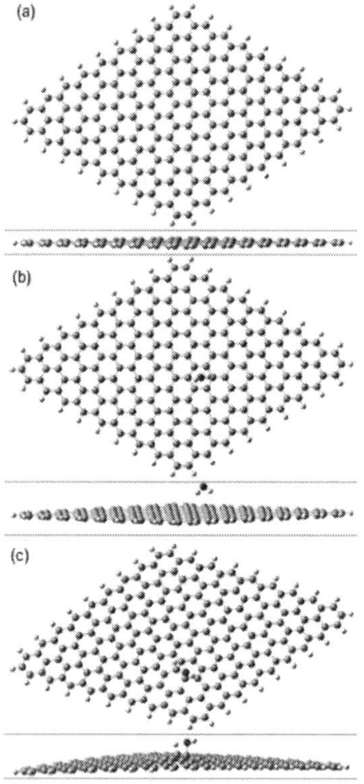

Figure 4. Optimised structures of molecules for 81 rings, a) graphene, b) physically adsorbed graphene and c) chemically adsorbed graphene.

4 Conclusion

Graphene has two types of interaction with water: physical and chemical. These interactions were studied theoretically with the DFT and TD-DFT methods. All results indicate that the the surface conductivity of graphene decreases in the presence of water molecules.

5 Acknowledgments

This work is supported by the Scientific Research Projects Unit of Uludağ University, project numbers: KUAP(F)-2015/20 and KUAP(F)-2013/25.

References

[1] Zhu, Y.F., Dai, Q.Q., Zhao, M., Jiang, Q. Sci. Rep. **3**, 1524, (2013).

[2] Weber, C.M. et al., Small **6**, 184–189 (2009).

[3] Yang, N., Zhai, J., Wang, D., Chen, Y., Jiang, L. ACS Nano **4**, 887–894 (2010).

[4] Lee, C., Wei, X., Kysar, J. W., Hone, J. Science **321**, 385–388 (2008).

[5] Abdelsayed, V. et al., J. Phys. Chem. Lett. **1**, 2804–2809 (2010).

[6] Zhang, K., Dwivedi, V., Chi, C., Wu, J., J. Hazard. Mater. **182**, 162–168 (2010).

[7] Feng, X. et al., J. Am. Chem. Soc. **126**, 62–63 (2003).

[8] Wang, S., Zhang, Y., Abidi, N., Cabrales, L. Langmuir **25**, 11078–11081 (2009).

[9] Leenaerts, O., Partoens, B., Peeters, F.M. Phys. Rev. B **79**, 235440 (2009).

[10] Rafiee, J., Rafiee, M.A., Yu, Z.-Z., Koratkar, N. Adv. Mater. **22**, 2151–2154 (2010).

[11] Chen, H., Müller, M.B., Gilmore, K.J., Wallace, G.G., Li, D. Adv. Mater. **20**, 3557–3561 (2008).

[12] Bon, S. B., Piccinini, M.,Mariani, A., Kenny, J.M., Valentini, L. Diam. Relat. Mat. **20**, 871–874 (2011).

[13] Frisch, M.J., et al., Gaussian 03, Revision E.01, Gaussian, Inc., Wallingford, CT, (2004).

[14] Becke, A.D., J. Chem. Phys. **98** 5648-5652, (1993).

[15] Zhang, Y., Tang, Tsung-Ta, Girit, Caglar, Hao, Zhao, Martin, Michael C., Zettl, Alex, Crommie, Michael F., Shen, Y. Ron, Wang, Feng (11 June 2009). "Direct observation of a widely tunable bandgap in bilayer graphene". Nature **459** (7248): 820–823.

Noble gas cluster ions

Yunus Kaya[1,a], Yalçın Kalkan[2], and Rob Veenhof[2,3]

[1] *Department of Chemistry, Faculty of Arts and Sciences, Uludağ University, 16059 Bursa, Turkey.*
[2] *Department of Physics, Faculty of Arts and Sciences, Uludağ University, 16059 Bursa, Turkey.*
[3] *RD51, CERN.*

Abstract. In this work, a reaction mechanism of formation of noble gas (Ng) cluster ions has been theoretically investigated in detail. The kinetic studies of formation of $Xe^{+}Xe$ cluster in Xe, $Ar^{+}Ar$ cluster ions in Ar, and $Ne^{+}Ne$ cluster ions in Ne have been made as theoretically. The optimized structures in the ground state were calculated using the density functional theory (DFT) by the B3LYP method combined with the Stuttgart/Dresden effective core potential basis set (SDD). In addition, we calculated the rate constants of all cluster formations. The results are $1.15 \times 10^{-31}, 3.58 \times 10^{-31}, 0.23 \times 10^{-31} cm^{6}/s$, respectively for Neon, Argon, Xenon cluster ions.

1 Introduction

A general classification of clusters according to their binding strength into van der Waals clusters, molecular clusters, hydrogen bonded clusters, ionic clusters, valence clusters and metal clusters has been proposed by Jortner [1]. The current paper focuses on small aggregates, which are bound by weak van der Waals forces in the ionic state. Here, we consider the dimers with one van der Waals bond as the smallest cluster entities.

In recent decades, quantum chemistry calculations have been used for various chemical fields such as reaction pathway analysis and spectroscopic assignments due to the theoretical developments, especially accuracy improvement of functional in density functional theory (DFT), and high-speed parallel computers.

The kinetic studies of formation Noble gas (Ng) cluster ions, which are $Xe^{+}Xe$ cluster in Xe, $Ar^{+}Ar$ cluster ions in Ar, and $Ne^{+}Ne$ cluster ions in Ne have not been found as experimentally or theoretically in literature. Therefore, in this study, we report the DFT-based mechanistic studies and spectroscopic analyses on formations of these clusters. In addition, we have calculated the rate constants all cluster formations.

2 Computational methods

The experiments and their analysis described in the literature [2] are complemented by quantum chemical calculations. These calculations concern the equilibrium structures of the relevant clusters, their vibration frequencies and rotational constants. All calculations were conducted using density functional theory (DFT) as implemented in the GAUSSIAN 09 program package [3]. The molecular geometries and vibrational frequencies of the all molecules in the ground state have been calculated using the B3LYP [4] with the SDD basis set [5]. Vibrational frequencies were used for calculation of the Gibbs free energies.

[a] e-mail: ykaya@uludag.edu.tr

3 Result and discussion

In this study, a mechanism of the noble gas cluster formation has been theoretically investigated in detail. Firstly, the mechanism of the formation of cluster ions was suggested for the noble gaseous. Then, all suggested geometries were optimized by DFT/B3LYP method and SDD level. The optimized geometries are shown in figure 1. The Gibbs free and the relative energies for the formation reactions were calculated at same basis set in atmospheric pressure and at room temperature.

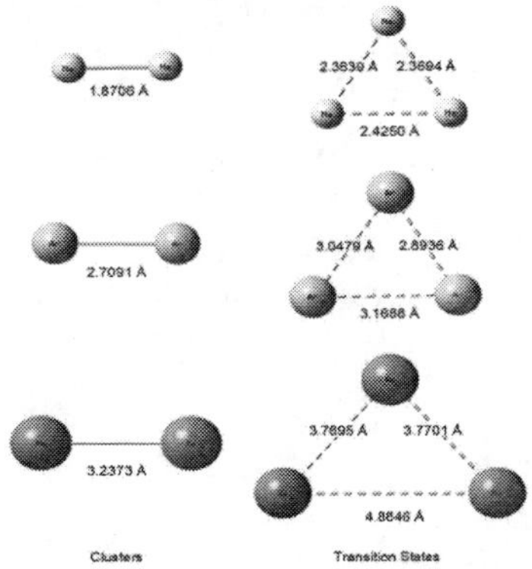

Figure 1. Optimized geometries of cluster ions and TS for Ng.

The Ng cluster ions can be formed in a three-body interaction at atmospheric pressure. The mechanism of the formation reaction of the Ng cluster ions involves one step, which are composed of triangular plane over transition state (TS).

According to the transition state theory, the rate constants are related to activation Gibbs free energies (ΔG^*). Therefore, analysis of these quantities in kinetic is of special importance. The activation Gibbs free energies related to TS are 249.6, 247.0 and 253.9 kJ/mole, respectively for Neon, Argon and Xenon cluster ions (Table 1).

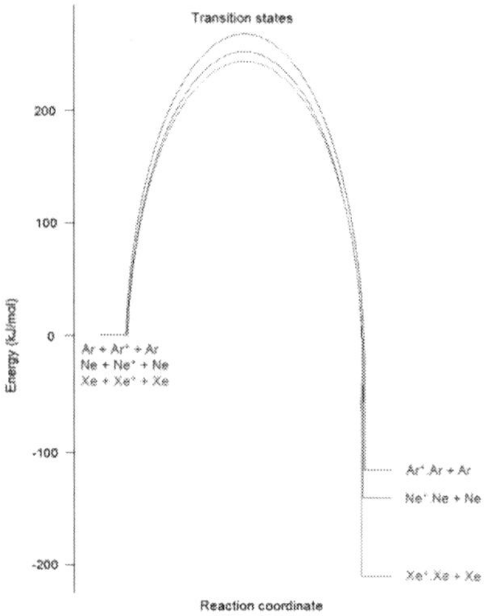

Figure 2. Enery profile for the formation of cluster ions of Ng.

Molecules	Energy (kJ/mole)	Freq. (cm^{-1})	$k x 10^{31}$ (exp.)[1] (cm^6/s)
Ne			
$Ne + Ne^+ + Ne$	0.0	-	
TS	249.6	−129.7	
$Ne^+ \cdot Ne + Ne$	−210.8	-	1.15(0.6)
Ar			
$Ar + Ar^+ + Ar$	0.0	-	
TS	247.0	−76.5	
$Ar^+ \cdot Ar + Ar$	−151.3	-	3.58(2.3)
Xe			
$Xe + Xe^+ + Xe$	0.0	-	
TS	253.9	−32.6	
$Xe^+ \cdot Xe + Xe$	−121.7	-	0.23

Table 1. Relative energy, negative frequency and rate constants of the structures of the reaction paths for formation of clusters ions for Ng.

Using traditional transition state theory:

$$k = \frac{k_b T}{h} \exp[-\frac{\Delta G^*}{RT}] \quad (1)$$

where k_b, h, R, and T are the Boltzmann constant, Plank constant, gas constant, and temperature, respectively. The values of theoretical rate constants are calculated from equation 1, which are equal to $k = 1.15 \times 10^{-31}$, 3.58×10^{-31}, 0.23×10^{-31} cm^6/s, respectively for Neon, Argon, Xenon cluster ions. The experimental values are 0.6×10^{-31} and $2.3 \times 10^{-31} cm^6/s$ for neon and argon cluster ions, respectively [2]. The calculated rate constants are in good agreement with experimental values. These all calculations show that the relative energy of TS for formation of argon cluster ion is smaller than TS for formation of neon cluster ion. Therefore, it was determined that the formation of argon cluster ion is faster than formation of neon cluster ion, although the neon is smaller than Argon [6]. The reason is that the cross section of argon may be higher than neon. In addition, the smallest rate constant was calculated in the formation of Xenon cluster ions. It is thought that the atomic size may be more effective than the cross section in the formation of cluster ions.

4 Conclusion

In this study, the kinetic studies of formation of Ng cluster ions have been calculated using density functional theory (DFT). The values of theoretical rate constants are calculated at $1.15 \times 10^{-31}, 3.58 \times 10^{-31}, 0.23 \times 10^{-31} cm^6/s$, respectively for Neon, Argon, Xenon cluster ions.

5 Acknowledgments

This work is supported by the Scientific Research Projects Unit of Uludağ University, project number: KUAP(F)-2015/20.

References

[1] Jortner, J., Bunsenges, B., Phys. Chem. **88** 188 (1984).
[2] Kalkan, Y., Arslandok, M., Cortez, A.F.V., Kaya, Y., Tapan, İ., Veenhof, R., JINST., DOI: 10.1088/1748-0221/10/07/P07004.
[3] Frisch, M.J., et al., Gaussian 03, Revision E.01, Gaussian, Inc., Wallingford, CT, (2004).
[4] Becke, A.D., J. Chem. Phys. **98** 5648-5652, (1993).
[5] Lv G., Wei F., Li Q., Shen Q., Jiang H., Zhou Y., Wang X. J. Nanosci. Nanotechnol. 10, 809-818 (2010).
[6] Smirnov, B. M., Cluster Ions and Van der Waals Molecules, Gordon and Breach Science Publishers (1992).

Speeding Up and Parallelizing the GARFIELD++

A. Sheharyar[1,a], O. Bouhali[1], and A. Castaneda[1]

[1] Texas A&M University at Qatar

Abstract. Garfield++ is a toolkit for the simulation of particle detectors that use gas and semi-conductors as sensitive medium. It takes enormous amount of time to complete the simulation of complex scenarios such as those involving high detector voltages, gases with large gains, or electric field meshes with large number of elements. We observed that most of the simulation time is being consumed in finding the correct element in the electric field mesh. We optimized the element search operation and achieved significant boost in the speed up. In addition, We added the parallel computing support in the toolkit to simulate multiple events simultaneously over multiple machines. In this paper, we present our approach of speeding up the computations and benchmark results.

1 Introduction

The use of Micro-Pattern Gas Detectors (MPGDs) for both applied and fundamental science has substantially increased during the last few years. For instance in the field of High Energy Physics (HEP) in which there is a constant demand for faster detection devices, MPGDs have become an excellent candidate for this task. In addition to a fast signal response, MPGDs provide high rate capability, unprecedented spatial resolution, large sensitivity area, operational stability and radiation hardness. The design of a MPGD can vary according to the experimental needs, two of the most popular ones are the Gas Electron Multiplier (GEM) [1] and the Micro-Mesh Gaseous Structure (MICROMEGAS) [2]. The role of MPGDs in HEP experiments is fundamental, as an example, the CMS experiment at the CERN Large Hadron Collider (LHC) is preparing for an upgrade of its muon detection system with the installation of new chambers based on GEM technology [3], given the large scale of such experiment, designing and producing large area gas detectors to preserve the desirable performance remains a challenge. Due to several experimental factors the performance of MPGDs can degrade, some of those include: changes in the detector geometry due to imperfections in the production or produced during the installation or operation, the choice of the sensitive material (which commonly consist on the mixture of two gases, one of them being a noble gas), the corresponding high voltage applied between the detector layers, among others. It can be costly and not feasible to study each one of these parameters in-situ, this is why Monte Carlo simulation represents a powerful tool. In order to study these parameters GARFIELD++ is used, which is the C++ version of the GARFIELD toolkit [4], this software can handle the simulation of detectors that use a gas mixture or a semi-conductor material as a sensitive medium.

[a]e-mail: ali.sheharyar@qatar.tamu.edu

1.1 Signal formation in a GEM detector

A GEM is a gaseous ionization detector, commonly used in particle physics. In the operation of a GEM detector a high energy particle (i.e. muons produced in the proton-proton collisions at the LHC) traverse the detector volume, due to interaction (collisions) with atoms and molecules in the sensitive medium (gas mixture) an electron is released via ionizing radiation, such electron is accelerated due to the presence of an uniform electric field, the electron is guided from the "drift" to the "amplification" region which consists of small semi-conical holes in a thin polymer sheet where an even more intense electric field is created, due to further interaction with other atoms and molecules more electrons are released producing an "avalanche" effect. Electrons in this avalanche are collected in the readout just after passing the amplification region (in the case of a single-GEM configuration) or are further amplified in a sequence of additional drift and amplification regions (which is the case for a triple-GEM detector configuration). Figure 1 (left) shows a typical electron avalanche simulated in GARFIELD++ using a triple-GEM detector configuration.

The generated electric field is modeled by an external package (ANSYS) which is a finite element solver. The 3D electric field modeling is described by a collection of tetrahedral elements building a mesh as it is shown in Figure 1 (right). The size of the elements in the mesh is a configurable in the simulation and it has a direct impact on the accuracy of the results and the time of the simulation, the finer the size of the elements the longer the simulation time, this fact combined with the complexity of the processes described in the simulation may lead to a total

 https://doi.org/10.1051/epjconf/201817406004

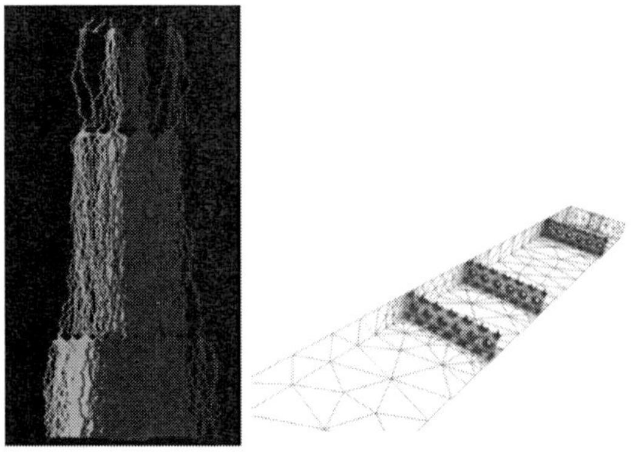

Figure 1. Left: Visualization of the simulated avalanche development for seven primary electrons in a triple-GEM chamber starting from the drift volume. Right: Electric field mesh generated by ANSYS for a triple-GEM detector

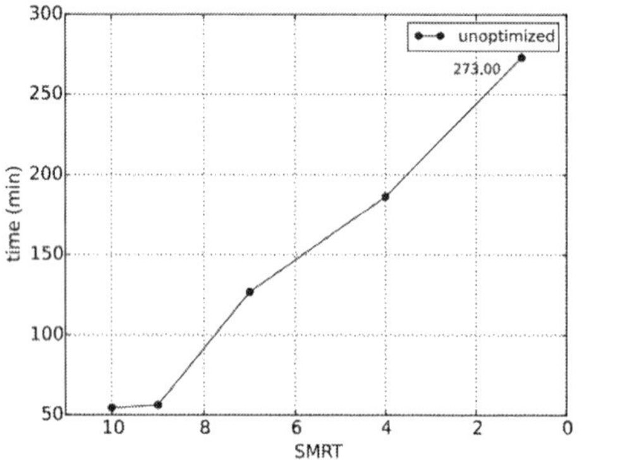

Figure 2. Simulation time in GARFIELD++ as a function of the size of the electric field mesh using a triple-GEM detector configuration.

simulation time of several days or even weeks, depending on the input parameters (high voltage, gas mixture, size of the mesh, number of events). Given the statistical nature of the related processes to get reliable results a sample of few thousand events should be collected, this fact makes the timely production of results very difficult to achieve. Figure 2 shows the simulation time as a function of the size of the electric field mesh for five different configurations and using as benchmark a single-GEM detector. It is clear how the simulation time drastically increases as the size of the mesh is reduced, such results were produced running GARFIELD++ in a serial mode, in fact, the two last points in the graph are extrapolations due to the jobs being aborted by exceeding the time limit commonly imposed by the cluster administrators. In this paper we propose several methods to optimize the algorithms and reduce the simulation time allowing for a fast generation of results without affecting the accuracy of the same.

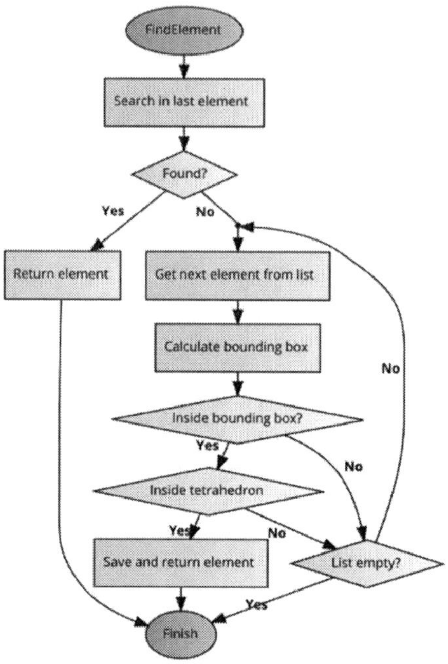

Figure 3. Workflow representing the different steps in the FindElement algorithm.

2 Optimization methods

The GARFIELD++ source code elements and operations were studied in order to find the one that takes the most of the simulation time. It was found that the so-called "FindElement" operation takes around 90% of the total execution time. The FindElement method basically search for elements in the tetrahedral mesh in order to find the position of the electron inside the electric field. As it is originally structured this method will search for compatible elements with the electron position, if such element is not found the search will start from the initial element in a sequential way, instead of searching for compatible elements around it, which clearly is a non-optimized implementation of this element searching algorithm. Figure 3 shows the workflow of the FindElement method which was later optimized to reduce the simulation time. Three optimization methods were studied and they are described in the following sub-sections.

2.1 Caching of bounding boxes

The original FindElement operation scans through all tetrahedrons (or elements) sequentially in order to find the corresponding tetrahedron that contains the given 3D point. This point-inside-tetrahedron test is an iterative and computationally expensive operation involving several matrix and vector operations. In order to speed up the computation, the FindElement operation calculates the axis-aligned bounding box (a cuboid) of the tetrahedron and checks if the point lies in it. The point-inside-cuboid operation is computationally inexpensive and simple as compared with the point-inside-tetrahedron operation. However, It was found that the bounding boxes were computed in every function call – each calculation requires 9 min and 9 max operations. In the optimized FindElement operation, the bounding boxes are computed when mesh is loaded initially when the simulation starts.

2.2 Element search using Tetrahedral Tree structure

As mentioned earlier, the FindElement operations scans through all tetrahedrons in a linear fashion to find the correct tetrahedron that contains a given 3D point. The order of scanning the tetrahedrons is arbitrary and depends on the way the mesh is stored on disk. In the optimized version, a spatial index is built based on the work by De Floriani et al. [5]. The tetrahedrons are arranged spatially using a *tetrahedral tree* that utilizes the *octree* as the underlying data structure. The *Octree* recursively subdivides the initial cubic domain (containing all tetrahedrons) into eight blocks. The octree has been selected instead of the kd-tree data structure as it produces tree of smaller depth hence reducing the number of tree nodes to be visited to reach the desired tetrahedron.

2.3 Search through neighbors

As the electron moves through the electric field or mesh, its new position is expected to be closer to the previous one. Based on this behavior, the non-optimized GARFIELD++ keeps track of the tetrahedron found in the last operation and runs the point-in-tetrahedron test. If the test is successful, then it skips the scanning through all elements and it simply returns the last tetrahedron. Otherwise, it scans through all elements sequentially. It was observed that if the new position does not lie in the last element, then it is most likely to be in one of the neighboring elements. The optimized FindElement operation search through the neighboring elements if the point-in-tetrahedron test fails for the last found element. Note that the neighbors of all tetrahedrons are precomputed initially when the simulation starts.

3 Results

The results of each of the optimization methods presented in section 2 are illustrated for a single- and triple-GEM configurations in Figure 4 and 5 respectively, in which the speed up factor is defined as the ratio of the simulation time using the standard GARFIELD++ to the simulation time using any of the optimization methods.

4 Introduction to a parallelized version of GARFIELD++ (pGARFIELD)

Based on a Message Passing Interface (MPI) architecture, the GARFIELD++ toolkit was adjusted to be used in parallel mode. Running with a master node which does random generation, the workload is distributed to and gathered back from the slave nodes. The performance was

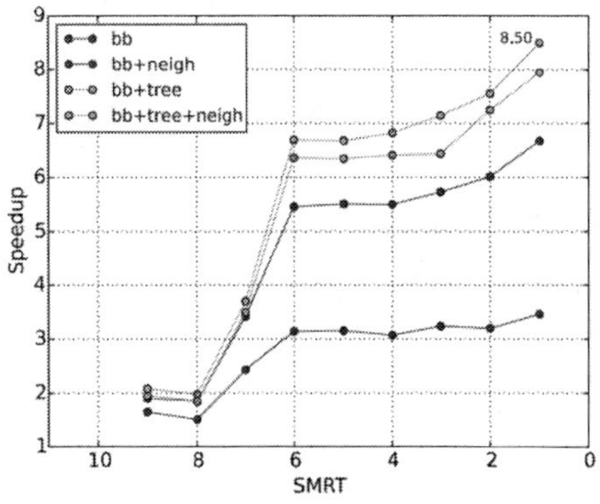

Figure 4. Speed up factor as a function of the size of the electric field mesh using a single-GEM detector configuration and comparing the performance of the different optimization methods

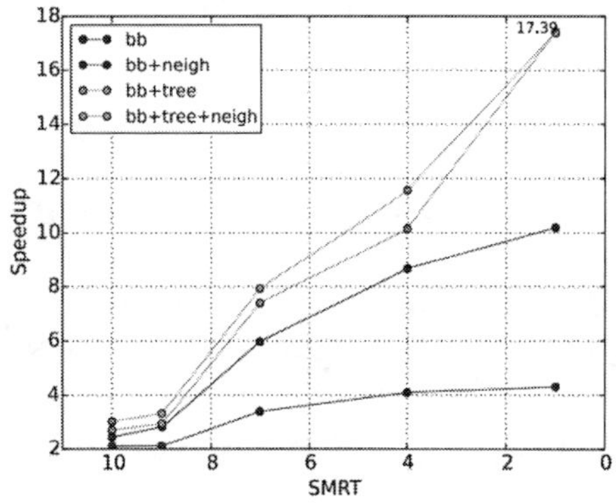

Figure 5. Speed up factor as a function of the size of the electric field mesh using a triple-GEM detector configuration and comparing the performance of the different optimization methods

studied on the HPC cluster at Texas A&M University at Qatar showing a good performance. The speed up factor reached is presented as a function of the number of nodes in Figure 6

5 Conclusions

Several optimization methods were studied to reduce the simulation time in GARFIELD++, using a single- and triple-GEM detector configurations results were obtained as a function of the size of the electric field mesh. It was shown that the simulation time is reduced by the use of each individual method, and it decreases drastically when a combination of them is applied. The largest reduction was observed by combining the three methods and using a triple-GEM detector configuration, in which, a reduction factor of 17 was achieved. This reduction was obtained without affecting the accuracy of the results. The algorithm improvements have been partially included in the official GARFIELD++ repository, while the remaining parts will be implemented soon. In addition to the optimization methods a parallelized version of GARFIELD++ (pGARFIELD) was introduced, allowing for an even further reduction in the simulation time.

6 Acknowledgements

This work was supported by the Qatar National Research Fund under the project NPRP-5-464-1-080

References

[1] F. Sauli, Nucl. Instrum. Meth. A **805**, 2-24 (2015). doi:10.1016/j.nima.2015.07.060

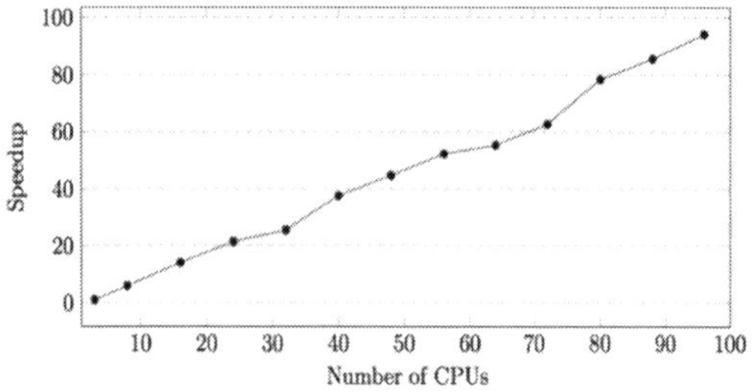

Figure 6. Speed up factor using a parallelized version of GARFIELD++ (pGARFIELD), the simulation time is reduced as the number of nodes is increased.

[2] G. Charpak, J. Derre, Y. Giomataris and P. Rebourgeard, Nucl. Instrum. Meth. A **478**, 26 (2002). doi:10.1016/S0168-9002(01)01713-2

[3] D Abbaneo, et al. JOURNAL OF INSTRUMENTATION **9**, 10 (2014). http://dx.doi.org/10.1088/1748-0221/9/10/C10036

[4] Rob Veenhof http://garfieldpp.web.cern.ch/garfieldpp

[5] De Floriani, Leila and Fellegara, Riccardo and Magillo, Paola, Proceedings of the 18th SIGSPATIAL international conference on advances in geographic information systems, 2010

https://doi.org/10.1051/epjconf/201817406005

Study of the spatial resolution of low-material GEM tracking detectors

V.N.Kudryavtsev[1,2], T.V.Maltsev[1,2], and L.I.Shekhtman[1,2,a]

[1] *Budker Institute of Nuclear Physics, 11.av.Lavrentiev, 630090, Novosibirsk, Russia*
[2] *Novosibirsk State University, 2 Pirogova str., 630090, Novosibirsk, Russia*

Abstract. The spatial resolution of GEM based tracking detectors has been simulated and measured. The simulation includes the GEANT4 based transport of high energy electrons with careful accounting for atomic relaxation processes including emission of fluorescent photons and Auger electrons and custom post-processing, including accounting for diffusion, gas amplification fluctuations, the distribution of signals on readout electrodes, electronics noise and a particular algorithm of the final coordinate calculation (center of gravity). The simulation demonstrates that a minimum of the spatial resolution of about 10 μm can be achieved with strip pitches from 250 μm to 300 μm. For larger pitches the resolution is quickly degrading reaching 80-100 μm at a pitch of 500 μm. The spatial resolution of low-material triple-GEM detectors for the DEUTRON facility at the VEPP-3 storage ring is measured at the extracted beam facility of the VEPP-4M collider. The amount of material in these detectors is reduced by etching the copper of the GEMs electrodes and using a readout structure on a thin kapton foil rather than on a glass fibre plate. The exact amount of material in one DEUTRON detector is measured by studying multiple scattering of 100 MeV electrons in it. The result of these measurements is X/X0 = 2.4×10^{-3} corresponding to a thickness of the copper layers of the GEM foils of 3 μm. The spatial resolution of one DEUTRON detector is measured with 500 MeV electrons and the measured value is equal to 35 $\pm$ 1 μm for orthogonal tracks.

1 Introduction

Triple-GEM tracking detectors can provide precise measurements of track positions in high rate environments [1]. These detectors can be produced with a low material budget that is important for precise tracking in medium and low energy particle and nuclear physics experiments [2]. Experimental data on the spatial resolution of the GEM detectors for relativistic charged particles lie in the range between 50 μm and 100μm (see for example [3] and [4]) depending on the particular type of the readout structure, gas mixture, electronics, etc. The spatial resolution of this kind of detectors is determined by a combination of many factors, such as: statistics and ranges of primary delta-electrons, transverse diffusion of electrons, configuration of the readout structure, noise of the readout electronics, fluctuations of the gas amplification, and the particular algorithm for the calculation of the track position. In order to determine the optimal set of parameters to get the best spatial resolution, we have started a systematic simulation and experimental study. The present paper presents preliminary results of the simulations of spatial resolution of the triple-GEM detectors with different layouts of the readout structure as well as experimental results, obtained with the detector produced for the DEUTRON facility at the VEPP-3 storage ring at the Budker INP [5].

[a] e-mail: L.I.Shekhtman@inp.nsk.su

2 Simulation studies

The simulation of the spatial resolution of the triple-GEM detectors was performed in two stages. At first, primary electrons with 1 GeV/c momentum perpendicular to the detector plane, and with randomly distributed initial coordinates in the detector plane are transported through the complete model of the detector, that includes the honeycomb cover with the drift electrode, the drift gap (3 mm), two transport gaps (1.5 mm), and the induction (2 mm) gas gap, three standard GEMs and the readout structure on a glass fibre plate (thick version of the detector, Fig. 1). In the simulation of the "thin" detector only 3 mm gas layer is left in the first stage. The energy depositions coming from other detector elements, such as the honeycomb with the drift electrode and the GEMs are removed. However in the second stage we take into account that the ionisation drifts through all GEMs and the induction gap, and distribute the signals using realistic readout strip plane configuration. After the recording of all energy depositions in the drift gap, the second stage is started that includes the introduction of the electron diffusion, gas gain fluctuations, the distribution of the signal between the readout strips, accounting for the electronic noise, and the calculation of the measured track position with the centre of gravity (COG) method.

Three types of the readout plane layouts were used in the simulation. The first layout contains parallel strips with their width equal to the pitch (no gaps between strips)

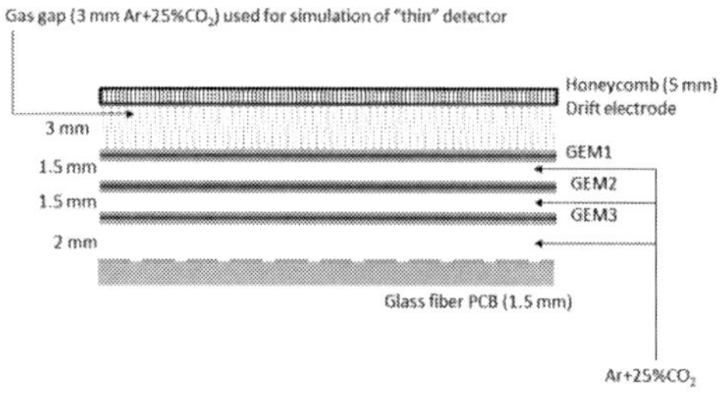

Figure 1. Schematic view of the model of "thick" detector. The "thin" detector was simulated with a single 3 mm layer of gas mixture (marked with pink background in the figure).

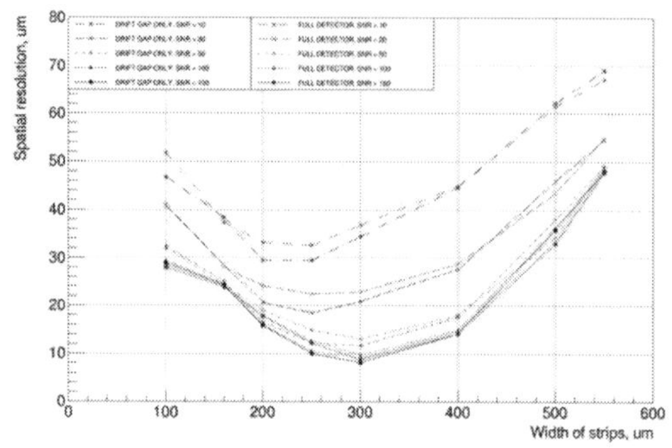

Figure 2. Comparison of spatial resolution of "low-material" detector (only 3mm drift gap, black points and curves) and regular detector with all the materials (red points and curves).

and each of these strips is read out. The second layout also contains parallel strips but only each second strip is read out. In the third layout that also contains parallel strips, only each third strip is read out.

These three layouts correspond to the cases where the full signal is induced on the strips (case 1), half of the signal is induced on the strips (case 2), and one third of the signal is induced on the strips (case 3). Case two and three correspond to the detectors with two layer readout planes with straight strips and stereo strips like in the detectors for the DEUTRON facility [6] and in the detectors for the tagging system of the KEDR experiment at the VEPP-4M [4]. In these detectors only part of the signal is induced on the straight strips used for the precise position measurement in one direction. The rest of the signal is induced at the stereo strips that are necessary for the position measurement in the perpendicular direction and were not used in the present study. In the detectors for the tagging system of the KEDR due to a mistake during manufacturing of the readout planes the signal induction is very non-uniform and only one third of the signal is in average induced on the straight strips.

At first the difference between the resolution of the "low material" detector and the thick detector was checked for the first layout of the readout plane. In Fig. 2 the resolution as a function of the strip pitch is presented for different values of the signal to noise ratio. The figure demonstrates that the effect of the low material on the spatial resolution is marginal and is only pronounced at strip pitches smaller than 300 μm. All further simulation results are obtained for the thick detector layout.

In Fig. 3 the resolution as a function of the strip pitch is shown for the readout plane of the third type and for several values of the signal to noise ratio. The experimental result from [4] for the detector of the tagging system of KEDR is also shown in the figure. The COG method used in this simulation for the calculation of the track position involved always three strips around the maximum of the charge distribution. Results of the simulations for the readout structure of the second type are presented in Fig. 4. The so called "flexible" COG method was used in this case, with the number of strips involved depending

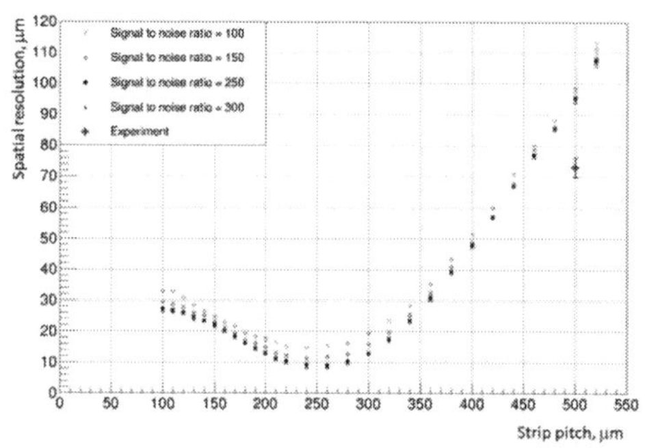

Figure 3. Spatial resolution as a function of the strip pitch for the readout structure of the KEDR TS detectors [4]. The experimental point is taken from [4].

on the relationship between the charge cluster size and the strip pitch. An increase of the number of strips involved in the COG calculation explains the sudden changes in the resolution around and below 100 μm strip pitch. From the figures we can see that the best resolution of 10–15 μm can be achieved for a strip pitch of 250 μm. For larger pitches the resolution degrades and the results of the simulation are worse than the experimental results (Fig. 3 and also see below).

3 Experimental results

The amount of material and the spatial resolution of the triple-GEM detectors for the DEUTRON photon tagging system (PTS) [6] (Fig. 5) were measured at the extracted electron beam facility at the VEPP-4M storage ring [7]. Three detectors were installed at the test bench. The first and the second were "thin" detectors for the DEUTRON PTS with the dimensions of the sensitive area of 160×40

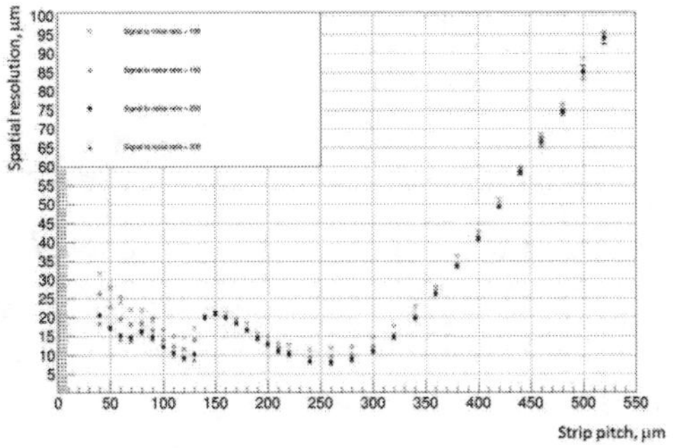

Figure 4. Spatial resolution as a function of the strip pitch for the readout structure of the DEUTRON PTS detectors [6] (see below).

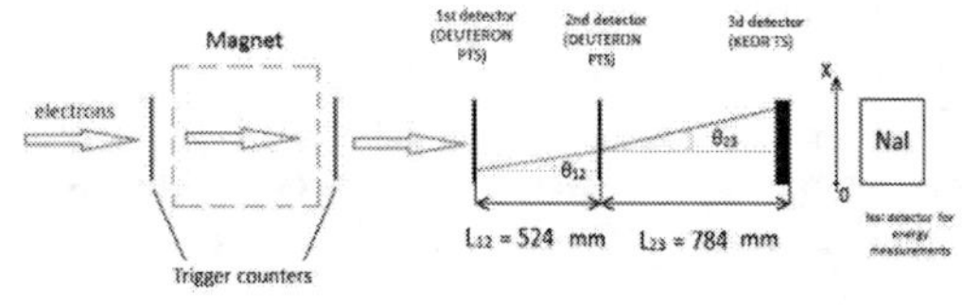

Figure 6. Schematic view of the set-up for the measurement of the amount of material in the detector.

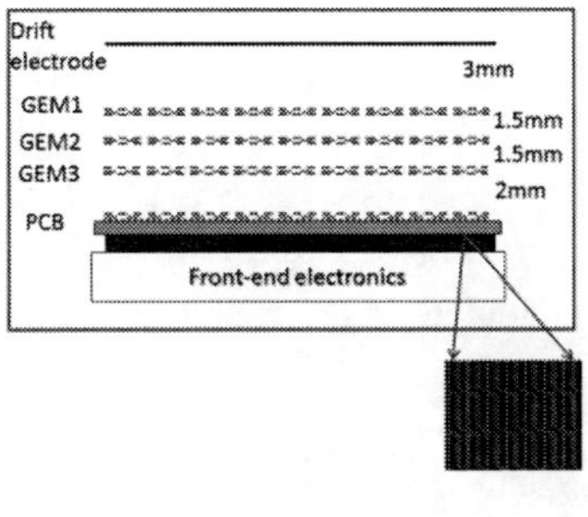

Figure 5. DEUTRON PTS detector structure and schematic view of the PCB design.

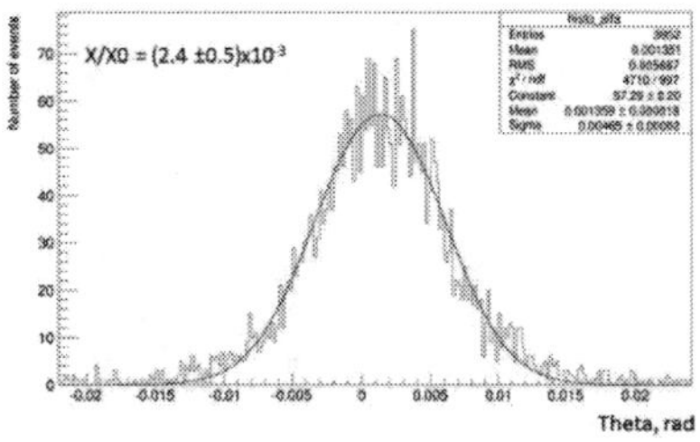

Figure 7. Angular distribution of tracks of 100 MeV electrons scattered in the 2nd detector (see Fig. 6).

mm^2 and the third was the detector for the KEDR TS with a sensitive area of 128×100 mm^2.

The readout strip structure of the DEUTRON PTS detectors is produced on a 50 μm thick Kapton foil and all copper layers on the GEM foils and the readout flexible PCB are reduced as much as possible to decrease the amount of material. The readout flex is produced in two layer technology and the strip pitch in each layer is equal to 0.5 mm. The strips of the stereo layer are shifted by 250 μm with respect to the strips of the straight layer. The amount of material in the detector was determined by measuring the multiple scattering of 100 MeV electrons in the experimental set-up shown in Fig. 6.

Due to the multiple scattering the track angles were changing after passing through the second detector in the setup. The distribution of the differences between the track angles after the second detector and before the second detector is shown in Fig. 7. Its width corresponds to an amount of material of ~2.4×10^{-3} X0. This value corresponds to a copper thickness on the GEM foils and the PCB of ~3 μm.

The spatial resolution of the 2nd detector was measured with a similar setup (Fig. 6), but the distances between the detectors were reduced to ~75 mm and the beam energy was increased to 500 MeV. The first and the third detector were used for tracking. Figure 8 demonstrates the resolution as a function of the track angle after correction for multiple scattering and for the limited resolution of the tracking detectors. The error bars in the figure represent the statistical errors. The curve in the figure is calculated using the formula of the quadratic sum of the resolution for orthogonal tracks and the track projection to the detector plane: $\sigma = \sqrt{\sigma_0^2 + (L * tg(\alpha))^2/12}$, where σ_0 is equal to 35 μm and L is 3 mm, the thickness of the drift gap.

4 Conclusions

Simulation studies show that an ultimate resolution of 10–15 μm can be achieved for orthogonal tracks of high energy electrons in triple-GEM detectors operating with a gas mixture of Ar–25%CO_2 with a strip readout structure at a pitch of 250 μm. At larger strip pitches the simulation shows a systematically worse resolution than the measurements that will require more studies.

The measurements demonstrate that a spatial resolution of the triple-GEM detector of 35±1 μm can be achieved for a strip pitch of 0.5 mm.

5 Acknowledgements

This study was supported by the Russian Science Foundation (project N 14-50-00080).

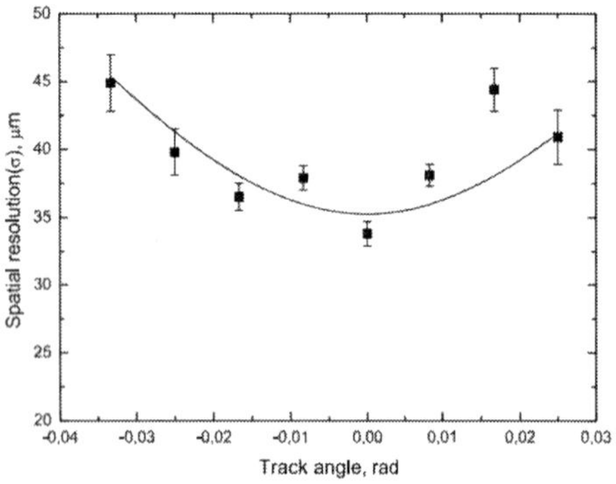

Figure 8. Spatial resolution as a function of the track angle.

References

[1] Sauli F., Sharma A. Micropattern Gaseous Detectors, Annu.Rev.Nucl.Part.Sci. 1999. **49**. P.341-88.
[2] A. Bondar, et.al, Nucl. Instrum. and Methods., **A556**, (2006), P.495-497.
[3] B.Ketzer, et.al., IEEE Transactions on Nuclear Science V.**49** II, N.5 (2002), P.2403-2410.
[4] V M Aulchenko, et.al., Journal of Instrum. **6**, P07001 (2011).
[5] Shestakov, Yu V., et al., PHYSICS OF PARTICLES AND NUCLEI, **45**, n.1 (2014), P.338-340.
[6] V N Kudryavtsev, et.al., Journal of Instrum., **9**, C09024 (2014).
[7] G N Abramov, et.al., Journal of Instrum., **9**, C08022 (2014).

Numerical Studies on Time Resolution of Micro-Pattern Gaseous Detectors

Purba Bhattacharya[1], Nayana Majumdar[2,a], Supratik Mukhopadhyay[2], and Sudeb Bhattacharya[3]

[1] *School of Physical Science, National Institute of Science Education and Research, Jatni, Khurda, India 752050*
[2] *Applied Nuclear Physics Division, Saha Institute of Nuclear Physics, 1/AF, Bidhannagar, Kolkata, India 700064*
[3] *Retired Professor, Applied Nuclear Physics Division, Saha Institute of Nuclear Physics, 1/AF, Bidhannagar, Kolkata, India 700064*

Abstract. The Micro-Pattern Gaseous Detectors offer excellent spatial and temporal resolution in harsh radiation environments of high-luminosity colliders. In this work, an attempt has been made to establish an algorithm for estimating the time resolution of different MPGDs. It has been estimated numerically on the basis of two aspects, statistics and distribution of primary electrons and their diffusion in gas medium, while ignoring their multiplication. The effect of detector design parameters, field configuration and the composition of gas mixture on the resolution have also been investigated. Finally, a modification in the numerical approach considering the threshold limit of detecting the signal has been done and tested for the RPC detector for its future implementation in case of MPGDs.

1 Introduction

Owing to the use of typical manufacturing techniques for microelectronics, the new genre of Micro-Pattern Gaseous Detectors (MPGDs) with high granularity and very small distances between the electrodes can offer high spatial and time resolutions and good counting rate capability [1]. The requirement of fast collection of data in various applications of the MPGDs has necessitated a thorough optimization of their time resolution through the modification of their design parameters and choice of gas mixture. In this context, the study of the time resolution of these detectors and its dependence on various parameters turns out to be an interesting aspect of MPGD development for many of the current and future applications.

The time resolution of a detector can be defined as the precision with which the detector can distinguish between two overlapping events in terms of time. It depends on the transit of elecrrons from their generation point to the collecting electrode. The spread on the duration of transit leads to a finite time resolution of the detector [2, 3]. The main two factors that contribute to the spread are the statistics and distribution of the primary electrons and their diffusion in gas medium.

From event to event, the electron is not produced at the same distance from the read-out plane of the detector under consideration. The general expression for the space distribution of the electron-ion pair j, closest to the read-out plane, when $\bar{N}_P$ is the average number of pairs produced, is obtained as follows [2, 3]:

$$A_j^{\bar{N}_P}(x) = \frac{x^{j-1}}{(j-1)!}{\bar{N}_P}^j e^{-\bar{N}_P x} \qquad (1)$$

[a] e-mail: nayana.majumdar@saha.ac.in

In particular, the distribution of the pair closer to one end of the detection volume, is given by

$$A_1^{\bar{N}_P}(x) = \bar{N}_P e^{-\bar{N}_P x} = \bar{N}_P e^{-\bar{N}_P u_e t} \qquad (2)$$

where u_e is the electron drift velocity. The distribution is shown in figure 1(a) with variance = $\frac{1}{\bar{N}_P u_e}$.

Again, due to the diffusion, the electrons produced at the same position in the gas arrive at the read-out plane at different times. The arrival time distribution, from a particular distance is Gaussian, as shown in figure 1(b). The mean of this distribution gives the mean arrival time and the variance is equal to $D\sqrt{z_{dist}}/u_e$. Here D is the diffusion coefficient, z_{dist} is the traveled distance. With varying distance from the read-out plane, the mean drift time, as well as the variance, changes accordingly.

Considering the above two factors, if the first cluster is always assumed to generate signal that is detected by the read-out, the temporal resolution can be defined as:

$$\sigma_T^2 = (\frac{1}{\bar{N}_P u_e})^2 + (\frac{D\sqrt{z_{dist}}}{u_e})^2 \qquad (3)$$

In the present work, a numerical simulation of the time resolution of a few MPGDs, is reported. A comprehensive study on the dependence of time resolution on detector design parameters and field configuration, has been made in addition.

2 Simulation Tools

The Garfield [4] simulation framework has been used in the following work. The 3D electrostatic field simulation

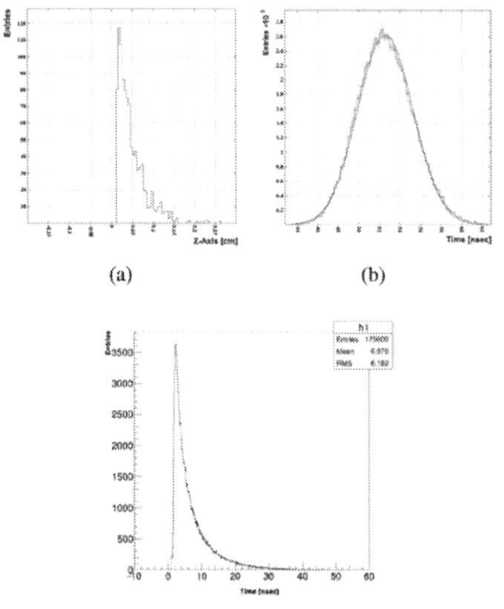

(a) (b) (c)

Figure 1. For a bulk Micromegas detector (a) minimum distance of an electron from the mesh plane, (b) the distribution of drift time of electron from a single point at fixed position above the mesh plane, (c) final time spectrum.

has been carried out using neBEM [5] toolkit. Besides neBEM, HEED [6] has been used for primary ionization calculation and Magboltz [7] for computing drift, diffusion, Townsend and attachment coefficients.

3 Simulation Models

In this work, Micromegas and GEM detectors have been opted as two cases of the MPGD genre for studying their temporal resolution. The design parameters of the bulk Micromegas detectors are compiled in table 1, whereas that for single and triple GEM detectors are listed in table 2.

Table 1. Design parameters of the bulk Micromegas detectors. Micromegas-wire diameter in all cases is 18 μm.

Amplification Gap (in μm)	Mesh Hole Pitch (in μm)
64	63
128	63
128	78
192	63

In our calculation, cosmic muon (1 − 3 GeV) tracks with different inclinations have been considered in the drift volume. In the first approximation, the resolution has been estimated on the basis of above two aspects while ignoring the electron multiplication. Rather, it has been assumed that the electrons which hit the readout plane first were multiplied adequately to produce a significant signal. For a particular track, the drift time of those primary electrons

Table 2. Design parameters of GEM-based detectors.

Polymer substrate	50 μm
Copper coating thickness	5 μm
Hole diameter (copper layer)	70 μm
Hole diameter (Polymer substrate)	50 μm
Hole to hole pitch	140 (S)
Drift Gap	3 mm
1st Transfer gap	1 mm
2nd Transfer gap	2 mm
Induction gap	1 mm

which hit the readout plane first to produce a considerable signal, has been recorded. Due the reasons mentioned above the time for the first hit varies from track to track and the final spectrum looks like as shown in figure 1(c), with a mean equal to the average drift time of the first hit and r.m.s (σ_T) equals to the time resolution.

Table 3. Design parameters of RPC detector.

Bakelite thickness	2 mm
Grapite coating thickness	20 μm
Copper strip thickness	200 μm
Gas gap	2 mm

In our calculation, some effects of electronics, such as shaping, noise etc., have not been considered. However, a modified simulation approach has been adopted where a threshold limit of detecting signal has been considered which is related to the gain variance of a detector.

4 Results

4.1 Bulk Micromegas

The temporal resolution of bulk Micromegas detector and its dependence on gas mixture, electrostatic field configuration and geometrical parameters have been simulated numerically [8]. Since the mean drift time of the first hit depends on the starting position of the electron, the inclination of the muon track plays an important role. The mean drift time for different inclined tracks and the corresponding resolutions have been plotted in figure 2. The electrons from the track which makes an angle of 5° with the XY-plane, are produced close to the micromesh plane and so these electrons traverse less path without being affected much by the diffusion. As a result the first hit time is less and also the resolution is better. With the increase of the inclination angle, the electrons have to travel much longer path which causes the worsening of the resolution and larger drift time.

The variation of the temporal resolution with drift field (E_{Drift}) has been plotted in figure 3(a) for the bulk Micromegas detector having amplification gap of 128 μm and pitch of 63 μm in Argon-Isobutane mixture with different mixing ratios. At the lower drift field, the larger transverse diffusion is responsible for worsening of resolution. At the higher drift field, due to poor funneling, the electrons traverse larger path and thus increase the drift time and temporal resolution [9]. For a particular drift field, the variation of resolution has been plotted with the change in amplification field, or with the mesh voltage (V_{Mesh}) in figure 3(b). The resolution is clearly improving because of better funneling and less transverse diffusion due to higher field in the amplification gap.

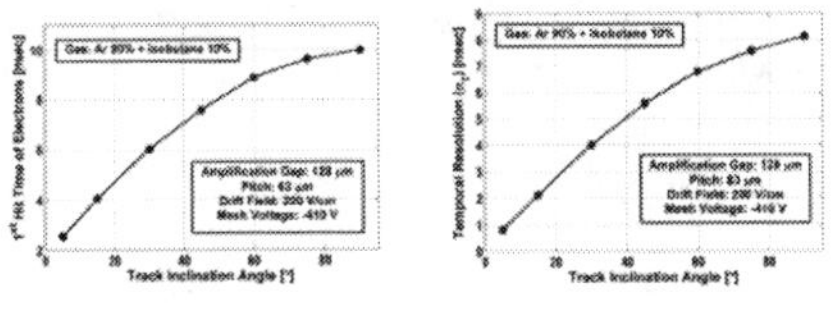

(a) (b)

Figure 2. Variation of (a) first hit time and (b) r.m.s. with track angle in Argon-Isobutane (90 : 10) mixture for bulk Micromegas detector having amplification gap of 128 μm and pitch of 63 μm.

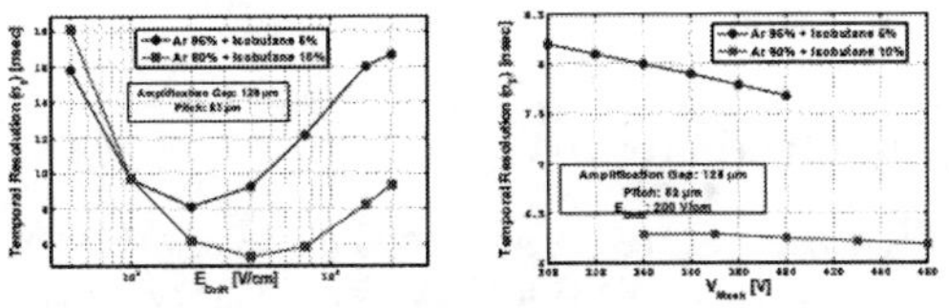

(a) (b)

Figure 3. Variation of r.m.s. with (a) E_{Drift}, (b) V_{Mesh} in different Argon-Isobutane mixtures for bulk Micromegas detector having amplification gap of 128 μm and pitch of 63 μm.

The effects of the variation of geometrical parameters such as variation of the amplification gap and mesh hole pitch have been also studied. The variation of the resolution with the drift field for several bulk Micromegas detectors has been estimated. However, no significant effect on the resolution has been observed except at higher drift fields where the detectors with larger pitch and smaller gap show comparatively better resolution (figure 4).

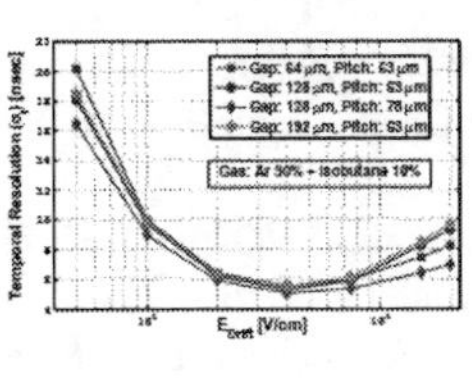

Figure 4. Variation of r.m.s. for bulk Micromegas detectors having different amplification gap and mesh hole pitch in Argon-Isobutane (90 : 10) mixture. Gap = 64 μm, pitch = 63 μm, mesh voltage = −330 V; gap = 128 μm, pitch = 63 μm, mesh voltage = −410 V; gap = 192 μm, pitch = 63 μm, mesh voltage = −540 V; gap = 128 μm, pitch = 78 μm, mesh voltage = −450 V.

4.2 Single GEM

To study the performance of GEM-based MPGDs, the temporal resolution of a single GEM detector and its dependence on the electrostatic configuration have been numerically simulated at first. The variation of temporal resolution with E_{drift}, V_{gem} and E_{ind} (induction field) has been carried out in Ar-CO_2 (70 : 30) mixture. Starting from the lower drift field, the resolution improves to a flat plateau. At lower drift fields, larger diffusion is responsible for a larger drift time and worsening of resolution (figure 5(a)). The variation with E_{ind} and V_{gem} reveals that the lower drift time and better temporal resolution can be obtained with higher induction field (figure 5(b)) and higher GEM voltage (figure 5(c)), respectively. At a higher induction field and high GEM voltage, the funneling of the electrons is such that the electrons have to travel smaller path to reach the readout plane. Also, at these higher field values, the transverse diffusion is low, whereas the drift velocity has an increasing trend. So the electrons take smaller time to reach the readout plane without much distortion.

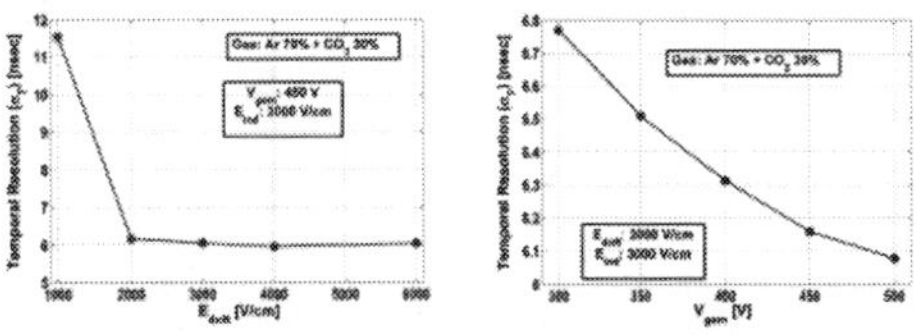

(a) (b)

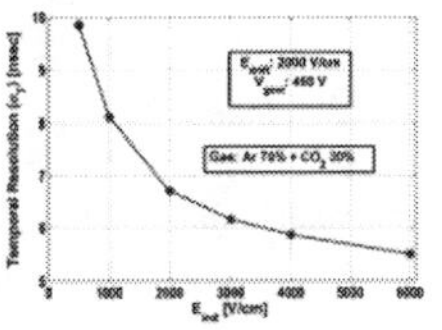

(c)

Figure 5. For a single GEM detector in Argon-CO_2 (70 : 30) mixture, the variation of temporal resolution with (a) E_{drift}, V_{GEM} = 450 V, E_{ind} = 3000 V/cm; (b) V_{GEM}, E_{drift} = 2000 V/cm, E_{ind} = 3000 V/cm; (c) E_{ind}, E_{drift} = 2000 V/cm, V_{GEM} = 450 V.

4.3 Triple GEM

The time spectrum of the triple GEM detector as a second case of GEM-based detector, is shown in figure 6(a) for two different Argon-based gas mixture. Due to higher drift velocity, the electrons in Ar-CO_2-CF_4 mixture take less time to hit the readout plane. Also, the lower transverse diffusion coefficient in this mixture helps to obtain a better temporal resolution. It may be noted here that the numerical estimates (~ 11 nsec for Ar-CO_2 and ~ 7 nsec for Ar-CO_2-CF_4) are quite close to experimentally measured values reported earlier [10–12].

The variation of the corresponding time resolution with applied high voltage for these two gas mixtures have been plotted in figure 6(b). As discussed in the case of

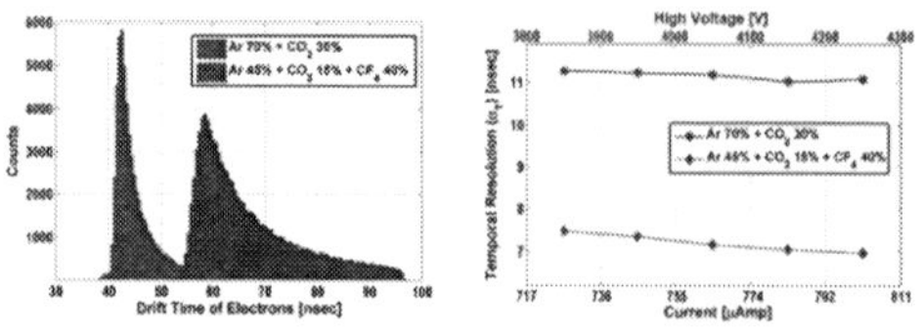

(a) (b)

Figure 6. For a triple GEM detector: (a) time spectrum, the variation of (b) temporal resolution with current and applied high voltage in different Argon-based gas mixture.

single GEM, an increase of the drift field, hole voltage and the induction field reduce the time taken by the electron to reach the readout plane, as well as improve the time resolution. For the triple GEM also, a larger value of the applied high voltage increases the field in the respective regions which in turn make the drift time less and the resolution better.

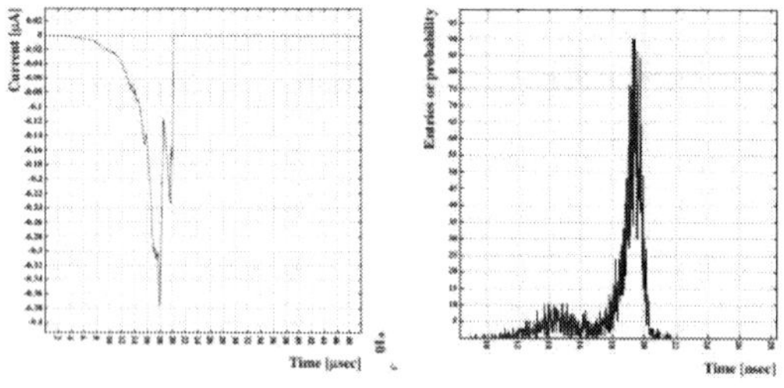

(a) (b)

Figure 7. For RPC, (a) induced signal; (b) time spectrum

5 Modified Simulation Approach

In the modified model, a threshold limit of detecting the signal has been considered which is related to the gain variance of the detector. At first, the new approach has been applied on a RPC detector. The detector dimension is listed in table 3. High voltages of ±5800 V have been applied in this detector to generate the signal in Freon-Isobutane-SF_6 (95 : 4.5 : 0.5) gas mixture. A typical induced signal for a cosmic muon track of 1 GeV is shown in figure 7(a). For the present calculation, 0.1 μA current at the rising edge of the signal has been considered as a lower threshold. For each track, the time to cross this threshold has been recorded and the final spectrum looks like as shown in figure 7(b). The r.m.s of this distribution has been found to 1.2 nsec which is very close to the experimental value. But this is only a preliminary calculation and further investigation is going on before drawing any firm conclusion. In future, this approach will be used to simulate the time resolution for the MPGDs.

6 Conclusion

In this paper, an attempt has been made to establish an algorithm for estimating the time resolution of an MPGD. A comprehensive numerical study on the dependence of time resolution on detector design parameters, field configuration and relative proportions of gas components has been made for a few MPGDs. The simulated results have been compared with available experimental data and the agreement between them is very encouraging. Please note that, gas composition used for Micromegas and GEM are made different to compare them with the available experimental data. Thus, a comparison between these two MPGDs is not possible here. The present work aims to accomplish a comprehensive characterization of the time resolution of the MPGDs on the basis of numerical as well as experimental measurements. In addition to further improvement in the numerical work, development of a test bench for studying the MPGDs invidually for their characteristics time resolution and its dependence on the design parameters and gas composition has been planned.

Acknowledgements

This work has partly been performed in the framework of the RD51 Collaboration. We wish to acknowledge the members of the RD51 Collaboration for their help and suggestions. We thank our respective Institutions for providing us with the necessary facilities.

References

[1] M. Hoch, Nucl. Instrum. Meth. vol. A **535** 1 (2004).
[2] F. Sauli, Yellow Report, CERN 77-09 (1977).
[3] M. Alfonsi, Ph. D Thesis, University of Rome.
[4] http://cern.ch/garfield.
[5] http://cern.ch/neBEM.
[6] http://cern.ch/heed.
[7] http://cern.ch/magboltz.
[8] P. Lengo, Proc. Sci. EPS-HOP 080 (2013).
[9] J. Bortfeldt, Diploma thesis, Ludwig-Maximilians University of Munich (2010).
[10] G. Bencivenni et al., Nucl. Instrum. Meth. A **494** 156 (2002).
[11] D. Heereman, Diploma Thesis, Ludwig-Maximilians University of Munich (2010).
[12] M. Alfonsi et al., Nucl. Instrum. Meth. Phy. Res. A **518** 106 (2004).

Numerical Study of Electrostatic Field Distortion on LPTPC End-Plates based on Bulk Micromegas Modules

Purba Bhattacharya[1], Deb Sankar Bhattacharya[2,4], Supratik Mukhopadhyay[2,a], Nayana Majumdar[2], Sudeb Bhattacharya[3], Paul Colas[4], and David Attié[4]

[1] *School of Physical Science, National Institute of Science Education and Research, Jatni, Khurda, India 752050*
[2] *Applied Nuclear Physics Division, Saha Institute of Nuclear Physics, 1/AF, Bidhannagar, Kolkata, India 700064*
[3] *Retired Professor, Applied Nuclear Physics Division, Saha Institute of Nuclear Physics, 1/AF, Bidhannagar, Kolkata, India 700064*
[4] *DSM/IRFU, CEA/Saclay, F-91191. Gif-sur-Yvette, CEDEX, France*

Abstract. The R&D activities for the linear collider TPC (LC-TPC) are currently working on the adoption of the micro pattern devices for the gaseous amplification stage. Several beam tests have been carried out at DESY with a 5 GeV electron beam in a 1 T superconducting magnet. We worked on a large prototype TPC with an end-plate that was built, for the first time, using seven resistive bulk Micromegas modules. During experiments, reduced signal sensitivity was observed at the boundary of these modules. Electrostatic field distortion near the module boundaries was considered to be the possible major reason behind these observations. In the present work, we will explore this hypothesis through numerical simulation. Our aim has been to understand the origin of distortions observed close to the edges of the test beam modules and to explore the possibility of using the Garfield simulation framework for investigating a phenomenon as complex as distortion.

1 Introduction

A Time Projection Chamber (TPC) [1] is envisaged just beyond the vertex detector of the proposed linear collider [2]. Because of the high flux environment, Micro-Pattern Gaseous Detectors (MPGDs) [3] have been proposed to build the end-plates of the TPC [4]. The end-plates are expected to be made of a number of MPGD modules, as shown in figure 1(a), because of the large area coverage and their geometry [5]. Since 2008, the LC-TPC collaboration has been involved in the design, development and test of a Large Prototype TPC (LP-TPC) with amplification systems based on different MPGDs and corresponding read-out schema.

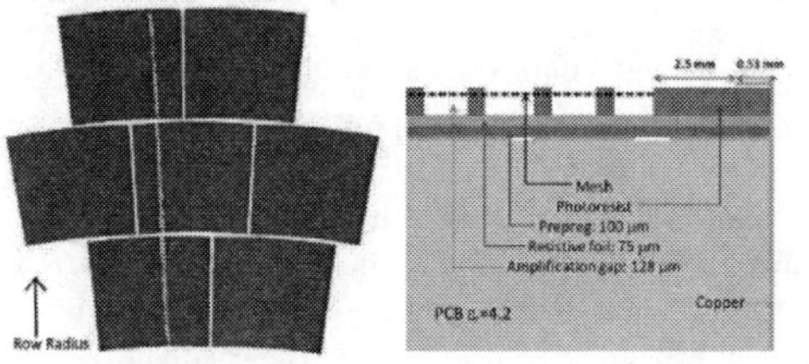

(a) (b)

Figure 1. (a) A reconstructed track on the TPC end-plate made of seven Micromegas modules; (b) schematic diagram of a region close to the edge of one of the module, side view.

[a] e-mail: supratik.mukhopadhyay@saha.ac.in

For the 2013 February test beam run, Micromegas modules with resistive coating were chosen [7]. These modules have keystone shape ($\sim$ 22 $\times$ 17 cm^2), which is similar to the shape foreseen for the ILD-TPC. The Micromegas in each of the modules has an amplification gap of 128 μm and a pitch of 63 μm. The resistive carbon-loaded Kapton®layer with a resistivity of about 3 MΩ/square was chosen as the resistive layer material. Each module had 1726 pads arranged in 24 rows by 72 pads. The pads were all grounded and were read out with the AFTER-based electronics designed for the T2K experiment [6]. The reconstruction of a typical track on theTPC end-plate, consisting of seven modules, is shown in figure 1(a). The close-up schematic diagram of the region near the module edge is shown in figure and 1(b) [7].

The DESY electron beam energy was varied from 1 – 5 GeV during the various phases of experiment. The working gaseous mixture in the TPC was Ar-CF_4-Isobutane (95 : 3 : 2), so-called T2K gas. Data related to cosmic rays, drift velocity estimation, Z scan, phi scan, drift field scan, variation of gain, effect of different magnetic field etc. were recorded and analyzed in detail [7]. During the analysis process, reduced signal sensitivity was observed near the edge of the module [9]. It was also found that the spatial resolutions of all the pads are consistent, except for those near the module edge [7, 8]. Figure 2 shows the distortion (residual) plot versus the row radius without or with magnetic field. The transfer between two detector modules is seen and in-between them, the distortion is larger and leads to 'S' shape as shown in figures 2(a)

(B = 0 T) and figure 2(b) (B = 1 T). It may be noted that the difference between the distortions is expected since in the presence of magnetic field, the transverse diffusion is known to be reduced and the distortion is likely to be influenced by $\vec{E} \times \vec{B}$ effect.

The most likely reason leading to this distortion is the electric field inhomogeneity introduced by the gap between a module and its neighbor. Similar conclusion was arrived at by studying the electric field configuration near the module boundaries with GEM amplification stages in [9] using a very simplified version of the actual experimental setup. In what follows, besides investigating the electric field at the module boundaries for an end-plate having Micromegas modules, we have numerically estimated the residuals for a relatively more realistic geometry and compared our estimates with the experimental data.

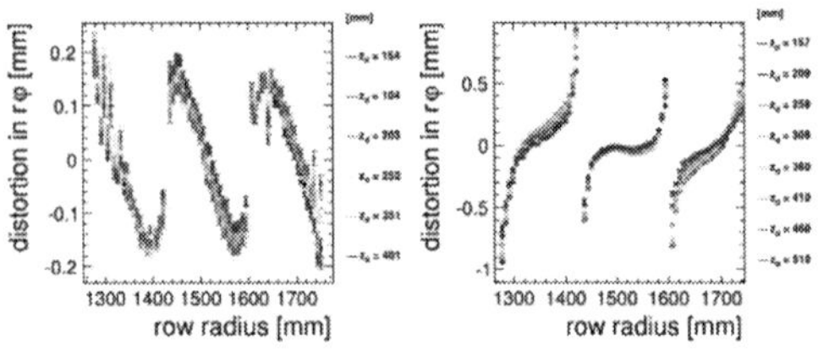

(a) (b)

Figure 2. Distortion (residual) versus row in (a) $B = 0$ T and (b) $B = 1$ T.

2 Simulation Tools

The Garfield [10] simulation framework has been used in the following work. Besides neBEM [11], which is used for electrostatic calculations, the Garfield framework provides interfaces to HEED [12] for primary ionization calculation and Magboltz [13] for computing drift, diffusion, Townsend and attachment coefficients.

3 Simulation Conditions

The model used for the simulation, although more realistic than the model used in [9], is simpler than that used in the real experimental setup and deals with a much smaller geometry in comparison to the actual system. By comparing the experimental condition, as depicted in figure 1, and the numerical model shown in figure 3, the following can be pointed out:

- In the place of seven, only three modules have been used.
- Each module is a rectangle of 4.0 cm × 3.7 cm along X-Y plane, in place of keystone shaped modules of 22 cm × 17 cm matching the r-phi placement of the LP-TPC.
- The actual experiment was performed using a resistive bulk Micromegas, whereas, for this simulation, the standard bulk Micromegas has been considered.
- Instead of cylinders, infinitely thin wire elements have been used to model the micromesh plane.
- A continuous grounded anode plane has been placed 128 μm below the mesh plane. A virtual pad structure has been evoked only when necessary.
- Instead of a TPC field cage, a drift plane has been placed 1 cm above the modules.
- As in the experiment, 3 mm thick photoresist has been used to support the modules.
- True to the experiment, a copper layer has been connected all around the module to the ground.
- Between the modules there is a gap of 1 mm which follows the experimental situation.

It should also be mentioned that, for the results presented here, the micromesh plane has been biased with a potential of −350 V whereas −550 V has been applied to the drift plane. Three different tracks, as shown in figure 3(a), have been considered.

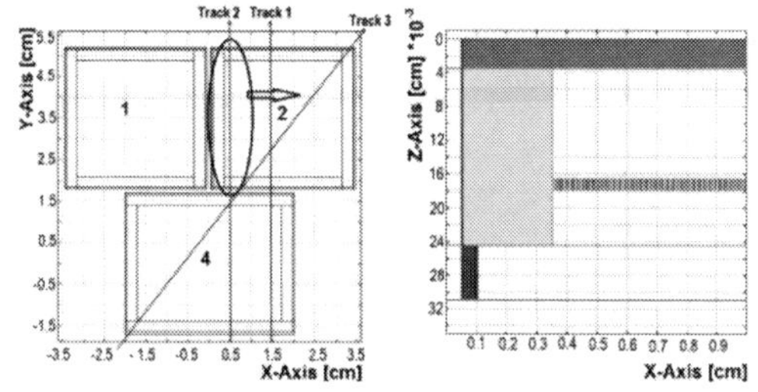

(a) (b)

Figure 3. (a) Top view of the simulated model, (b) closer (marked by the ellipse) side view of the edge of one module. Different tracks used in this simulation are shown in (a).

4 Results

4.1 Electrostatic Field

Figure 4(a) and figure 4(b) show the field contour along Track 1 and Track 2, respectively. In the vicinity of the gap between the modules, the field is affected by the presence of the additional ground line around the module and the dielectric material. It may be noted from figures 1(b) and 3(b) that the dielectric support is (2.5+0.55)mm, the ground line has 0.55mm length in the Y-direction, while the gap between the two modules is 1mm. So, after the micromesh ends on one module, there is a gap of 4.5mm before the micromesh in the next module begins. The gap lies in between 1.7cm to 1.8cm in the Y direction. The transverse component of the field, in the region between 1.2cm to 2.3cm is significantly large. Since Track 2 is close to the module edge in the X-direction as well, the field lines are found to be distorted further. This distortion of the field is very likely to affect the electrons drifting towards the read-out module. Similar estimates for GEM-based TPC end-plates [9] were made earlier, as mentioned

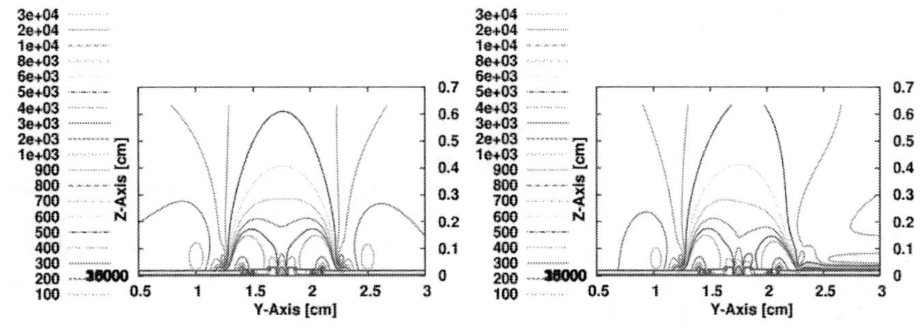

(a) (b)

Figure 4. Field contours on a YZ plane along Tracks (a) 1 and (b) 2.

above. Although overall features of the electrostatic field, electron drift and diffusion are the same, no direct comparison can be made since the geometries of the modules and the approximations involved are of a different nature.

4.2 Drift Lines

For the present calculation, T2K gas (Argon 95%, CF_4 2%, Isobutane 2%) has been considered. To study the effects of the field, electron tracks along Track 1 at a distance of 5 mm above the module have been considered. The electrons from these predetermined tracks are allowed to drift towards the read-out plane (figure 5(a)). At the edge of the module, the drift lines get distorted. It is observed from figure 5 that a significant number of electrons are lost on the additional ground and the dielectric material. From the drift lines, it is clear that the number of electrons at the read-out pads close to the module edge is less in comparison to that at the central part of the module. The presence of the magnetic field (1 T) reduces the diffusion as shown in figure 5(b), as expected. From a quantitative estimate of the efficiency carried out using a virtual 4mm long pad structure on the anode plane, it has been found that without magnetic field, the pad adjacent to the module edge has little charge collection, while the second pad has close to 60% collection. With magnetic field, the pad adjacent to the edge has a charge collection efficiency of around 40%, while the second pad is found to work almost at the full efficiency.

4.3 Calculation of Residue

In experiments, the residue is calculated to get an idea of the deviation of the measured hit position in the readout pad with respect to the position of the real track. Similarly, we have estimated residues for Track 3 for all the electrons that reach the anode plane by estimating the difference between the start and end points of these electrons. Since no pad structure has been considered, and the resistive nature of the anode plane has been ignored, the term intrinsic residue has been used to mean the numerical estimate. The patterns that emerge from these estimates follow the same trend as obtained through the analysis of real data.

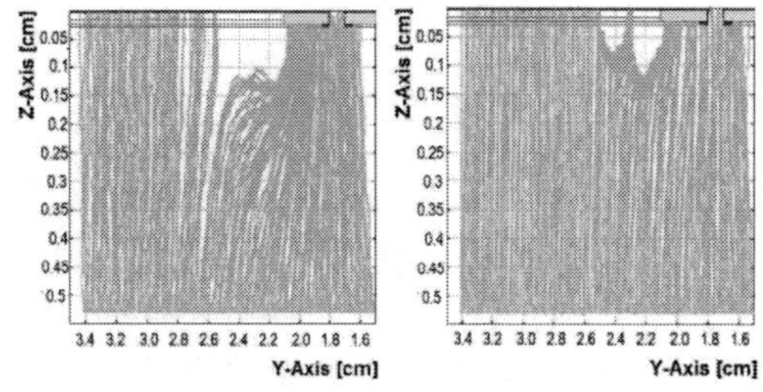

(a) (b)

Figure 5. Electron drift lines along Track 1 in (a) $B = 0$ T, (b) $B = 1$ T.

From the residue histograms (figures 4.3, 4.3), it can be observed that residues reduce significantly due to the presence of a magnetic field. This is easily explained by noting that magnetic field decreases diffusion since it acts as an additional constraint that makes the charged particle follow electric field lines. The X-residual plots for $B = 0$ T (figure 4.3 and $B = 1$ T (figure 4.3) reflects this fact once again. The effect of magnetic field is striking and the sign of the residues are found to change with and without magnetic field. Interestingly, this change of sign is also observed in the experimental data. It may be noted here that, though our simulation follows the experimental trend, the value of the distortion is larger in the numerical estimation, the maximum in residue being close to 1mm instead of around 0.5mm as found in the experimental data.

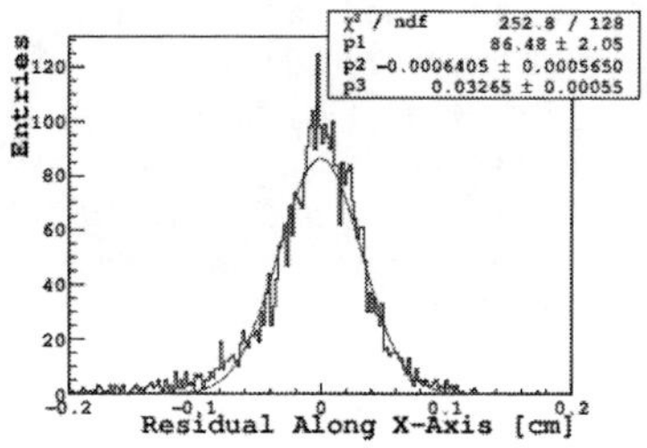

Figure 6. Histogram of residue alongX-Axis in $B = 0$ T.

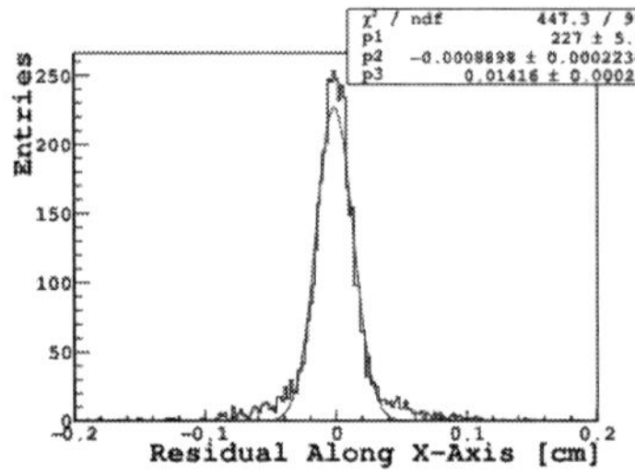

Figure 7. Histogram of residue along X-Axis in $B = 1$ T.

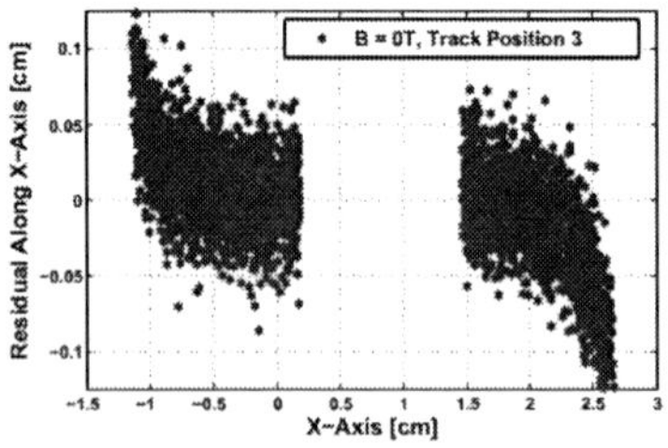

Figure 8. Residual along X-Axis in $B = 0$ T for Track 3.

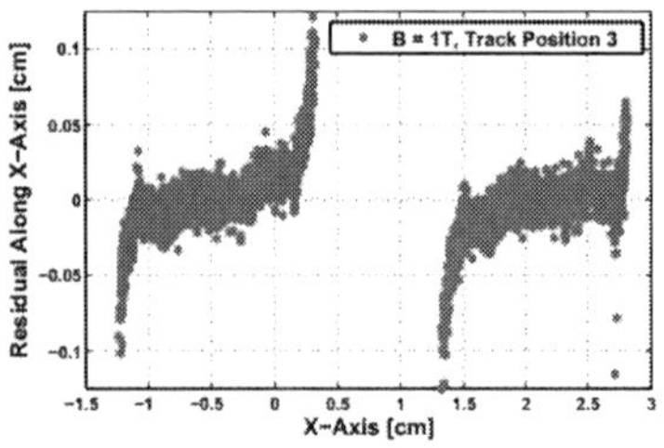

Figure 9. Residual along X-Axis in $B = 1$ T for Track 3.

The differences between the two estimates probably arise from the simplifications adopted in the numerical model. Besides the smallness of the modules, lack of resistive anode and the thin-wire assumptions are probably the most significant ones. Among these, the resistive anode in a real experiment is expected to improve reconstruction of the hit point and, thus, reduce the residue in the experimental data. The thin-wire assumption is likely to affect the transmission of electrons through the mesh. However, despite the short-comings arising out of these simplifying assumptions, this preliminary study is capable of bringing out the main features of the experimental observation.

5 Conclusion

Following the experimental activities and related data analysis, we have investigated the origin of the so-called distortions observed close to the module edges. We have been able to numerically simulate the observed patterns quite successfully, although we have not made any attempt to achieve quantitative agreement with the experimental data. Our computations indicate that the inhomogeneity of the electric field close to the module edges leads to a loss of efficiency of few pads close to the edge and also leads to the distortion in residue observed in the experimental data. The presence of magnetic field complicates the matter through the $E \times B$ effect, since the two fields are no more parallel (due to the fact that the E-field is distorted). The obtained qualitative agreements encourage us to continue with the study and, if possible, propose module design modifications that can alleviate the problem. At least, we should be able to suggest corrections that can be applied during the data analysis phase to take care of this issue. In future, the simulations will be made more realistic by making the end-plate geometry closer to the real one and by making the anode resistive. In addition, possible suitable design modifications leading to less inhomogeneity at the module edges will be investigated.

Acknowledgements

This work has partly been performed in the framework of the RD51 Collaboration. We wish to acknowledge the members of the RD51 Collaboration for their help and suggestions. We thank our collaborators from the ILC-TPC collaboration for their help and suggestions. We gratefully acknowledge IFCPAR/CEFIPRA (Project No. 4304-1) for partial financial support. Finally, we thank our respective Institutions for providing us with the necessary facilities.

References

[1] D. R. Nygren et al., Physics Today **31** 46 (1978).
[2] http://www.lctpc.org/
[3] M. Hoch, Nucl. Instrum. Method A **535** 1 (2004).
[4] P. Schade et al., Nucl. Instrum. Meth. A **628** 128 (2011).
[5] D. Attié, Jour. Instrum. **6** C01007 (2011).
[6] P. Baron et al., IEEE Trans. Nucl. Sci. **55** 1744 (2008).
[7] W. Wang, Ph. D Thesis, Univ. of Paris-Sud, (2013).
[8] D.S.Bhattacharya, Proc. Science (EPS-HEP2015) 277
[9] T. Behnke et al., arXiv:1309.1370v2 (2013).
[10] http://cern.ch/garfield.
[11] http://cern.ch/neBEM.
[12] http://cern.ch/heed.
[13] http://cern.ch/magboltz.

VMM - An ASIC for Micropattern Detectors

George Iakovidis[1,a] *on behalf of the ATLAS Muon Collaboration*

[1] *Brookhaven National Laboratory, 11973, Upton, NY, USA*

Abstract. The VMM is a custom Application Specific Integrated Circuit (ASIC) that can be used in a variety of charge interpolating tracking detectors. It is designed to be used with the resistive strip micromegas and sTGC detectors in the New Small Wheel upgrade of the ATLAS Muon spectrometer. The ASIC is designed at Brookhaven National Laboratory and fabricated in the 130 nm Global Foundries 8RF-DM process. It is packaged in a Ball Grid Array with outline dimensions of 21×21 mm^2. It integrates 64 channels, each providing charge amplification, discrimination, neighbour logic, amplitude and timing measurements, analog-to-digital conversions, and either direct output for trigger or multiplexed readout. The front-end amplifier can operate with a wide range of input capacitances, has adjustable polarity, gain and peaking time. The VMM1 and VMM2 are the first two versions of the VMM ASIC family fabricated in 2012 and 2014 respectively. The design, tests and qualification of the VMM1, VMM2 and roadmap to VMM3 are described.

1 Introduction

In order to handle efficiently the increased trigger rate that is expected from the high luminosity LHC performance after the long shutdown of 2019-2020 (Phase-I), the innermost station of the ATLAS muon end-cap system (Small Wheel, SW) will be replaced. The New Small Wheels (NSW)[1] will follow the same segmentation as the present detector system but will be equipped with two new detector technologies: resistive strip micromegas [2] and small-strip Thin Gap Chambers (sTGC) [8]. Both detectors have to operate under a high background radiation region (up to 15 kHz/cm^2 primarily from low energy photons and neutrons) while providing precise muon tracking information as well as Level-1 trigger primitives independently. This detector system of ~2.3 million readout channels (~2 million for micromegas and ~0.3 million for sTGC) adds highly challenging requirements for the front-end electronics. They should be able to operate in both input charge polarities due to different detector elements in an input capacitance range of 50 pF up to 1.5 nF while providing charge measurements up to 2 pC with a noise at ~ 1 fC level. Moreover, they should provide time measurements with high precision (~ 1 ns RMS) and trigger primitives, staying low power (< 1 W) and programmable. The VMM ASIC is developed at Brookhaven National Laboratory and will be used by both detector technologies in the NSW.

2 VMM1 ASIC

The VMM1 [3] [4] is the first version of the VMM ASIC family which was fabricated in 2012 in the 130 nm Global Foundries 8RF-DM process (former IBM 8RF-DM). It is packaged in 400 Ball Grid Array of 1 mm pitch. The block diagram of a channel and the common blocks is shown in Figure 1.

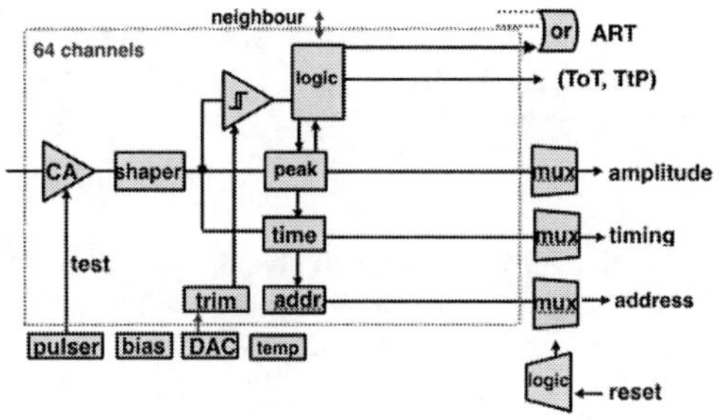

Figure 1. Architecture of VMM1 [3].

It is composed of 64 front-end channels each providing a low-noise charge amplifier, a shaper with baseline stabilizer, a discriminator with trimmer, a peak detector, a time detector, some logic, and a dedicated digital output for Time-over-Threshold (ToT) or Time-to-Peak (TtP) measurements. Shared among channels are the bias circuits, a temperature sensor, a test pulse generator, two 10-bit DACs for adjusting the threshold and test pulse amplitudes, a mixed-signal multiplexer, the control logic, and the Address in Real Time (ART) which consists of dedicated digital outputs (flag and address) for the first above-threshold event.

Figure 2 shows the theoretical and measured ENC as a function of the input capacitance as well as the measured timing resolution at 200 pF as function of the output signal amplitude for different values of peaking time. In this

[a]e-mail: giakovidis@bnl.gov

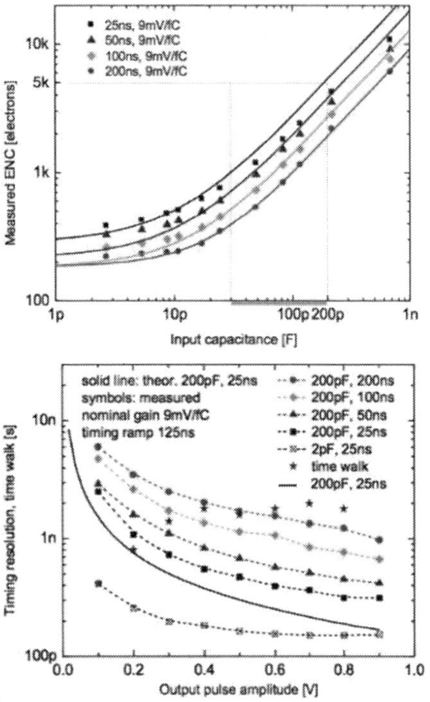

Figure 2. The measured Equivalent Noise Charge (ENC) as a function of the input capacitance (top). The timing resolution as a function of the output pulse amplitude (bottom) [3].

version of the ASIC, a non-negligible disagreement can be observed for the simulated and measured ENC at the lowest peaking time (25 ns) and large values of input capacitance due to an error in the adjustable lead-lag compensation of the front-end amplifier. Another issue was a considerable leakage current from the ElectroStatic Discharge (ESD) protection circuit at the input of the channel [3]. Moreover difference between the theoretical and the measured resolution at 25 ns was observed largely due to the actual peaking time. Though the timing resolution is at the level that satisfies the requirements of the Micromegas and sTGC detectors.

2.1 Testing of Micromegas detectors with VMM1 ASIC

The VMM1 was tested with eight identical $10 \times 10\,\mathrm{cm}^2$ resistive strip micromegas prototypes with 1D readout at the CERN beam facilities as shown in Figure 3. A spatial resolution of $\sim$ 65 µm , as shown in Figure 4, has been measured for perpendicular to the detector plane tracks as expected for these detectors [5]. Figure 5 shows the spatial resolution for particle tracks of incident angles 10° – 30°. The resolution was measured to be $\sim$ 180 – 120 µm respectively. Using the ART signals, the performance of the micromegas detectors was demonstrated to be below 1 mrad as shown in Figure 6. The details of the study above can be found in Ref. [6][7].

Figure 3. Setup of eight Micromegas detectors at CERN beam line readout by VMM1 [6].

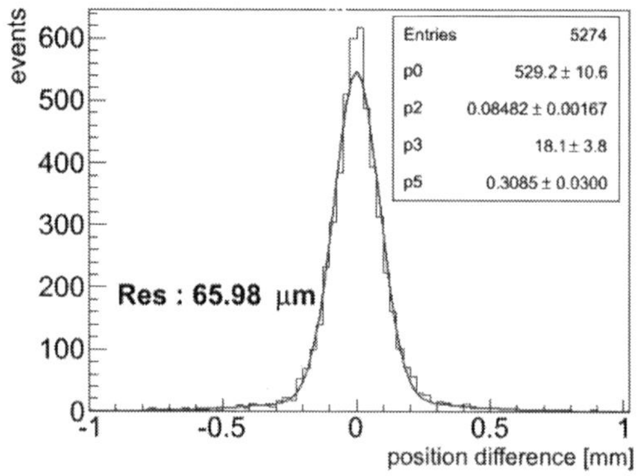

Figure 4. Residual distribution of two micromegas chambers with 400 µm strip pitch obtained with charge centroid method. The distribution is fitted with a double Gaussian function. An estimation of the position resolution is extracted by dividing the weighted Gaussian sigmas (p2, p4) with the Gaussian heights (p0, p3) and divide with $\sqrt{2}$. The resolution is demonstrated to be ~65 µm [6].

2.2 Testing of sTGC detectors with VMM1 ASIC

In 2014, the full-size sTGC prototype was tested with a high momentum pion beam at the Fermilab test beam facilities. Figure 7 shows the sTGC detector prototype in the middle of the silicon pixel telescope used for external tracking reference. Figure 8 shows the residual distribution with an intrinsic resolution of $\sim$ 44 µm (corrections applied). The details of this study can be found in Ref. [8].

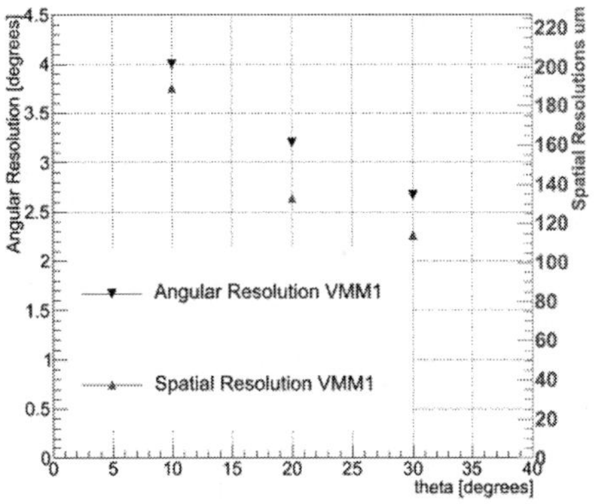

Figure 5. The spatial and angular resolution of a single micromegas chamber versus the incident track angle reconstructed using the μTPC method [6].

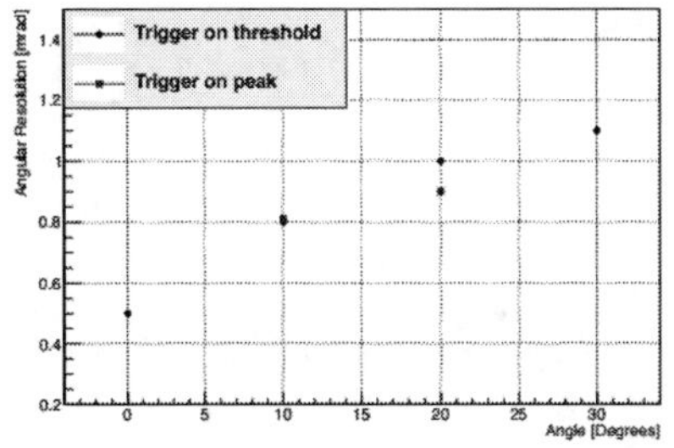

Figure 6. Angular resolution reconstructed with the eight micromegas telescope and the ART signals obtained for each of them. For 10° and 20° the ART was measured both in threshold and peak [6].

3 The VMM2 ASIC

The VMM1 ASIC was a successful first version of the VMM ASIC family. For the NSW needs though, a continuous fully digital readout is foreseen. The VMM2 is a deep redesign of the VMM1 correcting the drawbacks found as well as including a fully digital continuous readout mode. Including an order of magnitude transistors per channel more (80 k/ch), VMM2 is as well composed of 64 linear front-end channels. It was fabricated in 2014 in 130 nm Global Foundries 8RF-DM process (former IBM 8RF-DM) and packaged in 400 Ball Grid Array of 1 mm pitch. A block diagram of one of the identical channels is shown in Figure 9. The low-noise charge amplifier with adaptive feedback, test capacitor, and adjustable polarity (to process either positive or negative charge), was optimized for a capacitance of 200 pF and a peaking time of 25 ns. The input MOSFET is a p-channel with gate area of L × W = 180 nm × 10 mm (200 fingers, 50 μm each) biased at a drain current ID = 2 mA; this corresponds to an inversion coefficient IC ≈ 0.22, a transconductance $g_m \approx 50\,\mathrm{mS}$, and a gate capacitance $C_g \approx 11\,\mathrm{pF}$. The filter (shaper) is a third-order designed in delayed dissipative feedback DDF [4], has adjustable peaking time in four values (25, 50, 100, and 200 ns) and stabilized, band-gap referenced, baseline. The DDF architecture offers higher analog dynamic range, making possible a relatively high resolution at input capacitance much smaller than 200 pF. The gain is adjustable in eight values (0.5, 1, 3, 4.5, 6, 9, 12, 16 mV/fC).

Figure 7. The sTGC prototype detector, inside a copper Faraday cage, in the middle of the silicon pixel telescope [8].

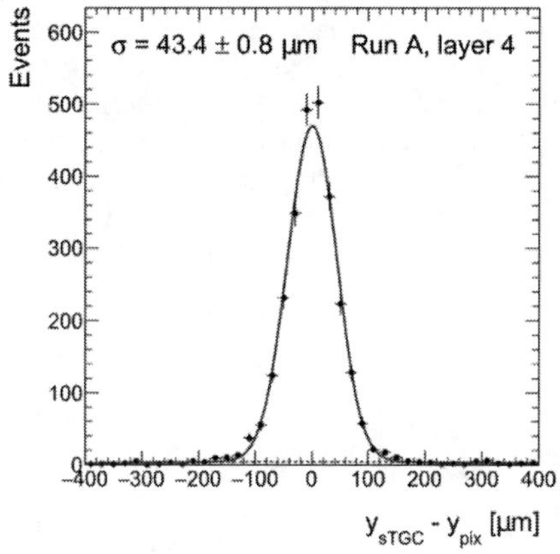

Figure 8. The residual distribution obtained with charge centroid method (corrections are applied) together with the result for the intrinsic resolution [8].

The peak detector measures the peak amplitude and stores it in an analog memory. The time detector measures the peak timing using a time-to-amplitude converter, i.e., a voltage ramp that starts at the time of the peak and stops at a clock cycle of the BC clock. The Time to Ampli-

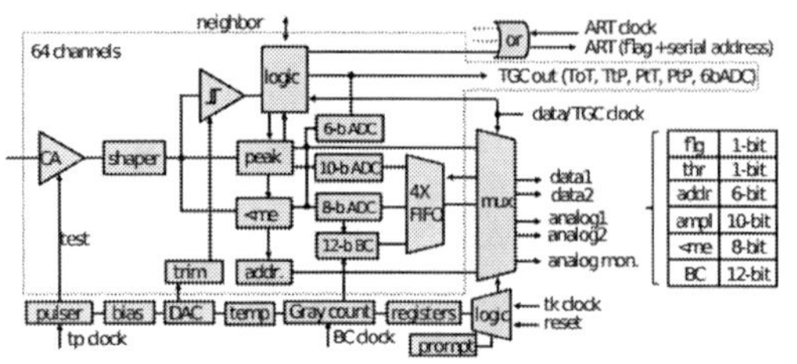

Figure 9. Architecture of VMM2 [10].

tude Converted (TAC) value is stored in an analog memory and the ramp duration is adjustable in four values (60 ns, 100 ns, 200 ns, 400 ns). The peak and time detectors are followed by a set of three low-power ADCs (a 6-bit, a 10-bit, and a 8-bit), characterized by a domino architecture [9] but of a new concept. For the time measurements an additional 12-bit Gray code time-stamp is provided, incremented by an external clock providing 20-bit timing information in total. For each channel, the amplitude measurement by the 6-bit ADC completes the conversion at 25 ns and it is driven as a direct output. These ADCs are enabled depending on the selected mode of operation. The output of the ASIC in digital continuous mode is composed of 38-bits as shown in Figure 9.

Some preliminary measurements of the VMM2 are shown in Figures 10, 11. The ASIC shows good performance with respect to the noise measurements, although the new circuitry shows also some drawbacks. The peak detector, followed by a voltage to current conversion circuitry, shows some bias due to overshoot which results in loosing the ADC precision. Moreover some accumulation in the ADCs was observed.

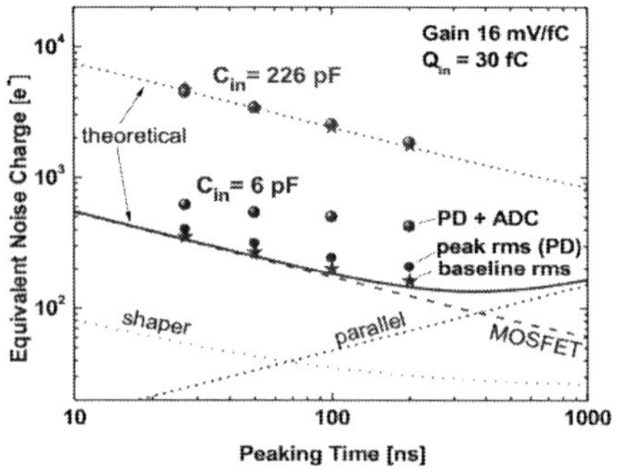

Figure 10. The measured ENC as a function of the peaking time.

The VMM2 ASIC is under testing with and without detectors in several labs and institutes around the world to measure its performance. Moreover all the drawbacks found until now are fully simulated and corrections have been provided for the next version.

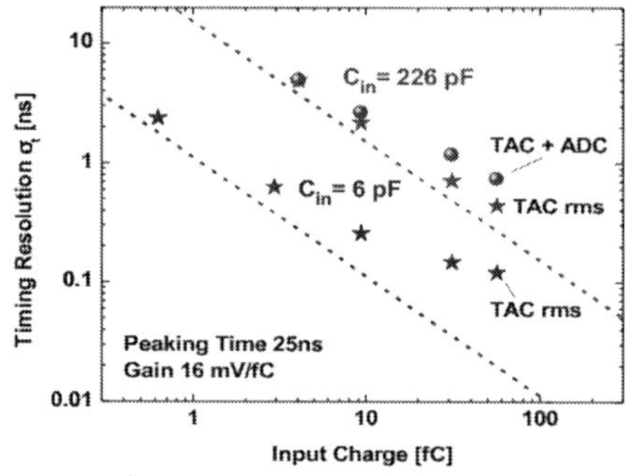

Figure 11. The timing resolution as a function the input charge.

4 Towards VMM3 - Final design

The next version of VMM, the VMM3, will be submitted for fabrication in January 2016. It will be packaged in the same BGA and will provide, in addition to the VMM2 drawbacks corrections, latency and Level-0 buffers necessary for the ATLAS readout. Moreover it will integrate features such as longer TAC ramps, SEU mitigation circuitry as well as handling of higher input capacitance of the order of 1.5 nF needed for the detector elements of NSW.

References

[1] ATLAS Collaboration, *New Small Wheel Technical Design Report*, ATLAS-TDR-020.

[2] T. Alexopoulos *et al.* Nuclear Instruments and Methods in Physics Research, **640**, 110 (2011).

[3] G. De Geronimo, *et al.* IEEE Transactions on Nuclear Science, **60**, 2314 (2013).

[4] G. De Geronimo and Li Shaorui , IEEE Transactions on Nuclear Science, **58**, 2382 (2011).

[5] G. Iakovidis, Journal of Instrumentation, **8**, C12007 (2013).

[6] ATLAS Collaboration, *Performance of the First Version of VMM Front-End ASIC with Resistive Micromegas Detectors*, ATL-UPGRADE-PUB-2014-001

[7] G. Iakovidis, *Research and Development in Micromegas Detector for the ATLAS Upgrade*, CERN-THESIS-2014-148.

[8] A. Abusleme, *et al.* arXiv:1509.06329, (2015).

[9] G. De Geronimo, *et al.* IEEE Transactions on Nuclear Science, **54**, 541 (2007).

[10] G. De Geronimo, *Readout Electronics for the ATLAS New Small Wheels (Phase 1)*, ACES 2014.

A MOdular System for Acquisition, Interface and Control (MOSAIC) of detectors and their related electronics for high energy physics experiment

G. De Robertis[1], G. Fanizzi[1], F. Loddo[1], V. Manzari[1], and M. Rizzi[1]

[1] *INFN Sezione di Bari, Bari, Italy*

Abstract. In this work the MOSAIC ("MOdular System for Acquisition, Interface and Control") board, designed for the readout and testing of the pixel modules for the silicon tracker upgrade of the ALICE (A Large Ion Collider Experiment) experiment at teh CERN LHC, is described. It is based on an Artix7 Field Programmable Gate Array device by Xilinx and is compliant with the six unit "Versa Modular Eurocard" standard (6U-VME) for easy housing in a standard VMEbus crate from which it takes only power supplies and cooling.

1 Introduction

This paper describes the main features and the preliminary tests of the electronics board, called MOSAIC, developed for the characterization of the detector building blocks of the new Inner Tracking System (ITS), which is a key element of the ALICE upgrade strategy for the Run 3 of LHC. ALICE [1] is the CERN LHC experiment designed to study the physics of strongly interacting matter at extreme conditions of energy density ($\sim$ 15 GeV/fm^3) and temperature over a large volume ($\sim$ 1000fm^3), and in particular the properties of the Quark Gluon Plasma (QGP), using nucleus-nucleus, proton-nucleus and proton-proton collisions. The ALICE long-term physics plans include a substantial upgrade of the central barrel detectors, which will be installed during the LHC Long Shutdown scheduled for the years 2019-20, in order to fully exploit the scientific potential of the LHC for fundamental studies of QCD [2]. The planned upgrades, which will enhance the ALICE vertexing and tracking capabilities at low p_T and allow data taking at substantially higher rates, include in particular a completely new ITS made of 7 cylindrical layers entirely based on an innovative active monolithic pixel sensor called ALice PIxel DEtector (ALPIDE) [3]. The ALPIDE chip is the full custom pixel sensor that will equip all modules of the new ITS [4].

2 Hardware Description

Mainly conceived for the testing of detectors and their related electronics for high energy physics experiments, the MOSAIC board is able to read data through ten high-speed serial links at rates up to 6.6 Gbps or using up to 126 slower channels, such as general-purpose Low Voltage Differential Signals LVDS lines.

The board (Fig. 1) is equipped with four programmable LEMO Input/Output complying with Nuclear Instrument Module standard and two FMC-LPC mezzanine slots in order to provide additional connectivity and flexibility.

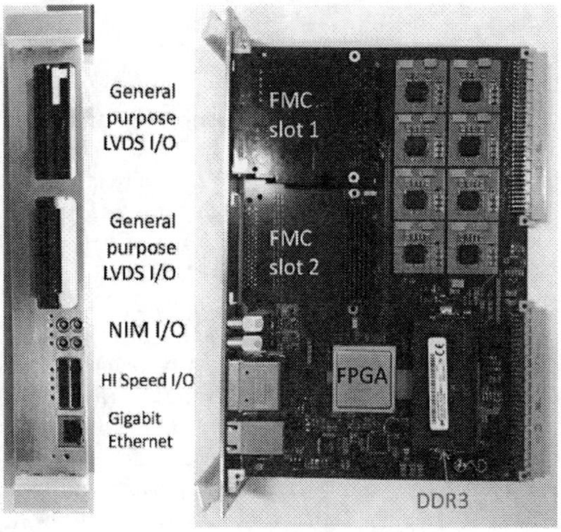

Figure 1. The MOSAIC board

The core of the board is the Xilinx FPGA model XC7A200T-2 FFG1156C [5] with the following relevant features:

- 215360 Logic Cells
- 730 Block RAM blocks (12.8 Mb)
- 10 Clock Management Tiles
- 10 I/O Banks
- 500 user I/O
- 16 Low-Power Gigabit Transceivers (up to 6.6 Gb/s)

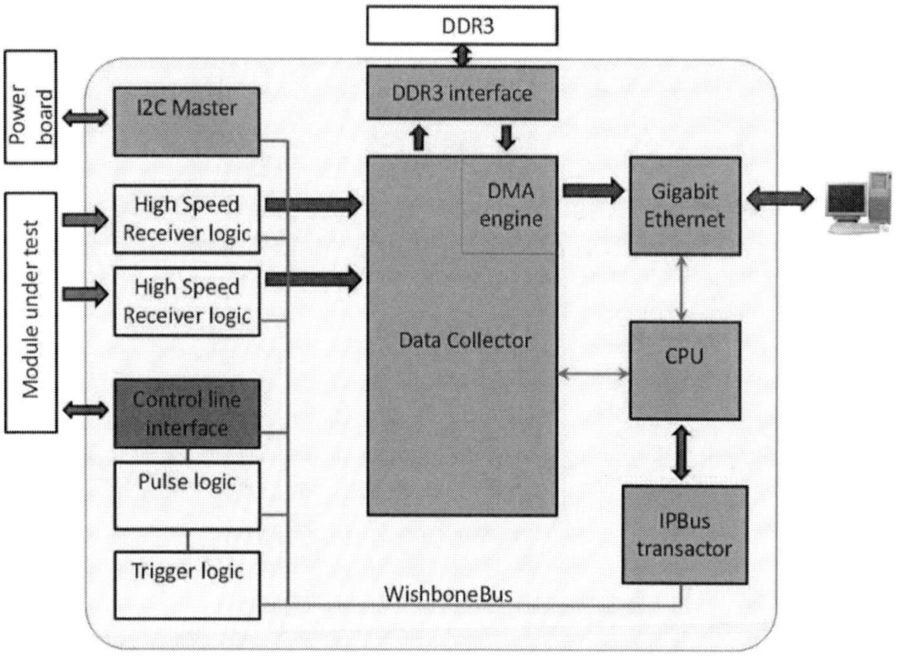

Figure 2. Block diagram of the FPGA firmware for the test of the ALICE new tracker modules

A SODIMM connector houses a 1 GB DDR3 memory module working at 400 MHz performing a nominal throughput of 6.4 GB/s. The board has been designed to support higher capacity modules, while the current version of the firmware can work only with this specific memory module.

Two Robinson Nugent P50E-068-P1-SR1-TG connectors are housed on the front panel, providing 33 LVDS signals for each connector plus the ground reference. Each LVDS couple is configurable as input or output and is protected against electrostatic discharges. The FPGA firmware can be designed to read data from these inputs at a maximum rate of 1.2 Gbps or to control ancillary electronic, like power control and monitoring interfaces.

To allow communication with high speed devices, a Samtec Eye Speed HDI6-035 connector is available on the front panel. Ten high speed transceivers, running up to 6.6 Gbps are connected to it. Moreover it provides two configurable LVDS couple of lines and two M-LVDS bidirectional differential lines, intended to control and monitor the front-end electronics.

All communication with the board for data readout, setup and status monitoring are performed through an Ethernet interface capable of running at 10, 100 and 1000 Mbps. The interface is able to configure itself automatically based on the connected device, so it can be connected directly to a PC without a network switch or a crossed cable.

The system can be expanded plugging up to two "FPGA Mezzanine Card" (FMC) Low pin count or one Double Width Card. Each FMC connector provides 34 differential pairs lines, configurable as 68 single ended lines, two clock differential lines and the connection to one high speed transceiver running up to 6.6 Gbps. The FMC slot is mechanically and electrically compliant to the ANSI/VITA 57.1 standard.

3 Firmware

The FPGA firmware has been designed to provide a common infrastructure and a set of modules specific for the external hardware that it will control or read. The hardware specific modules ("High Speed Receiver" and "Control line interface" in Fig. 2) are connected to the infrastructure in two ways:

a) using a custom interface designed to move data as fast as possible to DDR3 main memory for data acquisition

b) through a Wishbone bus for configuration and monitoring

The infrastructure is made by the blocks described in the following sections.

The Data Collector is devoted to move data from the hardware specific block buffers to the DDR3 memory. When a High Speed Receiver needs to move data from the buffer, it sets a read request flag and, as soon as possible, the Data Collector moves the requested amount of data to the DDR3 memory at a sustained rate of 5 GB/s. A header is inserted before each data block; it contains a detailed description of the block like its size, source, errors and additional flags from the data source.

The Gigabit Ethernet interface has been designed to allow high transfer rate with a very low CPU load. While the communication protocols are handled by the local CPU, all data are transferred by hardware using the DMA engine in the Data Controller module. Thanks to the hardware offloads implemented in the interface, like IP fragmentation and TCP checksum, the board is able to send data on the TCP connection at a transfer rate of 120 MB/s (the Gigabit Ethernet speed limit) with a CPU load of 30%.

The CPU implemented in the firmware is a 8-bit microprocessor running at 50 MHz with 12 kB of program memory and 2 kB of RAM, plus additional peripherals such as Interrupt logic, TIMER, I/O etc. The microprocessor has a Complex Instruction Set Computer (CISC) architecture, that has the advantage to allow the programmer to write complex programs even though its small size. The software running into the microprocessor needs less then 10 kB and implements the TCP/IP and UDP/IP protocols and accessory functions such as board startup diagnostic tests, firmware upgrade, IP configuration etc.

The IPBus Transactor translates the commands coming from the PC encapsulated in UDP/IP packets into Wishbone bus transactions. A single network packet can contain many bus read/write commands: the CPU forwards the request packet to the transactor that processes it and builds the answer packet containing transaction results.

The MOSAIC firmware further provides utility blocks such as the Pulse Logic, which implements a programmable periodic test-pulse/trigger generator to speed-up the characterization procedures, and the Trigger Logic to control and monitor the internal and external Trigger pulse production.

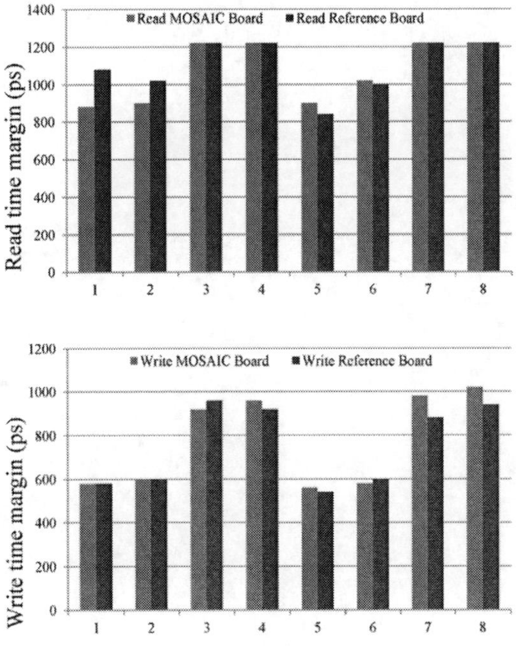

Figure 3. DDR time margin for read operations (top) DDR time margin for write operations(bottom)

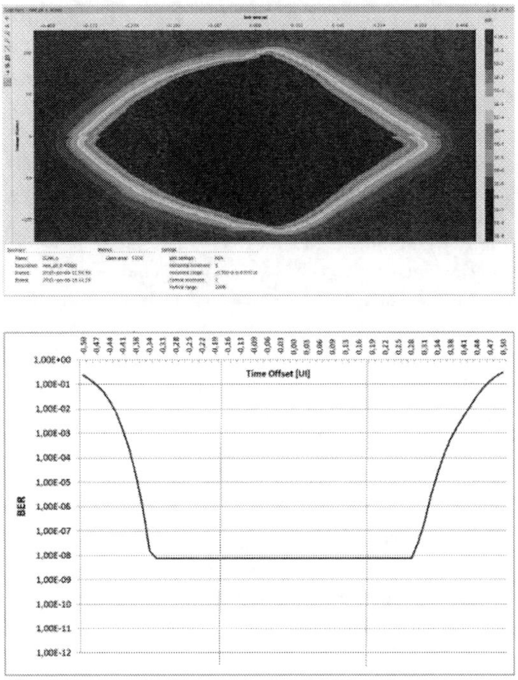

Figure 4. Eye scan for the transceiver at 6.4 Gbps (top) and bathtub diagram from the eye scan data (bottom). The eye scan shows the Bit Error Rate (BER) as a function of time (horizontal axis) and voltage offset (vertical axis) with respect to the ideal sampling point. Fixing the value of the voltage offset at 0, the recorded BER as a function of time expressed in UI give rise to the bathtub diagram in the bottom panel.

The whole structure in Fig. 2, as needed for the ALICE pixel module test, requires less than 10% of the FPGA resources, meaning that additional data processing can be potentially implemented in the hardware specific blocks and many external devices can be controlled with a single board.

4 Prototype test results

Three board prototypes have been tested to validate the design before final production. Two critical blocks have been identified: the DDR3 interface and the High Speed Transceivers.

The DDR3 interface requires special care in the PCB design due to the timing constraint, signal integrity and crosstalk for all signals connecting the FPGA to the SODIMM connector. The parameter suggested by Xilinx to measure the design quality is the time margin on sampling the signal on data and address busses. This measure has been performed on the MOSAIC board and on Xilinx AC701 evaluation board and the set of results are shown in Fig. 3, where the setup+hold margin is plotted for each byte lane of the bus for read and write cycles.

The performance of the High Speed data link has been evaluated connecting in loopback one of the FPGA transceivers through the board connector and a 2 m long cable. The Xilinx tool [6] has been used to perform the statistical eye scan and evaluate the Bit Error Rate (BER) as a function of the time offset of the sampling time from the ideal time (on the horizontal axis) and voltage offset (on the vertical axis). Data are plotted in the top picture of Fig. 4 where time is in Unit Interval (UI), i.e. the symbol period, and voltage in DAC codes, equivalent to 2 mV. BER for voltage offset equal to zero is shown on the bottom plot of Fig. 4 to obtain the bathtub diagram. The vertical bars at +/-0.175 UI are the maximum time error allowed by the receiver. From the diagram we can extrapolate the BER curve to estimate the eye shrink for an eye opening of 0.60 UI at 10E-15. The width and the height margins can be evaluated as 70% and 180%, respectively.

5 Conclusions

A custom board for readout and testing of detectors and their related electronics has been designed. From the tests carried out on the prototype, it fulfills all the design specifications. The board is currently used to perform the functional test of the modules of the new pixel tracker for the upgrade of the ALICE experiment. The firmware architecture provides the possibility to easily add new custom interfaces to other front-end modules.

References

[1] ALICE Collaboration, *The ALICE experiment at the CERN LHC, J. Inst. 3 S08002 (2008).*

[2] ALICE Collaboration, *Upgrade of the ALICE Experiment: Letter of Intent, J. Phys. G 41 (2014) 087001.*

[3] ALICE Collaboration, *Technical Design Report for the Upgrade of the ALICE Inner Tracking System, J. Phys. G: Nucl. Part. Phys. 41 (2014) 087002.*

[4] G. Aglieri et al., *Monolithic active pixel sensor development for the upgrade of the ALICE inner tracking system, JINST 8 (2013) C12041.*

[5] XILINX, *7 Series FPGAs Overview, www.xilinx.com*

[6] XILINX, *Serial Link Signal Integrity Analysis with IBIS-AMI Simulation and On-Chip Eye Scan for Low-Cost, High-Volume FPGA Transceivers, www.xilinx.com*

The TOTEM DAQ based on the Scalable Readout System (SRS)

Michele Quinto[1,a], Francesco S. Cafagna[1], Adrian Fiergolski[2], and Emilio Radicioni[1]

[1] *Istituto Nazionale di Fisica Nucleare (INFN) Sezione di Bari, Via E.Orabona - 4, Bari, Italy*
[2] *Warsaw University of Technology, Warsaw, Poland*

Abstract. The TOTEM (TOTal cross section, Elastic scattering and diffraction dissociation Measurement at the LHC) experiment at LHC, has been designed to measure the total proton-proton cross-section and study the elastic and diffractive scattering at the LHC energies. In order to cope with the increased machine luminosity and the higher statistic required by the extension of the TOTEM physics program, approved for the LHC's *Run Two* phase, the previous *VME* based data acquisition system has been replaced with a new one based on the Scalable Readout System. The system features an aggregated data throughput of $2GB/s$ towards the online storage system. This makes it possible to sustain a maximum trigger rate of $\sim 24kHz$, to be compared with the $1KHz$ rate of the previous system. The trigger rate is further improved by implementing zero-suppression and second-level hardware algorithms in the Scalable Readout System. The new system fulfils the requirements for an increased efficiency, providing higher bandwidth, and increasing the purity of the data recorded. Moreover full compatibility has been guaranteed with the legacy front-end hardware, as well as with the DAQ interface of the CMS experiment and with the LHC's Timing, Trigger and Control distribution system. In this contribution we describe in detail the architecture of full system and its performance measured during the commissioning phase at the LHC Interaction Point.

1 Introduction

The TOTEM (TOTal cross section, Elastic scattering and diffraction dissociation Measurement at the LHC) [1] experiment at LHC has been designed to measure the total proton-proton cross-section with a luminosity independent method, based on the optical theorem, and to study the elastic and diffractive scattering at the LHC energies.

To perform these measurements, TOTEM requires a good acceptance for particles produced at very small angles with respect to the beam. TOTEM's coverage in pseudo-rapidity spans the ranges $3.1 \leq |\eta| \leq 4.7$ and $5.3 \leq |\eta| \leq 6.5$ on both sides of the Interaction Point (IP); this is accomplished by two gas detector telescopes, named T1 and T2. The T1 and the T2 telescopes adopt respectively Cathode Strip Chambers (CSC) and Triple Gas Electron Multiplier (GEM) chambers which are able to detect inelastically produced charged particles. The inelastic telescopes are complemented by silicon detectors housed in special movable structures embedded in the beam-pipe, called Roman Pots (RP). Two stations, each composed of 6 RP, are placed at about 210m and 220m on both sides of the IP. The TOTEM's RP are designed to detect leading protons down to a few mm from the beam centre. The layout of the TOTEM apparatus is shown in Figure 1.

All TOTEM's detectors adopt the same front-end chip despite their different technologies. The chip, called VFAT2 [2], has 128 analogue input channels and digital serial readout interface.

[a] e-mail: michele.quinto@cern.ch

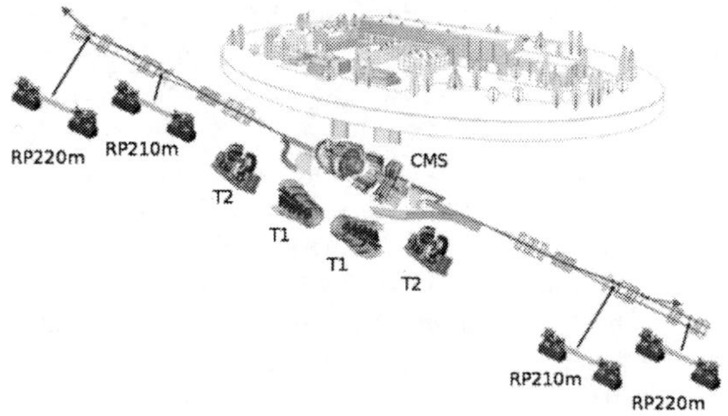

Figure 1: TOTEM experiment apparatus layout

TOTEM has already measured elastic, total and diffractive dissociation cross sections at the energies explored during the LHC's *Run One* phase. The future physics programme of TOTEM requires an increase of statistics by a factor 10 to 100. This goal has to be reached by minimizing the data taking time; for TOTEM, this means to take the maximum advantage, in terms of statistics, from the few special runs in which a machine optics configuration, reserved for TOTEM, is provided. In this framework a consolidation program [3] has been approved for TOTEM. The program includes the data acquisition system upgrade whose main requirement is to increase by more than one order of magnitude the experiment trigger rates.

Figure 2: FEC card with its Opto-FEC card and an OptoRx plugged mezzanine onto it.

2 Data acquisition system upgrade

In the legacy TOTEM DAQ architecture, optical receiver mezzanines (OptoRx), collecting data from detectors are hosted on Versa Module Eurocard (VME) boards. Each OptoRx is able to handle 12 Gigabit Optical Hybrid (GOH) [4] links running at $800Mb/s$. Having up to 16 VFAT readout chip connected to one GOH the amount of data processed by the OptoRx is $\sim 4.7KB$ per event. Table 1 shows the event raw data size in detail for each of the TOTEM's detector. The data throughput on the VME bus is the bottleneck of the system: the maximum VME transfer rate of $23MB/s$ translates into a maximum trigger rate of $1KHz$ for the experiment.

A new DAQ architecture, shown in Figure 3, was proposed to remove the system bottleneck by replacing the VME interface with the Scalable Readout System (SRS) [5] components which provide a faster and cost effective transmission medium. In the new system OptoRx modules are plugged onto a custom designed card, named Opto-FEC, which allows the connection with the SRS Front-End Concentrator (FEC) board. Figure 2 shows a FEC card and an Opto-FEC card connected via PCI edge connector. An OptoRx mezzanine is plugged onto the Opto-FEC.

Data received from the OptoRx are processed in the FEC and formatted using the User Datagram Protocol (UDP) protocol. Up to 16 FECs are read out and controlled by 4 PCs via of point-to-point connections. A Scalable Readout Unit (SRU) is connected to all FECs by means of Data, Trigger, Clock and Control (DTCC) links [6] and acts as system master fulfilling the following tasks:

- receive data from the LHC Timing, Trigger and Control (TTC) system
- distribute the machine clock, the Level One trigger signal and the fast-commands (Resynch, Bunch Crossing zero) to each FEC and to each OptoRx
- receives the Trigger Throttling System (TTS) data from each FEC

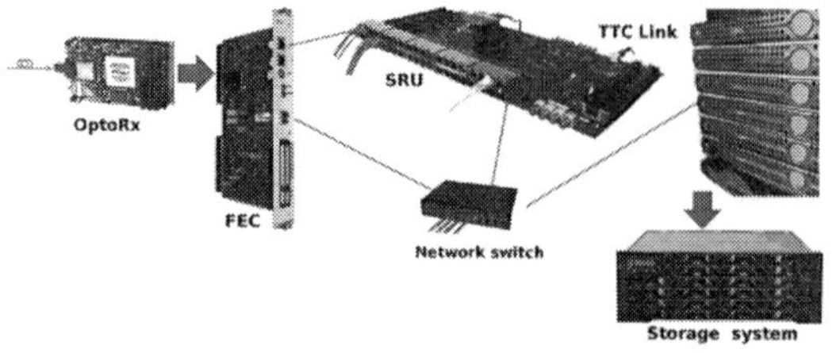

Figure 3: TOTEM DAQ architecture scheme using the SRS.

- merge 16 TTS connections into a single one and send it back to the TTC system

Detector	VFAT per ORx	N. of ORx	Ev. Size [kB]
T1	69	4	18.7
T2	170	4	18.7
RP 147m	123	4	18.7
RP 220m	123	4	18.7

Table 1: TOTEM Detectors raw data frame size.

2.1 System design and verification

The SRS firmware for TOTEM has been developed using System Verilog [7]. This allowed to integrate hardware description and verification in the same standard language. Although System Verilog is relatively new and Electronic Design Automation (EDA) tools supporting it are not fully mature, the language gradually gains attention in the industry thanks to its compactness and syntax structures which translate into more reusable and less error prone code. The usage of the SystemVerilog has implied a complete re-factory of the firmware shipped with the SRS system.

The architecture of the new FEC firmware is presented in Figure 4. It consists of two blocks: System Unit and Application Unit. The System Unit provides a set of common interfaces and services to application specific part which is included in the Application Unit. Such a distinction allows independent and parallel development of modules in the System Unit and in the data processing part enclosed in the Application Unit. The interconnections between all modules are implemented using standard buses from the Advanced Microcontroller Bus Architecture (AMBA) family [8]:

- Advanced High-performance Bus (AHB) for dedicated for memory-mapped modules such as registers and peripherals;
- Advanced eXtensible Interface 4 Stream (AXI4-Stream) unidirectional data push interconnection for modules exchanging stream of packets.

Although the firmware is mainly developed in SystemVerilog, the design still uses legacy, verified code. In

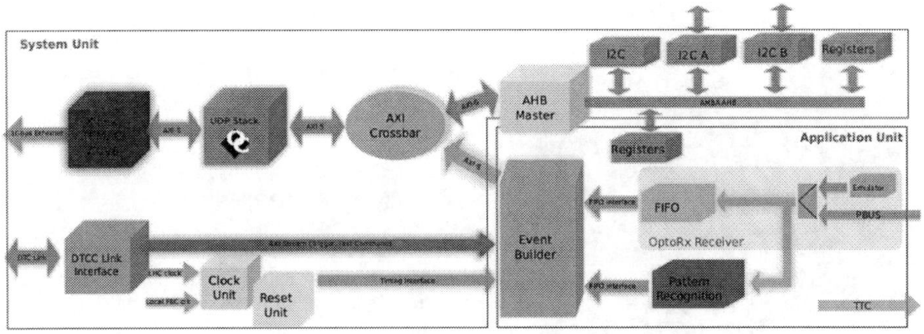

Figure 4: The FEC firmware architecture block diagram.

particular the User Datagram Protocol (UDP) stack and Inter-Integrated Circuit (I2C) master components were taken from the *opencores.org* open source projects community and have been adapted to the application.

A first SRS based demonstrator was build and tested before the end of LHC's *Run One*. The new DAQ platform was proved to run stably for several hours reaching $\sim 24Khz$ average trigger rate. Such limitation is set by the Gigabit Ethernet bandwidth that saturates at about $120MB/s$ when jumbo packet of $4.7kB$ are transferred.

2.2 Hardware data processing

In order to improve the rate limitation of $24kHz$, and with the aim of increasing not only the trigger rates but also the quality of acquired data samples, data processing algorithms have been studied and implemented on Field Programmable Gate Arrays (FPGA). The algorithms are based on *cluster reconstruction* and *pattern recognition* techniques tailored to the Roman Pot silicon detector layout. The cluster reconstruction consists in translating the VFAT channels hit pattern into a list of clusters. Each cluster is a 2-bytes information: the first byte contains the cluster starting strip position, the second byte contains the number of adjacent strips that are active. After having studied data collected by TOTEM during *Run One*, under beam conditions similar to the ones expected in *Run Two*, the cluster reconstruction was proved to be effective in reducing the event size by up to one order of magnitude. Figure 5 shows the histogram of the number of clusters in 10 RP detector planes extracted from a real data sample. The distribution shows that most of the events have less than $\sim$ 24 clusters leading to a reduction of the event size of a factor from 5 to 10.

Pattern recognition algorithms have been studied as well in order to further improve data purity by introducing second level trigger event selection. These algorithms, based on pattern-recognition techniques, are able to identify particle track segments by looking at coincidences between detector planes. The reconstruction of track segments within detector planes is actually the first step to identify particle track candidates which are intercepted by more than one pot along the beam line, a similar pattern

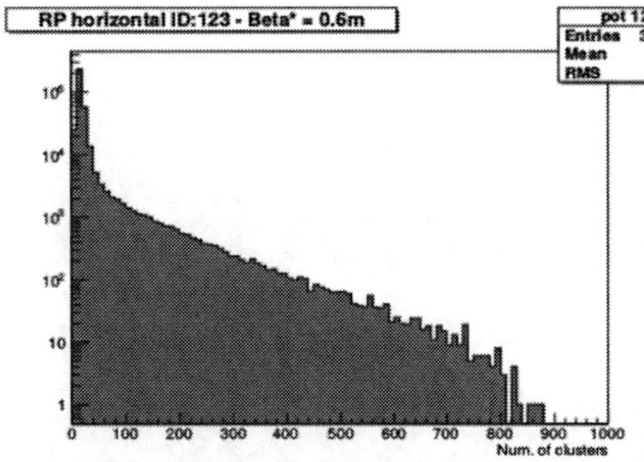

Figure 5: Histogram of the number of clusters per plane in the TOTEM RP detector.

recognition technique is implemented in the TOTEM offline analysis framework. The hardware algorithms were implemented in the FEC Virtex 5 FPGA and were proved to be as efficient as the offline reconstruction algorithms. A comparative analysis between software and hardware algorithms was possible thanks to the usage of advanced firmware simulation techniques which exploit SystemVerilog in verification environment. Universal Verification Methodology (UVM) [9] test-benches were built to simulate the full DAQ chain from the VFAT output up to the FEC board using real data as test vectors. This allowed both to test extensively design functionalities and to tune the hardware selection algorithm on real data samples.

The online algorithms will be further extended in the prospect of implementing an online event selection at the SRU level in which tracking information from the full Roman Pot detector can be combined.

3 System commissioning and operation

The full system has been installed and commissioned in the experimental area during August 2015. Figure 6 shows part of the SRS installation. The system is composed of 16 FEC modules and one SRU. Every FEC module is equipped with one OptoRx receiving data from 3 full RP detectors containing about 120 VFATs. Up to 4 FECs

Figure 6: TOTEM DAQ installation of two SRS crates and one SRU module at IP5.

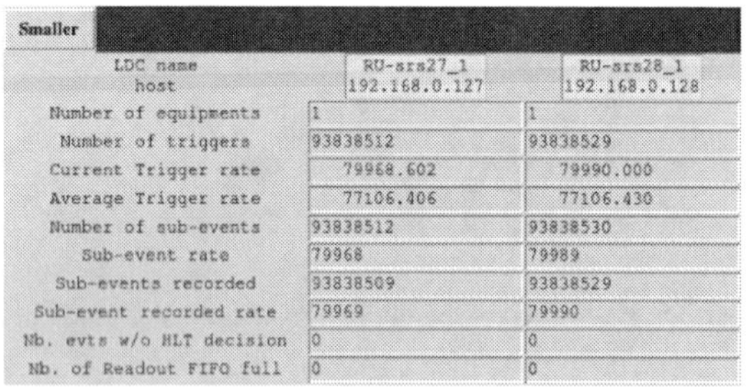

Smaller		
LDC name host	RU-srs27_1 192.168.0.127	RU-srs28_1 192.168.0.128
Number of equipments	1	1
Number of triggers	93838512	93838529
Current Trigger rate	79968.602	79990.000
Average Trigger rate	77106.406	77106.430
Number of sub-events	93838512	93838530
Sub-event rate	79968	79989
Sub-events recorded	93838509	93838529
Sub-event recorded rate	79969	79990
Nb. evts w/o HLT decision	0	0
Nb. of Readout FIFO full	0	0

Figure 7: Screen-shot of DATE software GUI. Three LDC status displays showing the trigger rates achieved reading out 12 FECs.

are read out by a standard PC, running the DATE software developed by the ALICE Collaboration [10]. The readout PCs stream the data to 12 event builder processes running on 5 different storage servers. Each server has a storage capacity of $\sim 70TB$ and can achieve up to $1GB/s$ data throughput in write mode. This configuration allows TOTEM to run continuously for a few days at full rate.

The system is being operated during the TOTEM's special runs, it has allowed the experiment to reach $\sim 77kHz$ (on average) trigger rate, which is about 2 order of magnitude higher than the performance of the TOTEM legacy DAQ. The Figure 7 shows a screen-shot of the DATE data acquisition software GUI during data taking operation. The two columns show the mean performance of two out of sixteen Local Data Collector (LDC) readout processes; each software instance reads out data from a single FEC board.

4 Acknowledgements

We would like to thank the RD51 collaboration and in particular H.Muller and J.Toledo for their valuable support in getting our hands on the SRS system and their help providing the SRS components. We are indebted to R.Liuzzi and to the CAD Service of INFN Bari unit for their contribution to the project. We also would like to acknowledge the Istituto Nazionale di Fisica Nucleare (INFN) for supporting this work.

References

[1] G.Anelli et al., *The TOTEM Experiment at the CERN Large Hadron Collider*, **2008 JINST 3 S08007**.

[2] P.Aspell et al., *VFAT2: A front-end system on chip providing fast trigger information, digitized data storage and formatting for charge readout of multi-channel silicon and gas particle detectors*, **TWEPP-07, p.292**

[3] Collaboration TOTEM, Timing Measurements in the Vertical Roman Pots of the TOTEM Experiment, **Technical Report CERN-LHCC-2014-020**, Sep 2014.

[4] Grahl, James F, Optical Data Links in CMS ECAL, *https://cds.cern.ch/record/814079*

[5] S.Martoiu et al., *Development of the scalable readout system for micro-pattern gas detector and other application*, **2013 JINST 8 C03015**.

[6] J.Toledo et al., *The Front-End Concentrator card for the RD51 Scalable Readout System*, **2011 JINST 6 C11028**.

[7] IEEE Standard for SystemVerilog — Unified Hardware Design, Specification, and Verification Language. 2013, *doi:10.1109/IEEESTD.2013.6469140*

[8] AMBA Open Specifications, *http://www.arm.com/products/system-ip/amba/amba-open-specifications.php*

[9] Universal Verification Methodology (UVM), *http://www.accellera.org/downloads/standards/uvm*

[10] ALICE Technical Design Report of the Trigger Data Acquisition High-Level Trigger and Control System, **CERN-LHCC-2003-062**

AUTHOR INDEX

AUTHOR INDEX

AUTHOR INDEX

AUTHOR INDEX